Frying Technology

Covering all the recently adapted developments, challenges, and other healthy approaches in the process of frying, this book provides the details of various frying technologies and discusses its operations and machinery in depth. Emphasis is placed on healthy prospects, nutritional values, and the emerging threats (e.g., acrylamide, acrolein, oxidation, rancidity, and other hydroperoxides) of the frying process and the effective ways to minimize them.

Key Features

- Provides a complete guide to production and consumption of fried foods along with discussions on packaging and labeling with global perspectives
- Discusses textural, sensory, and nutritional profiles of fried, baked, and puffed foods
- Explains the impact of frying on macromolecular constituents, fats/oils, starches, and proteins

A cohesive exploration of food-frying technology, this book appeals to students, academicians, researchers, and professionals in the fields of nutrition and food sciences.

Frying Technology

Recent Development, Challenges, and Prospects

Edited by

Ajay Singh, Sajad Ahmad Wani, and Pradyuman Kumar

CRC Press is an imprint of the
Taylor & Francis Group, an **informa** business

First edition published 2024
by CRC Press
6000 Broken Sound Parkway NW, Suite 300, Boca Raton, FL 33487-2742

and by CRC Press
4 Park Square, Milton Park, Abingdon, Oxon, OX14 4RN

CRC Press is an imprint of Taylor & Francis Group, LLC

ISBN: 9781032348735 (hbk)
ISBN: 9781032358994 (pbk)
ISBN: 9781003329244 (ebk)

DOI: 10.1201/9781003329244

Typeset in Century Old Style
by codeMantra

Contents

Preface

The content of this book is designed to be suitable for undergraduate, graduate, and postgraduate students and professionals in oil and fried food industries. This book also caters to the needs of doctoral- and post-doctoral-level researchers working on fats and oil frying.

This book has 14 chapters each dealing with different aspects of frying, viz. introduction to global trends, economics of fried foods, upgradation of fabrication to mechanized technology, quality of fats and oils used for frying, and packaging and labeling of fried foods. Moreover, details of vacuum frying, alternative frying, safety aspects of fried foods, and chemistry of polar compound generation are furnished, as well as nutritional facts and threats, along with biodiesel elaboration, are also considered. Common attributes of fried foods such as structural changes, flavor development, and shelf-life extension are elaborated. In industry, a person involved in a frying process must have a thorough knowledge and practical experience to understand physical and chemical properties of the frying oil and the effect of the frying process on oil quality. An oil with a minimum amount of damage can deliver fried foods with high flavor stability. Development of flavor in fried foods is a process with a complex phenomenon of temperature–time interactions that govern many sensorial, physical, textural, and chemical attributes of eatables. Labeling policies of different organizations (e.g., European Union, US Department of Agriculture, Food and Drug Administration, and Food Safety Standards Authority of India) in reference to restricted use of reused cooking oil (RUCO) also emphasize healthy processing of both vegetarian and non-vegetarian foods. Thus, used cooking oils are reused as biodiesel keeping the safety of the environment in mind.

PREFACE

The present effort to compile this book is much needed due to adoption of sedentary lifestyle by people across all income groups, sex, and age levels. Binge eating habits, improper food processing, and wrong selection of raw materials, along with adoption of unlawful tactics, are the mainstream reasons for the increase in the population of unhealthy indiviudals. This leads us to the search for better practices among frying technologies through this collection of literature from the experts in the field.

Editors

Dr. Ajay Singh is an Assistant Professor and Head of the Department of Food Technology of Mata Gujri College (an autonomous body, Estd. 1957), Punjab, India. He earned his Bachelor of Science in Life Sciences from Kurukshetra University, Kurukshetra, India, Master of Science in Food Science & Technology from CDLU, Sirsa, India, and Doctorate in Food Engineering and Technology from SLIET, Longowal, Punjab, India. He has qualified for ICAR NET-2014. He also participated in a NDRI and IIT Kharagpur collaborated project on Impact of High-Pressure Processing (HPP) on Physic-Chemical Characterization of Buffalo Milk. He also acquired Rheology diploma; Rotation and Oscillation from Anton Paar, Austria. He has attended and participated in various conferences and workshops of national and international repute. He is a DST-sponsored speaker for an oral talk at an international conference in Chicago, USA. He has also made many reviews, conducted research, and published popular articles in national and international journals to his credit along with a list of online MOOCS courses (IIT Kanpur & COL, Canada) awarded with distinction. He demonstrated a TEQIP-sponsored workshop as a resource person in Kurukshetra University, Kurukshetra. He was also awarded with the second prize in a MOFPI-sponsored international conference at JNU, New Delhi, India. In recent past (2022), he showcased the mentorship role for two students for R&D practices among traditional Indian baked stuffs, fully sponsored through RONDO, Switzerland. He is presently associated as an editor with three book assignments at CRC Press, one book of NIPA, New Delhi, and one book of Food Science Quiz Compendium from Brillion Publishers, New Delhi.

EDITORS

Dr. Sajad Ahmad Wani is currently working as a D.S. Kothari Post-Doctoral Fellow in the Department of Food Technology, SKUAST-Kashmir, India. He has completed his Master's degree in Food Technology at IUST, Awantipora, J&K, India and PhD at Sant Longowal Institute of Engineering and Technology, Punjab, India. He was qualified in the National Eligibility Test (NET) conducted by the Indian Council of Agricultural Research (ICAR) in December 2015. He has published more than 35 research/review articles, 10 book chapters, and 1 book. He is also writing articles for the popular magazine *Food & Beverage News* (India's first magazine for food and beverage industry). He has attended 50 national and 20 international conferences, seminars, and workshops across the world. Dr. Wani has participated in various faculty development programs. Dr. Wani serves as an Editorial Board Member of many national and international journals. He is Potential Peer Reviewer of reputed international journals related to food science and technology published by the likes of Elsevier, Taylor & Francis, Wiley, Springer, etc. He is also a member of various associations such as IFT, IFERP, International Association for Agricultural Sustainability, AFSTI, Asia Society of Researchers, and Asian Council of Science Editors. Dr. Wani is the recipient of prestigious "D.S. Kothari Post-Doctoral Fellowship" and "Maulana Azad National Fellowship" (MANF-2013–14) from University Grant Commission, New Delhi, India.

Professor Pradyuman Kumar is a Professor in the Department of Food Engineering and Technology, Sant Longowal Institute of Engineering and Technology, Longowal, Punjab, India. Dr. Kumar has teaching and research experience of more than 25years. He has guided 9 Ph. D. and 27 M. Tech and is currently guiding 4 Ph. D. His area of research and teaching interests include food processing, formulation, product development, characterization and optimization. He has published more than 100 research/review articles and many book chapters and books. He is also writing articles for the popular magazine *Food & Beverage News* (India's first magazine for food and beverage industry). He has attended more than 50 national and international conferences, seminars, and workshops across the world. Dr. Kumar serves as an Editorial Board Member on many national and international journals. He is a Potential Peer Reviewer of reputed international journals related to food science and technology from renowned publishing houses such as Elsevier, Taylor & Francis, Wiley, Springer, etc. He is also a member of various associations such as AFSTI. He authored 16 chapters for various books published by CRC Press, Elsevier, and Springer. Presently, Professor Kumar is associated as an editor for two edited books with CRC Press and one book of Food Science Quiz Compendium with Brillion Publishers, New Delhi. He is serving as Guest Editor for three journals namely *Foods*, MDPI publisher, *Frontiers in Bioinformatics* and *Frontiers in Sustainable Food Systems*, Frontiers publisher. He received best paper awards in various international/national conferences and recipient of Young Scientist award for 2005 and Dr. JS Pruthi award for 2019 from AFST, Mysore.

Contributors

Aaliya Ali
Faculty of Biotechnology and Applied Sciences
Shoolini University of Biotechnology and Management Sciences
Solan, India

Prajya Arya
Department of Food Engineering and Technology
Sant Longowal Institute of Engineering and Technology (Deemed to be University)
Longowal, India

Ritee Basu
Centre for Bio Separation Technology (CBST)
Vellore Institute of Technology (VIT)
Vellore, India

Zarina Begum
Department of Biotechnology
Shaheed Udham Singh College of Engineering and Technology
Mohali, India

Tejaswi Boyapati
Department of Food Technology
Vignan Foundation of Science, Technology and Research (Deemed to be University)
Guntur, India and
Department of Food Technology
Koneru Lakshmaiah Education Foundation (KL – Deemed to be University)
Guntur, India

Anjali Chaturvedi
Sai Institute of Paramedical and Allied Sciences
HNB Garhwal University
Dehradun, India

CONTRIBUTORS

Chandrakant Genu Dalbhagat
Department of Food Process Engineering
National Institute of Technology
Rourkela, Odisha, India

Sukanya Dasgupta
Centre for Bio Separation Technology (CBST)
Vellore Institute of Technology (VIT)
Vellore, India

A. Ganga
Dairy Microbiology Division
ICAR-National Dairy Research Institute (NDRI)
Karnal, India

Shikha Gupta
SSD Girls College Bathinda
Punjabi University Patiala
Punjab, India

Gurpreet Kaur
Department of Zoology
Mata Gujri College
Fatehgarh Sahib, India

Parneet Kaur
Faculty of Biotechnology and Applied Sciences
Shoolini University of Biotechnology and Management Sciences
Solan, India

Johra Khan
Department of Medical Laboratory Sciences
College of Applied Medical Sciences, Majmaah University
Al Majmaah, Saudi Arabia

Anusha Kishore
Dairy Chemistry Division
ICAR-National Dairy Research Institute
Karnal, Haryana, India

Saurabh Kulshreshtha
Faculty of Biotechnology and Applied Sciences
Shoolini University of Biotechnology and Management Sciences
Solan, India

Pradyuman Kumar
Department of Food Engineering and Technology
Sant Longowal Institute of Engineering and Technology (Deemed to be University)
Longowal, India

Shubham Mandliya
Department of Agriculture and Food Engineering
Indian Institute of Technology Kharagpur
West Bengal, India

Subhadip Manik
Dairy Technology Division
ICAR-National Dairy Research Institute
Karnal, India

Anupriya Mazumder
Food Science and Technology Department
St. Joseph's College
Bengaluru, India

Sourav Misra
Mechanical Processing Division
ICAR-National Institute of
Natural Fibre Engineering and Technology
Kolkata, West Bengal, India

Abhirup Mitra
Department of Agriculture
Invertis University
Bareily, India

Bharti Mittu
Central Instrumental Lab
National Institute of
Pharmaceutical Education and Research
SAS Nagar, Mohali

Chander Mohan
Dairy Technology Division
ICAR-National Dairy Research Institute
Karnal, India

Mohona Munshi
Department of Food Technology
Vignan Foundation of Science, Technology and Research (Deemed to be University)
Guntur, India

Ayesha Noor
Centre for Bio Separation Technology (CBST)
Vellore Institute of Technology (VIT)
Vellore, India

Jessica Pandohee
Telethon Kids Institute
Nedlands, Australia

Chirasmita Panigrahi
Chemical Engineering Department
IIT Madras
Chennai, Tamilnadu

Indira Dey Paul
Department of Food Technology
Haldia Institute of Technology
West Bengal, India

K. Pushpalatha
Dairy Microbiology Division
ICAR-National Dairy Research Institute (NDRI)
Karnal, India

Winny Routray
Department of Food Process Engineering
NIT Rourkela
Odisha, India

Mukul Sain
Dairy Engineering Division
ICAR-National Dairy Research Institute (NDRI)
Karnal, India

Mubasshira Shaikh
St. Xavier's College (Autonomous)
Mumbai, India

Renu Sharma
Department of Chemistry
Akal Degree College Mastuana Sahib
Punjab, India

Richa Singh
ICAR-NDRI Karnal
Haryana, India

CONTRIBUTORS

Sandeep Kumar Singh
Department of Agriculture
Mata Gujri College, Fatehgarh Sahib
Punjab, India

Priya Sundarrajan
Department of Life Science & Biochemistry
St. Xavier's College (Autonomous)
Mumbai, India

Surbhi
Biotechnology Department
Punjabi University
Patiala, Punjab, India

P. S. Sushama
Dairy Chemistry Division
ICAR-National Dairy Research Institute (NDRI)
Karnal, Haryana, India

Anjali Thakur
Department of Agriculture and Food Engineering
Indian Institute of Technology Kharagpur
West Bengal, India

Sheetal Thakur
University Centre for R & D
Department of Biotechnology
Chandigarh University
Mohali, Punjab, India

Ira Tripathi
Department of Home Science
Kashi Naresh Govt. PG College
Gyanpur, Bhadohi, UP, India

Siddharth Vishwakarma
Department of Agriculture and Food Engineering
Indian Institute of Technology Kharagpur
West Bengal, India

Chapter 1

Food Frying

Global Trends and Economics

Bharti Mittu

National Institute of Pharmaceutical Education and Research (NIPER)

Shikha Gupta

Punjabi University Patiala

Zarina Begum

Shaheed Udham Singh College of Engineering and Technology

DOI: 10.1201/9781003329244-1

1.1 Introduction

1.1.1 What Is Frying?

Frying is a process of cooking food in oil or any other fat. In pan frying, food products are generally flipped over once or twice using a spatula to confirm that they are well cooked, while they are "tossed in the pan" in the case of sautéing. A huge variety of fried foods are now being marketed in large quantities. Frying is a unique and very famous technique since ancient times. It has been used for preparing different kinds of dishes, especially in celebrations. Popularity and cheap cost of large-scale frying lead to the growth in the number of fried food staples in the late 20th century fast-food industry. Frying is continuous transfer of mass and heat, which results in two counterfluxes: firstly, transfer of moisture from the food into the hot oil, and secondly, transfer of hot oil into the food, giving the final product its unique textural and organoleptic characteristics (Arafat, 2014).

Researchers have been looking for techniques aimed to overcome the limitations associated with deep fat-frying practices, mainly high oil uptake and compromised food quality. Some recently developed frying technologies have come forth where vacuum is used alone or along with microwave or ultrasonic methods. This chapter discusses these advanced alternative frying technologies which not only improve the quality of products but also help reduce energy costs. These alternative frying technologies include air frying, radiant frying, ultrasound microwave-assisted vacuum frying (USMVF), microwave frying, vacuum frying, microwave-assisted vacuum frying (MVF), high-pressure deep frying, and pulsed-spouted microwave vacuum frying (Devi et al., 2021). These novel techniques significantly reduce the oil content of food products and also provide them with improved textural and organoleptic qualities in comparison to conventional frying practices. Furthermore, novel post-frying treatments of food products can be applied to the frying process to get a synergistic impact for maintaining quality (Zhang et al., 2020).

1.1.2 History of Frying

Frying is thought to have first appeared in the kitchens of Asian countries around 2500 BCE. The earliest evidence of frying technique in the western world was traced from a 17th-century painting depicting an elderly lady frying an egg, which demonstrates the birth of frying (captured while doing, Figure 1.1) (Tannahill, 1995). Since then, it has been regularly used by in the western food industry.

Figure 1.1 A 17th-century painting showing frying process

1.1.3 Deep Fat Frying

The process of cooking food materials partly or fully immersing in oil at some point or total duration of cooking at atmospheric pressure (760 mm Hg or 1,013 kPa absolute) is termed as deep fat frying. Frying food materials in oil not only cooks the food products but also provides a crunchy texture to them, for example, French fries, potato chips, and fried chicken (Moreira et al., 1997). The final product is generally golden-brown in color having oil content varying from 8% to 25%.

In the deep fat-frying process, the food is immersed in oil completely. The deep frying in oil shows drastic changes in the level of moisture from the food, giving height to hydrolytic revision. In a basic deep fat fryer, hot oil and food products are placed in a chamber. The speed and effectiveness of the frying process rely on the amount of heat and the overall quality of the oil including decomposition of triglycerides and alteration in thermal and physical properties such as color and thickness of oil (Moreira et al., 1999). The frying temperature normally ranges from 160°C to 190°C. Cooking oils (e.g., olive oil, canola oil, corn oil, sunflower oil, peanut oil, and soybean oil) not only act as a heat transfer medium, but they also penetrate into the product and give taste.

1.2 Global Trends in the Pretreatment of Frying Products

Frying is a centuries-old process that is still in use for preparing various food products of our regular diet. The increase in consumer health awareness has motivated more research in the development of healthy and safe fried products. Recent strategies to attain this goal have been focused on reduction in oil uptake by fried products and increasing the stability of the frying medium. Another approach is the development of alternative frying technologies such as pressure frying, microwave frying, vacuum frying, and radiant frying or their combinations. Among all the alternative frying technologies, only vacuum frying has received acceptance and commercial utilization from the food industry. Pressure frying and radiant frying have also shown some promising potential, but they need further exploration (Pankaj and Keener, 2017). The various fried items, their increased consumption, and market trends are presented in Table 1.1.

Table 1.1 Trends Emerging from the Fryer

List of Fried Items	*% Increase in Consumption and Trend in the Market*	*Interpretation*
Sweet potato fries	18.5%; Famous in almost all parts of the world	The rising popularity of pre- Fryer adds an extra flavor dimension to fries.
Breaded and battered fried chicken	12.7%; Famous in Menu Monitor.	Breaded and battered fried chicken are restaurant staples, but some creative chefs made the outer coating a shortcut that yields crispy skin and a cleaner product. Plenty of vegetables, too, made more attractive and healthier for the customer.
Fried cheese and cheese sticks	2.3%; Famous in all restaurants.	Fried cheese reflects an ever-expanding global palate that includes not only traditional mozzarella and provolone but also Edam, cheddar, American and other ethnic varieties which boosted the market at glance.
Scotch eggs	Trending largely in America	The classic recipe involves a runny- to hard-boiled egg.
World street food	Latest in Middle and Far Eastern origins	Items such as Japanese tonkatsu, sandwich or served with dips, Indian pakoras (fritters) and crispy samosas, and Filipino-style lumpia (spring rolls) and deep-fried bananas are most trending these days.

1.3 Formation of Acrylamide in Some Egyptian Foods due to Food Processing Conditions and Pre-Frying Treatments

Acrylamide is a colorless, odorless crystalline monomer having the formula C_3H_5NO ($CH_2=CH-CONH_2$) and a molecular weight of 71.08g. It dissolves in dichloromethane and methanol solvents. It converts into acrylic acid during hydrolization.

1.3.1 Uses of Acrylamide

It is used to form polyacrylamide which is used in the analyses of drinking and wastewater. It is used in the production of paper and the petroleum industry. It is used in the production of minerals and care of land and soil. It is also used in preparation of alloys in dentistry, formation of photographic films, and formation of varnish and dyes (EURAR, 2002).

1.3.2 Case Studies Related to Acrylamide (Health and Risk)

In 1997, a huge water leak occurred while building a tunnel in Sweden which led to the risk of acrylamide spread in the environment. Consequently, a significant number of dead fish and paralyzed animals were found near the establishment site. Monomeric acrylamide and *N*-methylolacrylamide were found in the walls of the tunnel, and the presence of these compounds in large amounts caused many health problems. Nerve-related symptoms were developed in several tunnel laborers similar to acrylamide toxicity (Eriksson, 2005; IARC, 1994).

In 2001, the Scientific Committee of Toxicity, Ecotoxicity, and the Environment found poisonous properties of acrylamide. The benefit of acrylamide as a food additive was first examined by Tareke et al. (2002) who found that giving fried foods to rats raises their hemoglobin level. It also helps in reducing sugars and amino acids, mainly asparagine. (It is an amino acid used in the biosynthesis of proteins.) In April 2002, some Swedish researchers reported that starch-rich foods (such as potatoes and cereal products) and coffee-related products contained high levels of acrylamide (Surdyk et al., 2004; Tareke et al., 2002). Table 1.2 presents acrylamide levels (μg/kg) in some Egyptian foods.

In 2005, Swedish National Food Administration declared that high levels of acrylamide can be caused by processing or cooking food at high temperatures (Zhang et al., 2005). It was seen that coffee “Bin Shaheen” with cardamom dark color had the highest acrylamide level (1,970.35 μg/kg), which was observed to be highly significant, while cooked rice had the lowest amount (2.05 μg/kg).

1.3.3 Reduction of Acrylamide Formation in Rice Using Different Pre-Frying Treatments

The core objective of this investigation was to examine the impact of pre-frying treatments on acrylamide formation in fried rice. According to prevention of acrylamide formation can be done by removing asparagine, reducing sugar, removing acrylamide after its formation,

Table 1.2 Acrylamide Levels (µg/kg) in Some Egyptian Food

Food Commodity	*Number of Samples*	*Acyrlamide Concentration (µg/Kg)*	*Range*
Potato with salt	5	25.97±5.41	17.64–49.53
Toast with olive	2	40.53±0.63	39.15–43.02
Coffee with cardamom dark color	3	1970.35±809.92	23.04–5473.02
Peanut	2	193.93±51.74	77.01–322.05
Falafel light and dark color	2	23.78±2.74	16.02–33.48
Fried noodles at different temperatures and times	3	192.93±35.88	61.59–312.21
Fried rice at different temperatures and times	3	142.25±15.9	101.04–208.8
Cooked rice	1	2.05±0.01	2.01–2.11

Source: EFSA (European Food Safety Authority), 2012.

or minimizing acrylamide formation by altering frying conditions such as heat, volume, and cooking time. The percentages of acrylamide reduction in different pre-frying treatments are recorded in Table 1.3.

1.4 Global Trends in Alternative Frying Technologies

1.4.1 Air Frying

Instead of submerging raw food ingredients in hot oil to cook them, air frying technology uses hot air containing oil droplets. In this technique, the food materials and the oil droplets in the hot air stream inside a chamber are maintained intact. The heat evenly distributes throughout the food, dehydrates the food ingredients, and gradually forms crust on the surface of the final product (Teruel et al., 2015a; Shaker, 2014). There is a substantial reduction in the time required for frying with greatly decreased oil and calorie contents. Apart from these health benefits, the technique is also associated with reduced fuel consumption and emissions of effluents and energy costs (Santos et al., 2017; Shaker, 2014). Air frying technology does not produce toxic compounds and also retains the sensory characteristics of the final product similar to deep fat frying (Zaghi et al., 2019). A major limitation of this technique is the compromised physical properties such as texture, flavor, color, and moisture content in the final food (Yu et al., 2020).

Table 1.3 Acrylamide Content (μg/kg) of Pretreated Rice with Different Treatments before Frying for 10 Minutes

Pretreatment of Rice	*Acrylamide Concentration (μg/kg)*	*% Reduction*
Control	1302.36±2.28	-
Water washing	351.94±1.4	72.97
Washing and soaking in water for 20 minutes	130.43±1.45	89.98
Soaking in citric acid (1%) soaking for 20 minutes	80.96±0.38	93.70
Soaking in acetic acid (1%) soaking for 20 minutes	69.70±0.82	94.65
Soaking in water "resulting from grape leaves soaking" for 20 minutes	99.18±1.63	92.38
Soaking in water "resulting from grape leaves boiling" for 20 minutes	104.91±0.78	91.91
Soaking in water "resulting from poached grape leaves soaking" for 20 minutes	143.01±1.18	89.01

Source: Ismial, Rehab, Askar, and Wafaa (2013).

1.4.1.1 Operating Principle

The system uses high-speed air circulation technology and consists of the following basic components: an air fryer chamber, a circulating fan, and a heating element. In an air fryer system, the food is cooked by scattering oil droplets through a hot air stream, and the product achieves organoleptic characteristics similar to that of conventional frying (Santos et al., 2017). During the process of air frying, oil droplets are scattered into the stream of hot air forming a mist, which comes in direct contact and surrounds the food ingredients. The food is constantly moved so that it remains in constant contact with the hot oily mist. The food is heated up, the internal moisture of the food is evaporated, and gradually a layer appears on the food surface (Joshy et al., 2020). The air is blown in high speed, and the circulating fan maintains a continuous movement of the air inside the chamber, which leads to uniform frying of the food within a short period of time.

1.4.1.2 Effect on Food Safety and Food Quality

Air-fried food products tend to be healthier as they are less oily and have lower levels of harmful substances. Sansano et al. demonstrated that an air fryer system reduces acrylamide content by about 90% in comparison to the ancient deep fat frying technique, and the acrylamide levels were further lowered if the food items (e.g., potatoes) were experimented with solutions of citric acid, nicotinic acid, 1% amino acid, and 2% sodium chloride (Sansano et al., 2015). Acrylamide mainly forms while frying starch-rich foods. Basuny et al. (2016) also supported this by providing evidence from air-fried potatoes and reported the reduction in the acrylamide levels of deep oil-fried potatoes from 290 to 78 ppm in air-fried ones. This may be due to the lower frying temperature used in air frying systems. Further, air-fried foods also exhibited lower polycyclic aromatic hydrocarbon contents which can be attributed to the requirement of less oil in this process (Kim et al., 2021; Lee et al., 2020). In comparison to the conventional frying, concentrations of oxidized fatty acids and free fatty acids were found to be significantly better in air-fried food products. Basuny et al. (2016) reported that the concentration of oxidized fatty acids was 0.06% and there is an increase in free fatty acids from 0.09% to 0.12% in air-fried foods. Air-fried food preserved large levels of vitamins with improved quality of oil since chemical degradation reactions are inhibited (Heredia et al., 2014).

Different workers have reported the sensorial and textural attributes of air-fried food (Santos et al., 2017; Wang et al., 2021). The final products exhibited pleasant taste, color, crispiness, and appearance and absorbed 90% less fat compared to deep frying process (Zaghi et al., 2019). The frying time, frying temperature, and microstructure of the food crust perform major roles in the frying process (Salamatullah et al., 2021). Appearance-wise, air-fried samples appear puffed and dry, but deep fat-fried samples displayed a greasy touch and oily coating. Furthermore, reduced starch gelatinization due to lower temperature results in a noticeable difference in texture. So, air frying technology provides better nutritional quality and healthier fried foods probably due to less oil absorption, degradation, and oxidation.

1.4.2 Vacuum Frying (VF)

This technology was found between the 1960s and 1970s and performed at pressures lower than atmospheric pressure (Wang et al., 2021). It has several advantages compared to deep fat frying such as the oil concentration of the final product is greatly decreased (Caro et al., 2020), the formation of toxic products due to oil degradation is almost nullified due to the lower heat and volume used (Granda et al., 2004), and finally the natural color, flavor, and quality of nutrition in the food are better saved (Da Silva and Moreira, 2008). So this technology is widely employed to dehydrate fruits (e.g., apples, bananas, jackfruit, kiwi, pear, strawberries), vegetables (e.g., carrots, mushroom, tomatoes, potatoes, sweet potato, onions) (Andrés-Bello et al., 2011), nuts and dried fruits (e.g., dates, peanuts), and some products of animal origin (e.g., fish fillets, chicken nuggets). This technique may give safer and stable fried foods with lower content of oil, thus improving the diet of consumers and reducing production costs (Sosa-Morales et al., 2022).

1.4.2.1 Operating Principle

All vacuum frying systems consist of three basic components: a vacuum pump (a liquid ring or oil sealed to keep the frying process in a vacuum state), a frying chamber (which contains an oil heater and a frying basket), and a cooler vapor condenser (to trap the steam and other volatiles evaporated during frying of the food) (Figure 1.2) (Pandey et al., 2020; Diamante et al., 2015). Before frying, the high-intensity vacuum pump is opened and oil is heated by gas burner or an electric heater below the frying basket. The food material is loaded into the basket. The pressure in the frying chamber is also lowered to the pre-determined level. A rotating lift rod lowers the basket into the hot oil as soon as it reaches the required temperature. The vapor cooling condenser then traps the water vapor released during the heating of the food through a vacuum pipe. After frying, the extra amount of oil present in the food product can be removed by centrifugation.

Literature studies revealed a rapid increase in the internal heat of the food during vacuum frying until it acquires the boiling point of water (under the required pressure). Till the evaporation of water, it remains at

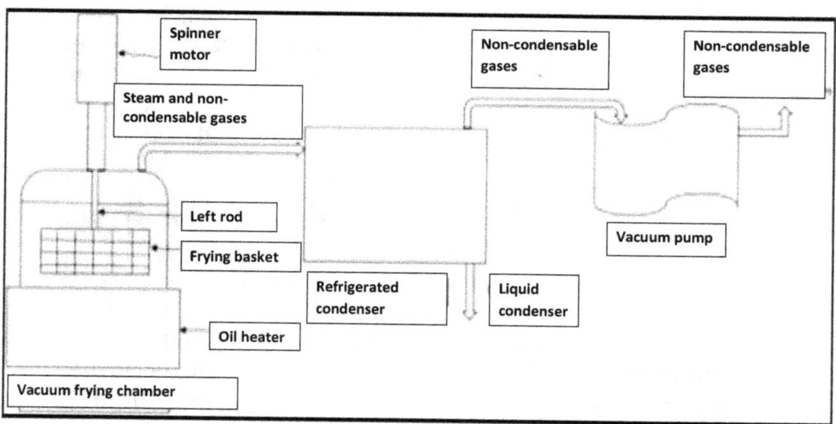

Figure 1.2 Vacuum frying system (Diamante et al., 2015).

this temperature and then steadily increase until it matches the temperature of the surrounding oil (Diamante et al., 2015). As the volume in the system increases, so does the boiling point of water. Under a vacuum, the moisture in fried food reaches its boiling point and completely evaporates at a lesser temperature. During frying, moisture is lost from the food in the form of vapors and oil enters so that vacuum-fried products have lesser oil uptake (Mir-Bel et al., 2012). Dueik and Bouchon (2011) stated that the operating pressures must be lower than 7 kPa (vacuum pressure used varies from 1.33 to 90 kPa), which causes a significant reduction in the boiling point of water, making frying to be done at temperatures lesser than 90°C (Dueik and Bouchon, 2011).

1.4.2.2 Effect on Food Safety and Food Quality

Vacuum-fried food products have reduced oil uptake. The researchers further demonstrated that the amount of oil absorbed into the fried food product was equal to the moisture loss from the final product (Andrés-Bello et al., 2010; Dueik and Bouchon, 2011). Vacuum-fried products have fewer open pores at lower pressure and frying temperatures, and hence less amount of volatilized moisture is trapped. The reduced temperatures correspond to the reduced cooling time of the final product, which greatly reduces the time for the oil to enter the voids (Soto et al., 2021).

Pandey and Moreira (2012) pointed out that the fat removal from potato chips could be maximized by reducing the frying time and/or by increasing the centrifuging time during the process. Moreover, the production of potentially harmful substances (e.g., acrylamide, trans-fatty acids, funan, polycyclic aromatic hydrocarbons, epoxy propionamide, and propylene) is considerably decreased due to the lower temperatures used. Researchers have attributed the reduction of toxic substances to decreased O_2 concentration in vacuum frying, which discourages the oxidation of oils and proteins (Karimi et al., 2017; Belkova et al., 2018). The funan content of vacuum-fried products was found to be reduced by 81%, 46%, and 34% under atmospheric pressures corresponding to frying temperatures of 96°C, 106°C, and 116°C, respectively (Mariotti-Celis et al., 2017).

1.4.3 Microwave-Assisted Vacuum Frying (MVF)

In this technique, the capacity of vacuum frying is increased by combining it with a microwave oven to increase the sensory and organoleptic features of the fried final food. The fried products have enhanced color, texture, and nutritional attributes (Albertos et al., 2016; Su et al., 2015). The most important advantage of this method is the low frying temperature and effective moisture diffusivity from food resulting in products with decreased oil uptake, less generation of toxic compounds, and better texture and flavor (Quan et al., 2014).

1.4.3.1 Operating Principle

The basic components include a frying chamber, a vacuum chamber, an oil tank, a vacuum pump, a heat sensor and controller, a microwave source, a circulation pump, and an electric motor (Su et al., 2017). Microwave is a form of electromagnetic radiation, and the ionic polarization mechanism is the main principle of microwave dielectric heating in foods; the internal heat generation leads to internal vapor generation (Sun et al., 2008). In MVF, heat is generated due to the fast rotation of polar molecules and ions within the food product under rapidly oscillating electric fields and transferred within the food sample (Al Faruq et al., 2018). The heat is rapidly transferred throughout the food leading to uniform heating of the product (Zhang et al., 2006). The food sample is also dehydrated fast due to the growth of a pressure gradient which significantly increases the rate of moisture transfer (Parikh and Takhar, 2016).

1.4.3.2 Effect on Food Safety and Food Quality

Researchers have reported that MVF requires lower temperatures and lesser time for effective dehydration of food samples. Sahin et al. stated that though the required temperature and moisture content of the final product were similar to traditional frying, the acrylamide content of MVF-fried potato chips was approximately 88% less than that of deep fat fried potato chips (Sahin et al., 2007). Similar reports were documented by Sansano et al., where the acrylamide content of fried foods was decreased with increasing microwave power (Sansano et al., 2018). However, acrylamide formation was increased with the up level of microwave power and frying time (Sahin et al., 2007). MVF-fried products have reduced acrylamide formation, provided there is reduced frying temperature and frying time. Moreover, due to the internal pressure of vapor, the heat movement is very fast, which can dry the food sample fast (Amrein et al., 2006). The flow of water in the vapor state may remove part of the acrylamide and its precursors. The drying of a product during cooking leads to a significant increase in the acrylamide content, and so, keeping the food moisture during frying leads to the reduction of formation of acrylamide.

MVF food products have enhanced nutrition and flavor, reduced oil content, and better color when compared to vacuum-fried ones. The uniform heating of MVF products leads to the outward movement of water, resulting in a faster heat transfer process increasing the rate of dehydration in foods and shortening the frying time (Zhang et al., 2006). Su et al. (2016, 2017) reported that the MVF potato chips have greatly reduced oil content (by about 22%–31%), which decreases with increased microwave power, which may be due to the increased porosity of fried foods (Su et al., 2016, 2017). Sahin et al. (2007) further reported change in the color of food during the frying time.

1.4.4 Power Ultrasound and Microwave-Assisted Vacuum Frying (USMVF)

The synergistic effects of microwave and ultrasound technology applied to vacuum frying have helped in increasing the energy efficiency and the quality of fried products. This technology has wide application in different fried products, including pumpkin chips (Teruel et al., 2015b), apple slices (Nelson et al., 2013), and mushroom chips (Su et al., 2018a). The major benefit of this method is improved quality parameters such as reduced oil uptake, shorter frying time, and improved nutritional attributes. USMVF increases the evaporation rate of moisture in the food sample resulting in increased dehydration efficiency. Caro et al. (2020) reported the greater preservation of protein content and bioactive compounds (phenolic and flavonoid) in USMVF-fried potato chips. Furthermore, Granda et al. (2004) showed the highest preservation of vitamin C and chlorophyll in USMVF-fried edamame. This technology demonstrated significant improvement in maintaining nutritional and sensorial attributes of food products during storage when compared to conventional deep-fried products (Munoz-Almagro et al., 2021).

1.4.4.1 Operating Principle

The basic components of USMVF are a microwave and ultrasound source, a vacuum pressure balance system, a vacuum chamber, a frying chamber, a circulation pump, an oil tank, a heating system, a controller, and an operation panel (Su et al., 2018b).

The microwave source helps heat all over the vacuum chamber evenly, while a series of ultrasonic sources located at the bottom of the frying container help in mixing properly.

High-energy ultrasonic waves (20–100 kHz) perform a series of compressions and rarefactions that can break up and mix the fluids they are passing through. In addition, cavitation, micro-streaming, and channeling effects can promote mass transfer and evaporation processes (Carcel et al., 2007), which can accelerate dehydration processes.

1.4.4.2 Effect on Food Safety and Food Quality

The power ultrasound and microwave-assisted vacuum frying (USMVF) combines the potential of ultrasound and microwave in vacuum frying to improve the quality of products (flavor and color) within a short frying time. In comparison to MVF, the final product exhibits effective moisture diffusivity and increased water evaporation rate, decreasing 20%–28% of the drying time compared with the MVF, mainly at lower frying temperatures. The textural parameters and nutritional attributes were also improved, though no significant reduction in oil uptake was evident in comparison to MVF. The application of ultrasound provides low oil absorption and better crispness of USMVF food products. Su et al. (2018a) also stated that the energy utilization rate in USMVF potato chips was decreased by 20.4%–24.7% (in comparison to MVF) depending on the frying temperature. The ultrasonic waves provide additional energy which helps in reducing the effective frying temperature required, which subsequently decreases the generation of harmful and toxic compounds like acrylamide. USMVF also shows effective dehydration of fruits and vegetables (Al Faruq et al., 2018). However, the ultrasonic waves generated by the machine are a major hurdle and pose health hazards (Azam et al., 2020).

1.4.5 High-Pressure Deep Frying

High-pressure deep frying is an upgraded technique similar to conventional deep frying, where the moisture release from the food sample generates pressure inside the fryer. Under pressure, the steam generation is comparatively reduced, and hence more moisture is retained in the food sample. This technology has been found effective in reducing the frying time with improved texture and flavor of the final product. Reports also showed that chicken fried using this technique softer and juicier due to better moisture retention and lower oil induction. Moreover, oil deterioration was reduced and hence, fewer harmful substances were generated in high-pressure deep frying. Under high pressure, the food is cooked up to a temperature of 120°C internally, compared to around 100°C during the regular frying method (Pankaj and Keener, 2017). Some researchers have reported that in the case of high-pressure frying of potato slabs, the heat transfer coefficient doubled at a pressure of 2 bar when compared to the normal atmospheric pressure frying (Erdoğdu and Dejmek, 2010).

1.4.6 Radiant Frying

This technology uses high-temperature radiant emitters to simulate the heat flux profile which foods generally experience during deep fat frying. Through alteration in emitter-sample geometry, power settings, and product exposure time, operators can tailor the radiant heat flux profile to that of oil immersion frying. According to Nelson et al. (2013), the radiant-fried chicken patties had 19% more water and 16% less oil in comparison to those produced by conventional deep fat frying technique. However, the final product fried using this technique was less satisfying in terms of sensory attributes and crispiness, and the radiant-fried products were darker in color and not as crispy as deep fat-fried samples (Nelson et al., 2013).

1.4.7 Electric Field Frying (EFF)

This technology enhances the dehydration of foods during frying by employing high-frequency, low-voltage electric fields. The electric field acts on the frying oil which substantially decreases the detrimental effect on the food. Researchers have shown a significant reduction in the rate of formation of toxic substances due to degradation of oil. In a study by Yang et al. (2004), electric field-fried shrimps displayed higher water content ($p<0.05$), reduced oil content ($p<0.05$), and decreased microstructure damage than traditional deep fat-fried shrimp. The fried shrimp also demonstrated uniform oil distribution and a more compact final structure (Yang et al., 2004).

1.5 Novel Post-Frying Techniques

A drastic change in the color of food was observed during the post-frying holding time, i.e. the time between the removal of fried foods from the fryer and their consumption. This shows that post-frying treatments are very important to meet the expectations of customers and sustain the maximum possible level of product quality after its preparation. The temperature of fried food rapidly starts to fall after its removal from the frying chambers, which shows drop of internal pressure and higher oil uptake (Su et al., 2016). Moreira et al. (1999) have researched the usage of oil during the frying and cooling of tortilla chips, which shows absorption of 20% of fat during the process of frying, and collection of 80% of the fat at the exterior part. Among this surface oil, about 64% of the entire fat content was absorbed by chips and 36% was collected at the outer part in the cooling phase.

A de-oiling device, called a centrifuge, is seen to be a potential technology for reducing uptake of oil in the post-frying step (where centrifugal force works perpendicular to the surface of chip and thus, separating the oil from the surface of chips). According to the findings of Yagua and Rosana (2011), the centrifugal de-oiling device is more useful in vacuum conditions than atmospheric conditions, and the authors estimated that the de-oiling device, before pressurization, eliminated approximately 87% of the oil from fried potato chips. Similarly, Moreira et al. (2009) also observed the impact of a de-oiling mechanism on the production of high-quality vacuum-fried potato chips by employing a centrifugal de-oiling device (750 rpm for 40 s). The findings indicated that non-centrifuged potato chips fried at 120°C for 360 seconds showed a 0.43 g/g of final oil uptake in comparison to 0.097 g/g in the case of centrifuged product.

Even the position and speed of the centrifugal de-oiling device affect uptake of oil in the post-frying stage. Sothornvit (2011) examined fried banana chips under vacuum conditions at two different centrifugal speeds (140 and 280 rpm) and found reduced oil uptakes by 17.22% and 17.31% for both centrifugal speeds, respectively. According to their study, the edible coated fried foods and high centrifugal speed lowered the oil uptake by 33.71% compared to control samples. Khalilian et al. (2020) also investigated the impact of two different centrifugal forces (perpendicular and parallel) on the oil content of eggplant during the post-frying stage. The authors found a regular decrease in oil uptake with increase in centrifugal force and time. The final fat was decreased by 40% at the perpendicular and 80% at the parallel positions of the centrifugal de-oiling device.

Post-frying drainage is another process of minimizing the oil uptake of fried foods. The fried product is kept in a basket above the soil surface in this process, permitting the oil to drain off the food surface under vacuum or atmospheric condition (Tarmizi and Niranjan, 2013). Several researches reported that immediate post-frying drainage followed by frying process prevented a significant oil absorption in the fried foods (He et al., 2013; Odunlami et al., 2021). Another study observed the impact of post-frying drainage time and vacuum-breaking velocity on the oil content of fried potatoes and concluded that lesser drainage time and low vacuum-breaking velocity led to higher oil uptake in the final products. Also, the long drainage times had lesser impact on oil uptake. However, breaking the vacuum fast can minimize the oil uptake up to 69.8% (Mir-Bel et al., 2009). The effects of different frying processes on foods are presented in Table 1.4.

1.5.1 Frying and Health

Loss of nutrients in cooked foods generally happens when they are subjected to high temperature oil heating and re-frying. There is decrease in concentration of amino acids due to re-frying and sidewise maillard browning also ends up that somehow contributes to loss of necessary amino acids among cooked food. There is a joint loss of vitamins like vitamin C (ascorbic acid), vitamin B (thiamine, riboflavin, vitamin B complex, and B6), vitamin A, and vitamin E, which alters the flavor and color of oil. Because of the loss of these nutrients, negative health effects might occur like mutations or gastrointestinal upset. The various food components changed during and after frying process are shown in Table 1.5.

1.5.2 Growth Rate of Different Categories of Food in the Market in 2003–2004

There are different stages of food frying indirectly related to our health. In 2004, health, convenience, and private label—in the same order—are the three chief carriers of the global food industry. Nielsen (2004a) examined 89 food categories from 59 countries and described that drinkable yogurts and soy beverages were the two topmost growth categories (Table 1.6). In a similar study in 2004, they were also among the fastest-growing categories. Six of the seven categories that enjoyed double-digit inflation presented health benefits and weight loss (Nielsen 2004b).

Table 1.4 Effects of Different Frying Processes on Foods

Sample	*Frying Process*	*Pre-Treatment*	*Post Frying*	*Findings*	*Reference*
Potato	Deep fat frying	HWB, ultra sound	Postfrying drainage and wiped off	Ultrasonic pretreatments reduced around 23.19% oil uptake depend on frying time	Mohammad Alinejhad and Dehghannya, 2018)
	Vacuum frying	Vacuum microwave drying	Centrifuged at 300 rpm for 1 minute	The rate of water loss and oil uptake in potato chips decreased due to pre-drying system	Song et al. (2007)
	Microwave frying	Hot water blanching	Cooled at room temperature	Less oil degradation and lower production of artifacts	Gharachorloo et al. (2017)
	Ultrasound microwave vacuum frying	Hot water blanching	Centrifuge with deoiling device	UMVF consumed 34.9%–48.3% less energy and absorbed 27.4%–32.3% less oil compared to MVF and VF	Su et al. (2018)
	Air frying	HWB and freeze drying	Cooled at room temperature	Air frying showed lower oil uptake and longer processing time compared to deep fat frying	Andres et al. (2013)

(Continued)

Table 1.4 (*Continued*) Effects of Different Frying Processes on Foods

Sample	*Frying Process*	*Pre-Treatment*	*Post Frying*	*Findings*	*Reference*
Apple	Vacuum frying	NaCl solution and sucrose solution	Centrifuged under vacuum at 350 rpm for 30 minutes	The oil uptake decreased with increasing temperature and time, while the moisture content and breaking force increased.	Shyu et al. (2001)
	Microwave vacuum frying	$CaCl_2$ solutions, steam blanched	Cooled at ambient temperature	The pretreated sample showed lower oil uptake and increased crispiness of chips.	Sham et al. (2001)
	Ultrasound microwave vacuum frying	Citric acid and HWB	Centrifuge with de-oiling device	Increased the textural crispiness and produced more desirable yellow colour	AlFaruq et al. (2019)
Mushroom	Vacuum frying	Freeze drying	Cooled at ambient temperature	It reduced up to 18.78% oil content compared to deep fat frying.	Deng et al. (2019)
Mango	Vacuum frying	Hot water blanching	Centrifuged to remove surface oil	Oil uptake due to the formation of vapour, which supposes to prevent oil from entering the pores.	Ayustaningwarno et al. (2020)

(*Continued*)

Table 1.4 (*Continued*) Effects of Different Frying Processes on Foods

Sample	*Frying Process*	*Pre-Treatment*	*Post Frying*	*Findings*	*Reference*
Carrots	Deep fat frying	Microwave drying	Cooled at room temperature	Pre-dried sample reduced 18% the oil content compared not to pretreat sample.	Karacabey et al. (2017)
Jackfruit	Vacuum frying	Hot water blanching	Centrifuged under vacuum at 500 rpm for 8 minutes	Retained bioactive compounds such as total phenolic, total flavonoids, and total carotenoids.	Maity et al. (2014)
French fries	Air frying Deep fat frying Microwave vacuum frying	Freeze drying	Cooled at ambient temperature	Reduced oil content by 45%–50%, but appearance, texture, flavor and odor characteristics were significantly different from deep fat fried.	Giovanell et al. (2017)
	Radiant frying	Freezing	Cooled at room temperature	Reduced oil uptake and overall acceptability of radiant frying French fries equivalent to deep frying	Lloyd et al. (2004)
Green beans, sweet potato	Deep fat frying Vacuum frying	Osmotic dehydration	Drainage by elevated on the baskets above the oil surface	Vacuum fried chips retained more natural colors and flavors while lowering the final oil content than traditional frying.	DaSilva and Moreira (2008)

Table 1.5 Components of Foods Changed after Frying

Components	*Changes during Frying*
Fat	Increased concentration and change in composition
Water	Significant loss
Sugars	Browning
Proteins	Alteration of the composition
Amino acids	Formation of undesirable flavoring substances
Vitamins	Moderate loss
Minerals	Small loss
Antioxidants	Moderate loss

Source: Miyagawa and others (1991).

Table 1.6 Category Growth Rate

Fastest Growing Category	*No. of Markets Growing/ Marketed*	*Category Growth Rate (2003–2004)*	*Category Growth Value (in Euros)*
Soya-based drinks	19 of 20	31%	244 million
Drinkable yogurts	37 of 40	19%	655 million
Eggs	13 of 13	16%	802 million
Cereal/fruit bars	26 of 30	14%	314 million
Sports/energy drinks	45 of 48	10%	438 million
Sugar substitutes	21 of 28	10%	438 million
Refrigerated complete meals	15 of 15	10%	487 million
Frozen fruit	14 of 14	9%	37 million
Refrigerated salad dressings	7 of 10	9%	21 million
Ready-to-drink non-carbonated beverages	46 of 55	8%	1.9 million
Cocoa/chocolate drinks	35 of 42	8%	189 million
Fresh ready-to-eat salads	9 of 10	8%	166 million
Frozen meat	19 of 25	7%	299 million

(*Continued*)

Table 1.6 (*Continued*) Category Growth Rate

Fastest Growing Category	*No. of Markets Growing/ Marketed*	*Category Growth Rate (2003–2004)*	*Category Growth Value (in Euros)*
Fresh vegetables	4 of 5	7%	640 million
Refrigerated deserts	24 of 27	7%	237 million
Cooking/edible oils	35 of 42	7%	640 million
Refrigerated meat	18 of 19	7%	2 billion
Shelf-stable cakes/gateaux	18 of 20	6%	259 million
Refrigerated fish/ seafood	8 of 8	6%	157 million
Frozen pizza	26 of 29	6%	323 million
Refrigerated soup	10 of 11	6%	25 million
Bottled water	42 of 53	6%	920 million
Shelf-stable fruit	20 of 26	6%	389 million
Toaster pastries	7 of 11	6%	56 million

Source: Nielson (2004a).

1.6 Conclusion

In this chapter, we have discussed frying types and their mechanisms. We have shown that oil uptake of fried products increases with increase in the intensity of surface compound and food roughness. Water, high temperature, oxygen, and the duration of frying process are the major contributors to these changes like rancidity of food, acrylamide formation on fried potatoes, etc. Some of the compounds formed like acrylamide and their effects on health have been discussed.

The current chapter evidences the feasibility of alternate frying technologies and their improved efficacy when different techniques are combined. The synergistic effects of the alternate frying techniques, either alone or in combination, greatly enhance the nutritional value and quality of the fried food samples. Nevertheless, a lot of work is still pending to overcome the disadvantages of these innovative frying technologies and make them commercially viable as well as healthier alternatives to the traditional frying techniques.

References

Al Faruq, A., Zhang, M., and Fan, D. "Modeling the dehydration and analysis of dielectric properties of ultrasound and microwave combined vacuum frying apple slices". *Dry. Technol.* 37 (2018): 409–423.

Albertos, I., Martin-Diana, A., Sanz, M., Barat, J., Diez, A., Jaime, I., and Rico, D. "Effect of high pressure processing or freezing technologies as pretreatment in vacuum fried carrot snacks". *Innov. Food Sci. Emerg. Technol.* 33 (2016): 115–122.

Amrein, T.M., Limacher, A., Conde-Petit, B., Amadò, R., and Escher, F. "Influence of thermal processing conditions on acrylamide generation and browning in a potato model system". *J. Agric. Food Chem.* 54 (2006): 5910–5916.

Andrés-Bello, A., García-Segovia, P., and MartínezMonzó, J. "Vacuum frying process of gilthead sea bream (Sparusaurata) fillets". *Innov. Food Sci. Emerg. Technol.* 11 (2010): 630–633.

Andrés-Bello, A., García-Segovia, P., and Martínez-Monzó, J. "Vacuum frying: An alternative to obtain high-quality dried products". *Food Eng. Rev.* 3 (2011): 63–78.

Arafat, S.M. "Air frying a new technique for produce of healthy fried potato strips". *J. Food Sci. Nutr.* 2, no. 4 (2014): 200–206.

Azam, S.M.R., Ma, H., Xu, B., Devi, S., Siddique, A.B., Stanley, S.L., Bhandari, B., and Zhu, J. "Efficacy of ultrasound treatment in the removal of pesticide residues from fresh vegetables: A review". *Trends Food Sci. Technol.* 97 (2020): 417–432.

Basuny, A.M.M., Oatibi, H.H.A., Amany, N., and Hala, N. "Effect of a novel technology (air and vacuum frying) on sensory evaluation and acrylamide generation in fried potato chips". *Banat. J. Biotechnol.* 1(4) (2016):058–065c.

Belkova, B., Hradecký, J., Hurkova, K., Forstova, V., Vaclavik, L., and Hajslova, J. "Impact of vacuum frying on quality of potato crisps and frying oil". *Food Chem.* 241 (2018): 51–59.

Blisard, N. "America's changing appetite: Food consumption and spending to 2020". Rural Economics Div., Econ. Res. Service, U.S. Dept. of Agriculture, Beltsville, MD (2004). nblisard@ers.usda.gov.

Carcel, J.A., Benedito, J., Rosselló, C., and Mulet, A. "Influence of ultrasound intensity on mass transfer in apple immersed in a sucrose solution". *J. Food Eng.* 78 (2007): 472–479.

Caro, C.A.D., Sampayo, R.S.P., Acevedo, C.D., Montero, C.P., and Martelo, R.J. "Mass transfer and colour analysis during vacuum frying of Colombian coastal carimañola". *Int. J. Food Sci.* 2020 (2004): 9816204.

Da Silva, P.F., and Moreira, R.G. "Vacuum frying of high-quality fruit and vegetable-based snacks". *LWT* 41 (2008): 1758–1767.

Devi, S., Zhang, M., Ju, R., and Bhandari, B. "Recent development of innovative methods for efficient frying technology". *Crit. Rev. Food Sci. Nutr.* 61, no. 22 (2021): 3709–3724.

Diamante, L.M., Shi, S., Hellmann, A., and Busch, J. "Vacuum frying foods: Products, process and optimization". *Int. Food Res. J.* 22 (2015): 15–22.

Dueik, V., and Bouchon, P. "Development of healthy low-fat snacks: Understanding the mechanisms of quality changes during atmospheric vacuum frying". *Food Rev. Int.* 27 (2011): 408–432.

Erdoğdu, F., and Dejmek, P. "Determination of heat transfer coefficient during high pressure frying of potatoes". *J. Food Eng.* 96, no. 4 (2010): 528–532.

Eriksson, S. "Acrylamide in food products: Identification, formation and analytical methodology". PhD thesis. Department of Environmental Chemistry, Stockholm University, Stockholm, Sweden (2005).

EFSA (European Food Safety Authority), 2012. doi 10.2903/sp.efsa.2013. EN-418.

EURAR (European Union Risk Assessment Report). "Acrylarnid". Luxemburg, European Union (2002): 210.

Fikry, M., Khalifa, I., Sami, R., Khojah, E., Ismail, K. A., & Dabbour, M. (2021). Optimization of the frying temperature and time for preparation of healthy falafel using air frying technology. Foods, 10(11), 2567.

Granda, C., Moreira, R.G., and Tichy, S.E. "Reduction of acrylamide formation in potato chips by low-temperature vacuum frying". *J. Food Sci.* 69 (2004): E405–E411.

He, D.B Xu, F., Hua,T.C., Song X Y., Oil absorption mechanism of fried food during cooling process Journal of Food Process Engineering, 36 (2013), pp. 412–417.

Heredia, A., Castelló, M.L., Argüelles, A., and Andrés, A. "Evolution of mechanical and optical properties of French fries obtained by hot air-frying". *LWT* 57 (2014): 755–760.

Innova. Droplets the refreshing trend for 2005. Innova Newsletter, (2005) Feb., 34–35.

Joshy, C.G., Ratheesh, G., Ninan, G., Kumar, K.A., and Ravishankar, C.N. "Optimizing air-frying process conditions for the development of healthy fish snack using response surface methodology under correlated observations". *J. Food Sci. Technol.* 57 (2020): 2651–2658.

Karimi, S., Wawire, M., and Mathooko, F.M. "Impact of frying practices and frying conditions on the quality and safety of frying oils used by street vendors and restaurants in Nairobi, Kenya". *J. Food Compos. Anal.* 62 (2017): 239–244.

Khalilian, s Mba, OI, Ngadi., OI., g-Frying of eggplant (Solanum melongena L.) Journal of Food Engineering (2020), Article 110358.

Kim, H.J., Cho, J., and Jang, A. "Effect of charcoal type on the formation of polycyclic aromatic hydrocarbons in grilled meats". *Food Chem.* 343 (2021): 128453.

Lee, J.S., Han, J.W., Jung, M., Lee, K.W., and Chung, M.S. "Effects of thawing and frying methods on the formation of acrylamide and polycyclic aromatic hydrocarbons in chicken meat". *Foods* 9 (2020): 573.

Mariotti-Celis, M.S., Cortés, P., Dueik, V., Bouchon, P., and Pedreschi, F. "Application of vacuum frying as a furan and acrylamide mitigation technology in potato chips". *Food Bioprocess. Technol.* 10 (2017): 2092–2099.

Mir-Bel, J., Oria,R., Salvador, ML., Influence of the vacuum break conditions on oil uptake during potato post-frying cooling Journal of Food Engineering, 95 (2009), pp. 416–422.

Mir-Bel, J., Oria, R., Salvador, M.L., and Solano, M.L.S. "Influence of temperature on heat transfer coefficient during moderate vacuum deep-fat frying". *J. Food Eng.* 113 (2012): 167–176.

Miyagawa K, Hirai K, Takezoe R. 1991. Tocopherol and fluorescence levels in deep-frying oil and their measurement for oil assessment. J Am Oil Chem Soc 68: 163–6.

Moreira, R.G., Castell-Perez, E.M., Barrufet, M.A., (1999). Deep-fat Frying: Fundamentals and Application. Aspen Publication, Gaithersburg, MD.

Moreira, R.G., Sun, X. & Chen, Y. (1997). Factors affecting oil uptake in tortilla chips in deep-fat frying. Journal of Food Engineering, 31, 485–498.

Munoz-Almagro, N., Morales-Soriano, E., Villamiel, M., and Condezo-Hoyos, L. "Hybrid high-intensity ultrasound and microwave treatment: A review on its effect on quality and bioactivity of foods". *Ultrason. Sonochem.* 80 (2021): 105835.

Nelson, L.V., Keener, K.M., Kaczay, K.R., Banerjee, P., Jensen, J.L., and Liceaga, A. "Comparison of the Fry Less 100 K radiant fryer to oil immersion frying". *LWT* 53 (2013): 473–479.

Nielsen, A.C. What's hot around the globe: Insights on growth in food and beverage. ACNielsen Global Services, Schaumburg, IL (2004a). www.acnielsen.com.

Nielsen, A.C. ACNielsen's hot 10—Frozen foods department—Largest sales in supermarkets Y/E 9/4/04. ACNielsen Global Services, New York (2004b).

Nielsen, A.C. ACNielsen's hot 10: Fresh produce department. Largest increases in supermarkets—Y/E 10/30/04. ACNielsen USA, Schaumburg, IL (2005).

Odunlami, Y., Sobukola, O., Adebowale, A., Sanni, S., Sanni,L., Ajayi, F., Faloye, O., Tomslin K., Effect of Ingredient combination and post frying centrifugation on oil uptake and associated quality attributes of a fried snack Journal of Culinary Science & Technology (2021), pp. 1–19.

Pandey, A., and Moreira, R.G. "Batch vacuum frying system analysis for potato chips". *J. Food Process Eng.* 35 (2012): 863–873.

Pandey, A.K., Ravi, N., and Chauhan, O.P. "Quality attributes of vacuum fried fruits and vegetables: A review". *J. Food Meas. Charact.* 14 (2020): 1543–1556.

Pankaj, S.K., and Keener, K.M. "A review and research trends in alternate frying technologies". *Curr. Opin. Food Sci.* 16 (2017): 74–79.

Parikh, A., and Takhar, P.S. "Comparison of microwave and conventional frying on quality attributes and fat content of potatoes". *J. Food Sci.* 81 (2016): E2743–E2755.

Perlick, A. 2005.Organics' chemistry. Restaurants & Institutions, Feb. 15, pp. 38–40, 42, 44, 46, 48, 50.www.rimag.com.

Quan, X., Zhang, M., Zhang, W., and Adhikari, B. "Effect of microwave-assisted vacuum frying on the quality of potato chips". *Dry. Technol.* 32 (2014): 1812–1819.

Sahin, S., Sumnu, G., and Oztop, M.H. "Effect of osmotic pretreatment and microwave frying on acrylamide formation in potato strips". *J. Sci. Food Agric.* 87 (2007): 2830–2836.

Salamatullah, A., Ahmed, M., Alkaltham, M., Hayat, K., Aloumi, N., Al-Dossari, A., Al-Harbi, L., and Arzoo, S. "Effect of air-frying on the bioactive properties of eggplant (Solanummelongena L.)". *Processes* 9 (2021): 435.

Sansano, M., Juan-Borrás, M., Escriche, I., and Andrés, A. "A Heredia: Effect of pretreatments and air-frying, a novel technology, on acrylamide generation in fried potatoes". *J. Food Sci.* 80, no. 5 (2015): T1120–T1128.

Sansano, M., Reyes, R.D.L., Andrés, A.M., and Heredia, A. "Effect of microwave frying on acrylamide generation, mass transfer, color, and texture in french fries". *Food Bioprocess. Technol.* 11 (2018): 1934–1939.

Santos, C.S.P., Cunha, S.C., and Casal, S. "Deep or air frying? A comparative study with different vegetable oils". *Eur. J. Lipid Sci. Technol.* 119 (2017): 1600375.

Shaker, M.A. "Air frying a new technique for produce of healthy fried potato strips". *J. Food Nutr. Sci.* 2, no. 4 (2014): 200–206.

Sosa-Morales, M.E., Solares-Alvarado, A.P., Aguilera-Bocanegra, S.P., Muñoz-Roa, J.F., and Cardoso-Ugarte, G.A. "Reviewing the effects of vacuum frying on frying medium and fried foods properties" (2022). https://doi.org/10.1111/ijfs.15572.

Soto, M., Pérez, A.M., Servent, A., Vaillant, F., and Achir, N. "Monitoring and modelling of physicochemical properties of papaya chips during vacuum frying to control their sensory attributes and nutritional value". *J. Food Eng.* 299 (2021): 110514.

Su, Y., Zhang, M., and Zhang, W. "Effect of low temperature on the microwave-assisted vacuum frying of potato chips". *Dry. Technol.* 34 (2015): 227–234.

Su, Y., Zhang, M., Zhang, W., Adhikari, B., and Yang, Z. "Application of novel microwave-assisted vacuum frying to reduce the oil uptake and improve the quality of potato chips". *LWT* 73 (2016): 490–497.

Su, Y., Zhang, M., Zhang, W., Liu, C., and Bhandari, B. "Low oil content potato chips produced by infrared vacuum pre-drying and microwave-assisted vacuum frying". *Dry. Technol.* 36 (2017): 294–306.

Su, Y., Zhang, M., Adhikari, B., Mujumdar, A.S., and Zhang, W. "Improving the energy efficiency and the quality of fried products using a novel vacuum frying assisted by combined ultrasound and microwave technology". *Innov. Food Sci. Emerg. Technol.* 50 (2018a): 148–159.

Su, Y., Zhang, M., Bhandari, B., and Zhang, W. "Enhancement of water removing and the quality of fried purple-fleshed sweet potato in the vacuum frying by combined power ultrasound and microwave technology". *Ultrason. Sonochem.* 44 (2018b): 368–379.

Sun, J., Wang, W., and Yue, Q. "Review on microwave-matter interaction fundamentals and efficient microwave-associated heating strategies". *Materials* 9 (2008): 231.

Surdyk N, Rosén J, Andersson R, Aman P. Effects of asparagine, fructose, and baking conditions on acrylamide content in yeast-leavened wheat bread. J Agric Food Chem. 2004 Apr 7;52(7):2047–51.

Tannahill, R. *Food in History*. Three Rivers Press (1995): 75. amazon.in.

Tarmizi, A.H.A. Niranjan, K. Post-frying oil drainage from potato chips and French fries: a comparative study of atmospheric and vacuum drainage Food and bioprocess technology, 6 (2013), pp. 489–497.

Tareke E, Rydberg P, Karlsson P, Eriksson S, Törnqvist M. Analysis of acrylamide, a carcinogen formed in heated foodstuffs. J Agric Food Chem. 2002 Aug 14;50(17):4998–5006.

Technomic. Consumers rate satisfaction with healthful restaurant chain offerings—Technomic Inc.'s Ahead of the Curve Survey. Press release, Jan. 17. Technomic Services, Inc., Chicago, IL (2005b).

Technomic. Breakfast survey. Technomic Inc., Chicago, IL (2005c).

Teruel, M.R., Gordon, M., Linares, M.B., Garrido, M.D., Ahromrit, A., and Niranjan, K. "A comparative study of the characteristics of French fries produced by deep fat frying and air frying". *J. Food Sci.* 80, no. 2 (2015a): 349–358.

Teruel, M.D.R., Gordon, M., Linares, M.B., Garrido, M.D., Ahromrit, A., and Niranjan, K.A. "Comparative study of the characteristics of french fries produced by deep fat frying and air frying". *J. Food Sci.* 80 (2015b): E349–E358.

Wang, Y., Wu, X., McClements, D.J., Chen, L., Miao, M., and Jin, Z. "Effect of new frying technology on starchy food quality". *Foods* 10 (2021): 1852.

Yang, D., Wu, G., Lu, Y., Jin, W.Q., et al., "Comparative analysis of the effects of novel electric field frying and conventional frying on the quality of frying oil and oil absorption of fried shrimps". *Food Control* 128, (2004): 1.

Yu, X., Li, L., Xue, J., Wang, J., Song, G., Zhang, Y., and Shen, Q. "Effect of air-frying conditions on the quality attributes and lipidomic characteristics of surimi during processing". *Innov. Food Sci. Emerg. Technol.* 60 (2020): 102305.

Zaghi, A.N., Barbalho, S.M., Guiguer, E.L., and Otoboni, A.M. "Frying process: From conventional to air frying technology". *Food Rev. Int.* 35, no. 8 (2019): 763–777.

Zhang, M., Tang, J., Mujumdar, A., and Wang, S. "Trends in microwave-related drying of fruits and vegetables". *Trends Food Sci. Technol.* 17 (2006): 524–534.

Zhang, X., Zhang, M., and Adhikari, B. "Recent developments in frying technologies applied to fresh foods". *Trends Food Sci. Technol.* 98, no. 1 (2020): 68–81. https://doi.org/10.1016/j.tifs.2020.02.007.

Zhang Y, Zhang G, Zhang Y. Occurrence and analytical methods of acrylamide in heat-treated foods. Review and recent developments. J Chromatogr A. 2005 May 20;1075(1-2):1–21.

Chapter 2

Fried Food Products

Their Classification, Significance, and Impact on Nutrition and Health

Prajya Arya and Pradyuman Kumar

Sant Longowal Institute of Engineering and Technology (Deemed-to-be University)

DOI: 10.1201/9781003329244-2

2.1 Introduction

Frying is the method of immersing raw or semi-processed food in hot liquid fat or oil in a pan over a set temperature of 150°C–190°C (Britannica, 2022). Frying is the most used technique in the food-processing industry and catering worldwide for dispensing fried food products. Fried foods are highly enriched in aromatic flavors, crispy crust, and microporous structural modifications with more oil and less moisture. Fried foods are produced from the various food matrices by using heat and mass transfer mechanisms equipped with frying equipment (Su et al., 2022). Frying of various foods is accomplished by three basic phases: (i) physical (water evaporation, oil uptake), (ii) chemical (amino acids, starch, and reducing sugars), and (iii) structural (alteration in the surface matrix of the fried foods by developing microporous crispy crust) (Eichenlaub and Koh, 2015). The involvement of various modern frying techniques made the process much easier and feasible. The recent involvement of techniques such as pulsed electric field frying (Dourado et al., 2019), infrared frying (Rahimi et al., 2018), vacuum frying, microwave frying, ultrasound frying, radiant frying, air frying, and their combinations aids in the agile frying process (Devi et al., 2021).

Fried food products are gaining popularity worldwide due to their rich aromatic taste and better sensory acceptability. The variety of foods that are fried include vegetables, meat and meat products, and different batters. Vegetables that are fried include taro, green beans, squash, celery bulb, red beet, green bean, eggplant (Da Silva and Moreira, 2008), and especially potato as in French fries (Kita, 2014). Meat products that are fried include chicken, mutton, eggs, and fish. The type, consistency, and raw materials used in the preparation of batter aid in the frying process (Rahimi, 2015).

2.2 Frying

Frying of the most food products depend on the type of their source material and their interaction with hot oil that was at the temperature range of 160°C–190°C, which hinders with crust formation of fried food products (Zhang et al., 2019). The frying process indulges with various stages of frying like raw, intermediate, and finished products (Pacheco et al., 2020). Fried foods majorly classified in two categories (a) from vegetable source like potato, yam, cassava (Alvis et al., 2008), sweet potato (Pacheco et al., 2020), green onions (Zhang et al., 2019), fried vegetables (Nguyen et al., 2022), mathri (Chauhan et al., 2022) and (b) animal source like chicken (Li et al., 2022), mutton (Bai et al., 2022), fish (Qin et al., 2022), and crab

(Cao et al., 2022), grass carp (Ji et al., 2022), omelet (de Oliveira et al., 2022) and many more. Food materials are generally undergo a quick heat and mass transfer process that takes place between the food materials and hot oil, which causes alternations in moisture content, crust surface, taste, and texture of the fried food (Gouyo et al., 2021).

During the frying process, the vast changes in physical and chemical characteristics depend on the starch gelatinization of the large granules, distinctive taste component formation, swelling, shrinkage, and crunchy crust formation (Oke et al., 2018). The forthcoming physical changes cause alterations in the structure of fried food crust at macro and micro levels (Al Faruq et al., 2022). The availability of fried foods keeps rising in the market as they attract consumers due to their unique taste and flavor generated by frying.

Frying food materials includes different processes including water removal, capillary pressure, adherence, draining oil, and vapor condensation that leads to the vacuum effect that characterizes oil uptake of fried foods (Ngadi et al., 2008). Frying process is affected by the type of oil used, pre-processing of the food materials, and post-frying conditions (Ziaiifar et al., 2010). The other stages of the frying process are water replacement, cooling phase, and addition of surface-active agents, which play a crucial role in the fulfillment of frying process (Dana and Saguy, 2006), accompanied by osmotic dehydration and vaporization (Guillermo et al., 2021). In the stage of water replacement, large pores are created due to evaporation of water. In the cooling phase, there is quick oil uptake due to alternations in surface characteristics of food materials that are produced due to the viscosity of the oil. After frying, the fried food products are taken out of the fryer, and then the food begins to cool down which causes vapor condensation and creates a sudden internal pressure drop. The pressure drop creates an environment for oil to remain over the fried food surface and drew inside the food due to the consequent vacuum effect. In the last and third stage, surfactants are generated due to prolonged frying which reduce the contact angle and the interfacial tension leading to considerable oil uptake. This leads to production of mono- and diglycerides and polar compounds along with changes in surface characteristics that aid in the suppression of interfacial tension and enhance the contact between frying oil and food, which enhances oil uptake (Dehghannya and Ngadi, 2021). Due to prolonged frying, there is a reduction in the viscosity of oil due to polymerization which results in adherence of oil to the product's surface. After taking the fried food out of the fryer, the oil uptake increases due to decreased drainage of oil from the surface of the fried food product because of increased viscosity (Dana and Saguy, 2006). The pictorial representation of the frying process is shown in Figure 2.1.

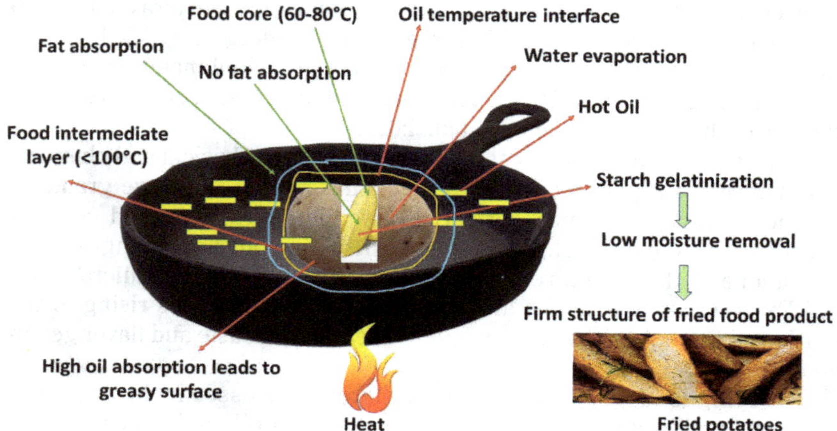

Figure 2.1 The pictorial representation of the frying process.

2.3 Classification of Frying

The categories of frying range from old traditions to recent technologies due to huge variations in methods adopted and techniques used. Different types of frying are discussed in the following part of the chapter. The majorly used frying techniques such as deep fat frying, air frying, pan frying, and vacuum frying are shown in Figure 2.2.

2.3.1 Deep Fat Frying

Deep fat frying originated in the Mediterranean region (Varela, 1988). Deep fat-fried products are rich in sensory acceptability, palatability, affordability, availability, and accessibility (Brannan and Pettit, 2015) but quite rich in oil content (Liberty et al., 2019). The amount of oil uptake of the fried product depicts the formation of crust which leads to product functionalities that reveal the structure of fried food (Liberty et al., 2019). It is the process of sealing and modifying the food surface by immersing it in hot oil to retain flavor, juiciness, and crispy crust and to check the proper cooking of the product (Ngadi et al., 2008). Deep fat frying is influenced by the mechanism of heat and mass transfer which takes place between the composition of the product to be fried and hot oil by including thermal conductivity, thermal diffusivity, density, and specific heat that cause major changes in fried products. The effect of hot oil initiates the temperature gradient that causes a convective heat transfer coefficient that occurs

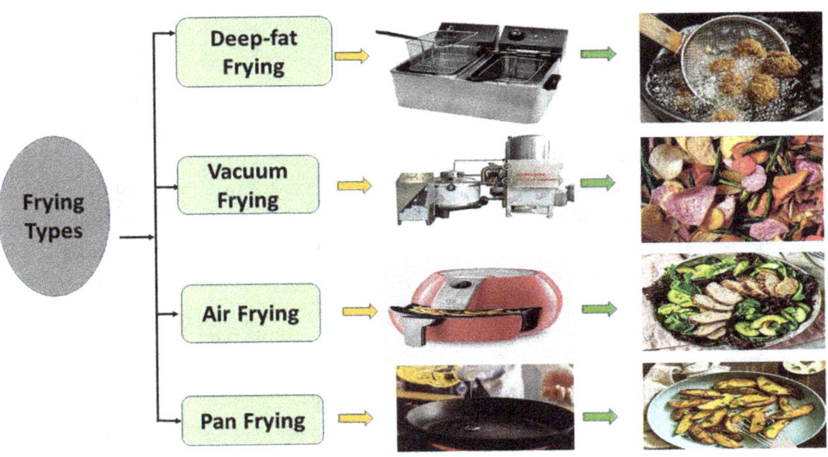

Figure 2.2 Pictorial representation of different types of frying.

between the food and oil interface. The bubbling that occurs in hot oil represents the evaporation of moisture from food products toward oil which causes a shift of textural and structural properties of the product (Singh and Mermelstein, 1995). During frying, the thickness of the food product plays a crucial role in determining product quality by observing its shape and moisture content, and the temperature inside the fried food product was hindered by the boiling point of moisture present in food. Depending on the thickness of the food product, its moisture content varies from medium to low, produced more structurally intact fried food and crispy crust caused by an outgo of internal temperature from the boiling point of water during frying (Farkas et al., 1996).

2.3.2 Air Frying

Air drying is a method of dehydrating food with hot air and spraying oil droplets inside the frying chamber which causes crust frying with minimal oil content and less frying time as compared to conventional frying (Andrés et al., 2013). Air frying is the process of producing fried products by blowing hot air around the food material rather than dipping the product in hot air (Teruel et al., 2014). It aids in the reduction of oil uptake as compared to traditionally frying (Zhu et al., 2021). In air frying, the advantages are less destruction of volatile compounds, retaining oil-soluble vitamins by omitting oil as heating medium, and betterment of fried food quality by avoiding damages caused by hot oil (Heredia et al., 2014).

2.3.3 Pan Frying

It is a method of frying a small amount of food products using a small amount of oil in a frying pan and is also called as shallow frying (Chiou et al., 2009). Pan frying is recognized as home frying that aids in maintaining the same moisture level that retained in other frying methods with continuous stirring until fried. Pan frying is used for obtaining proper crust of food products like fish, meat, and vegetables. It requires constant flipping of food products to obtain the crust color and texture, and to maintain appropriate degree of cooking by monitoring the heat and mass transfer that affects crust formation (Ikediala et al., 1996). Pan frying is accomplished in two parts: (i) the frying part in which the temperature of the product is raised to the ambient temperature of the boiling point of water followed by little oil absorption by capillary action (Moreira, 2014), and (ii) the cooling part in which the oil is absorbed on the surface of fried foods via micropores. The change in pressure is due to a fall in temperature and enhanced capillary pressure (Moreira and Barrufet 1998).

2.3.4 Vacuum Frying

Vacuum frying is one of the upcoming techniques that aids in food processing under a low sub-atmospheric pressure of up to 50Torr. Vacuum frying causes depression in boiling oil followed by a decrease in moisture content of the food products, resulting in less oil-containing fried food products because of normal atmospheric frying (Yagua and Moreira, 2011). Vacuum frying can be used with microwave- and ultrasound-assisted frying of food products (e.g., potato chips), which caused less oil absorption in the final fried products (Su et al., 2018). Microwave frying is based on the concept of high-frequency dipole– dipole interaction and ionic polarization of charged groups that generate heat and cooks food materials (Jumras et al., 2020). The microwave-assisted vacuum frying (MWVF) was conducted over banana chips which revealed 16%–20% less oil uptake, less processing time, accelerated frying rate, and better color retention (Devi et al., 2018). The use of ultrasound-assisted microwave frying (UAVF) aids in enhancing the frying process when taking place at low temperatures with substantially low amounts of oxygen required. UAVF provides quick heat and mass transfer, better retention of natural color, higher product yield, and better shelf life of fried products (Al Faruq et al., 2019). The effect of UAVF was observed in preparation of *Pleurotus eryngii* chips by monitoring pore size. The UAVF altered the pore size of the chips, which caused reduction in oil uptake by osmosis-induced micropore formation on the chips' surface (Ren et al., 2022).

Vacuum frying acts as an alternative for a thermal input control system that occurred at lower temperatures and causes reduction in the Maillard reaction and formation of acrylamides (Belkova et al., 2018). The increased consumption of potato chips fried in atmospheric frying leads to formation of acrylamides, while using vacuum frying, the acrylamide formation is reduced to a greater extent from 763±91.1 μg/kg in to 358±2.5 μg/kg (EFSA, 2015). Vacuum frying can be a better alternative to conventional frying methods due to its quick processing, less absorption, and production of better texture, color, and sensory attributes (Su et al., 2018). The vacuum frying of apple slices resulted in quick moisture removal, better retention of color, and better texture and crispness of the final product (Al Faruq et al., 2019). The roles of different frying techniques in the end quality of fried products with recently used frying methods are shown in Table 2.1.

Table 2.1 Role of Different Frying Techniques in the End Quality of Fried Products with Recently Used Frying Methods

Raw Material	*Method of Frying*	*Frying Conditions*	*Product Characteristics*	*Sensory Attributes*	*References*
Chicken nuggets	Vacuum frying	Time: 0, 2, 4, 6, 8 minutes Temp.: 130°C, 140°C, 150°C	Reduced oil uptake as compared to deep fat frying at 2 minutes with less moisture content	Better organoleptic properties	Teruel et al. (2014)
Orange peel	Vacuum frying	Time: 5, 10, 15, 20, 25, 30 minutes Temp.: 80°C, 85°C, 90°C, 95°C, 100°C	Reduction in hardness with the rise in temperature was observed at 95°C at 25 minutes	Better retention of color texture and acceptability enhancement after vacuum frying	Hien and Nguyet (2021)
Fried fish skin snacks	Vacuum frying	Time: 4, 8, 12, 16, 20, 24 minutes Temp.: 120°C	Longer time in frying, but good texture with 28% less oil content	Puffy, crispy, and yellowish color with better sensory attributes	Fang et al. (2021)

(Continued)

Table 2.1 (*Continued*) Role of Different Frying Techniques in the End Quality of Fried Products with Recently Used Frying Methods

Raw Material	*Method of Frying*	*Frying Conditions*	*Product Characteristics*	*Sensory Attributes*	*References*
Silver carp surimi chips	Vacuum frying	Time: 2.5 minutes Temp.: 118°C	A much higher thickness causes a reduction in the crispiness of the sample	Low oil uptake, optimal thickness, and acceptable sensory attributes	Hu et al. (2019)
Fish fillets	Microwave-assisted vacuum frying	Time: 18–36 minutes Power: 800, 900, 1,000 W	Quick moisture removal at 1,000 W with brighter product, well-distributed pores, thinner fiber interval	Maintain better quality of product, good appearance, and texture of fish fillet for consumer acceptance	Shi et al. (2019)
Pumpkin slices	Ultrasound-assisted vacuum frying	Power: 600, 800, 1,000 W Temp.: 90°C	Lower oil absorption, better crispness of the fried slices, maintains firm cellular structure of internal pumpkin slices	Not much effect on the color properties and better acceptability	Huang et al. (2018)
Purple flesh sweet potato slices	Microwave-assisted vacuum frying	Power: 900 W Temp.: 90°C Time: 15 minutes	Moisture reduction observed in the form of GAB model	Aids in maintaining better quality of fried foods	Fan et al. (2019)
Chicken drumsticks	Deep fat frying	Time: 3–4 minutes Temp.: 177°C	A less oil uptake in the 15% chicken protein sample by	Better retention of *L*, *a**, and *b** values by using 15%	Ananey-Obiri et al. (2020)

(*Continued*)

Table 2.1 (*Continued*) Role of Different Frying Techniques in the End Quality of Fried Products with Recently Used Frying Methods

Raw Material	*Method of Frying*	*Frying Conditions*	*Product Characteristics*	*Sensory Attributes*	*References*
			1.77% and higher moisture content by 57%	chicken protein sample	
French fries	Deep fat frying	Temp.: 180°C	A small indication of sound peaks, linear distance, and less sound pressure was observed	The less crispy and less hard structure of fries	Gouyo et al. (2020)
French fries	Air frying	Temp.: 140°C, 180°C, 200°C	Higher indications of sound peaks, linear distance, and maximum sound pressure were related to the more crispy and hard structure of the fries	Less oil uptake and more acceptable crust of fried	Gouyo et al. (2020)
Youtiao bread	Electrostatic frying	Time: 4 minutes Temp.: 180°C	Less oil uptake as compared to traditional and vacuum frying. Less acrylamide 667.37 ppb as compared to vacuum frying 1,084.98 ppb	Higher browning index that provides more acceptability from a consumer point of view	Shyu et al. (2021)

2.4 Classification of Fried Food Products

There are a range of fried food products well recognized worldwide such as French fries, nuggets, samosa, poori, mathri, cutlets, kababs, fritters, spring rolls, wonton, emping, churro, calas, boondi, bonda, luchi, medu-vada, hush puppies, and jalebi. The majorly consumed fried food products are discussed in the following part.

2.4.1 International Fried Food Products

2.4.1.1 French Fries

Potatoes are a staple necessity of today's smart world that constitute and consume many fast foods. The lion's share of potato production is utilized in processing channels that produce more than 60% of marketed potato-based food products. The potato-based frozen fries account for more than half of the total export volume of potatoes in the United States (Sadeghi et al., 2021). French fries are the most popular potato-based fried food products among all age groups due to their distinctive texture and mouthfeel (Gouyo et al., 2021). French fries consist of a specific crunchy dry crust solid layer with a moist and soft core (Van Koerten et al., 2015). The texture of French fries plays a crucial role in maintaining the peculiar taste and attracting consumers as the crispy and crunchy crust is imparting unique sensory attributes (Salvador et al., 2009). The texture of French fries is based on the variety of potato used, oil type used, process optimized followed by the finger spinning inspection from wall to core that provide access for the external and internal texture of the fries. It is checked by breaking the fries into two pieces for visual inspection (Waxman et al., 2019). There are different types of probes available in the market for texture analysis of French fries like a cylindrical probe, V-shape probe, puncture test with needle probe, shear test with Kramer shear cell five blade probe, and cut test with guillotine probe (Li et al., 2020).

French fries were prepared with steam-peeled potatoes with a thickness of 1 cm strips, followed by hot water blanching performed at 175°F for 6 minutes. The potato strips were dried in a hot-air oven at 180°F for 3 minutes. The oil was heated to 375°F and potato strips were par-fried for 40 seconds, and then the fries were kept in frozen condition at −4°F until further processing (Waxman et al., 2019). In the process of preparation of French fries, the formation of the crust is the outcome of heat and mass occurring during the interaction between a moist potato surface and hot oil (Pedreschi, 2009). The structure of fries varies based on the temperature and time of frying, which provides different types of crust hardness and color variation (Ziaiifar et al., 2010). The crust of French fries varies in the form of porosity, pore distribution, and pore size and shape (Bouchon, 2009), and it leads to the sensory appeal of the product (Van Dalen et al., 2007).

2.4.1.2 Fish and Chicken Nuggets

There is a growth in the demand for coated, enrobed, battered, and fried products in the food restaurant market in recent decades. Nuggets satisfy this demand by its sensory acceptance, inexpensiveness, and better storage life (Rahimi et al., 2018).

The increase consumption of meat products opened the gateways for the development of various fried meat products such as nuggets. Nuggets are produced after the processing of base raw materials like fish, beef, and chicken (Yogesh et al., 2013). The acceptance of nuggets depends on the processing, type of raw material, and overall sensory appeal. Nuggets should have high nutritional value, good textural properties, good taste with juiciness and tenderness, flavor, and low cholesterol (Behrends et al., 2005; Calkins and Hodgen, 2007). Fish nuggets are prepared by applying different coating materials on minced fish that provide a firm base for technological functionalities such as water-holding capacity, viscosity, gelling properties, fat-binding capacity, and texturization (Jayasinghe et al., 2013). Richness of fish nuggets majorly depends on the batter type, the oil used in frying, size of the fish piece, coating material, batter pickup, and the technology adopted for frying (Kang and Chen, 2015). There are some common fish varieties used in preparation of fish nuggets like Tilapia (Jayasinghe et al., 2013), fresh cassava croaker (*Pseudotolithus senegalensis*) (Oppong et al., 2022), live grass carp (Wu et al., 2022), frozen fish (*Coryphaena hippurus*) (Chen et al., 2009), trash fish (Amalia et al., 2016), and many more.

Chicken nuggets are deboned chicken meat that is enrobed in batter followed by frying and generally occupy a huge part of the food market in most developed countries (deShazo et al., 2013). Chicken nuggets are commonly produced using broiler chicken meat (Nurlela et al., 2018) due its high availability, tenderness, and high protein content (Sabikun et al., 2019), followed by its other advantages over spent hen meat (Bhosale et al., 2011). In the preparation of chicken nuggets, major factors that play crucial roles are batter preparation, processing of nuggets, emulsion stability, texture analysis, water loss, oil absorption capacity, color parameters, and sensory attributes (Sabikun et al., 2021). The different fried products are represented in Table 2.2 with their nutritional aspects.

Table 2.2 The Elaborated Representation of Various Fried Food Products with Their Different Nutritional Aspects

Raw Material	*Food Product*	*Properties*	*Microelements*	*References*
Cowpea	Boondi	AA (%): 314.22	Ca: 1.38 mg/100g Fe: 5.01 mg/100g	Bhati and Raghuvanshi (2020)
	Sev	AA (%): 274.5	Ca: 1.75 mg/100g Fe: 7.43 mg/100g	
	Enrobed nuts	AA (%): 208.17	Ca: 1.71 mg/100g Fe: 6.07 mg/100g	
	Fried nuts	AA (%): 212.88	Ca: 2.03 mg/100g Fe: 9.14 mg/100g	
Pokora (fritters)	Control	AA (%): 27.04 FFA (%): 0.15	Fe: 4.18 mg/100g Zn: 1.00 mg/100g Ca: 192.75 mg/100g P: 113.89 mg/100g	Diksha and Modgil (2020b)
	Street vendor	AA (%): 34.55 FFA (%): 1.13	Fe: 6.18 mg/100g Zn: 1.16 mg/100g Ca: 206.50 mg/100g P: 154.23 mg/100g	
	Value added	AA (%): 57.10 FFA (%): 0.12	Fe: 4.18 mg/100g Zn: 1.00 mg/100g Ca: 192.75 mg/100g P: 113.89 mg/100g	

(Continued)

Table 2.2 (*Continued*) The Elaborated Representation of Various Fried Food Products with Their Different Nutritional Aspects

Raw Material	*Food Product*	*Properties*	*Microelements*	*References*
Mathri	Oleogel (carnauba wax + soybean oil)	Carnauba wax (5%, 10%, 15%)	Oil uptake 5%: 27.7% less 10%: 22% less 15%: 19.3% less Biting strength 5%: 1,974.8 g 10%: 1,092.7 g 15%: 3,168.1 g	Chauhan et al. (2022)
Poori	Rice bran oil (1–6 frying cycles)	AA (%): 53.5%–37.9% Color: Red: 2–3 units (heating) & 2–4.5 units (frying) Yellow: 15–20 units (heating) & 15–25 units (frying)	Free fatty acid content: 0.3%–0.44%–0.48% Peroxide value: 2.2–3.6 meqO_2/kg (heating) 3.0–4.0 meqO_2/kg (frying) Total polar material: 3.7%–3.9% (heating) & 3.9%–4.4% (frying)	Debnath et al. (2012)
French fries	Apple pomace-based antioxidants with mustard oil	AA (%): 76%	Phenolic content: 5.58 ± 0.23 mg GAE/g Flavonoid content: 5.42 ± 0.45 mg GAE/g	Manzoor et al. (2022)
Papad	Extruded black gram flour (BGF) Control	Diameter (cm): Before: 11.03 After: 12.00	Oil content (%): 17.12 Texture crispness (N): 2.31	Ananthanarayan et al. (2018)
	Papad + 3% pakadkhar	Diameter (cm): Before: 11.08 After: 12.47	Oil content (%): 18.55 Texture crispness (N): 2.62	

(*Continued*)

Table 2.2 (*Continued*) The Elaborated Representation of Various Fried Food Products with Their Different Nutritional Aspects

Raw Material	*Food Product*	*Properties*	*Microelements*	*References*
	25% BGF+1% pakadkhar	Diameter (cm): Before: 11.07 After: 12.56	Oil content (%): 13.55 Texture crispness (N): 2.69	
	50% BGF+1% pakadkhar	Diameter (cm): Before: 11.04 After: 13.12	Oil content (%): 13.13 Texture crispness (N): 2.82	
	Market sample	Diameter (cm): Before: 11.05 After: 13.54	Oil content (%): 15.00 Texture crispness (N): 2.95	
Chicken kabab	Control	Firmness (kg/cm^3): 1.49 Moisture content (%): 65.72	Toughness (kg s): 7.64 Protein content (%): 16.43	Hussain et al. (2020)
	Butylated hydroxytoluene	Firmness (kg/cm^3): 1.22 Moisture content (%): 65.83	Toughness (kg s): 5.92 Protein content (%): 16.23	
	Shatavari root powder (SRP) 1%	Firmness (kg/cm^3): 1.59 Moisture content (%): 65.17	Toughness (kg s): 7.88 Protein content (%): 16.95	
	Shatavari root powder aqueous extract (SARE) 2%	Firmness (kg/cm^3): 1.29 Moisture content (%): 65.00	Toughness (kg s): 6.77 Protein content (%): 17.41	

(*Continued*)

Table 2.2 (*Continued*) The Elaborated Representation of Various Fried Food Products with Their Different Nutritional Aspects

Raw Material	*Food Product*	*Properties*	*Microelements*	*References*
Potato chips	Refined soybean oil	Breaking strength (kg) & frying interval (hour) 0: 26.50 12: 25.50 24: 25.00	Breaking strength (kg) & frying interval (hour) 36: 22.74 48: 23.10 60: 23.37	Rani and Chauhan (1995)
	Refined groundnut oil	Breaking strength (kg) & frying interval (hour) 0: 36.60 12: 34.20 24: 32.20	Breaking strength (kg) & frying interval (hour) 36: 31.00 48: 31.80 60: 29.80	
	Hydrogenated vegetable fat oil	Breaking strength (kg) & frying interval (hour) 0: 25.90 12: 24.00 24: 23.80	Breaking strength (kg) & frying interval (hour) 36: 22.80 48: 21.10 60: 19.00	

AA, antioxidant activity; FFA, free fatty acids; Ca, calcium; Fe, iron; Zn, zinc; P, phosphorous.

2.4.2 Indian Fried Products

2.4.2.1 Samosa

Samosa is the most consumed fried or baked dish that is prepared with a refined wheat flour stuffed with savory fillings like spiced potatoes, onions, peas, meat, and lentils. It is generally cone in shape, and its shape varies depending on the region (Diksha and Rajni, 2020a). The samosa sprang up in Central Asia and the Middle East, the Indian subcontinent, and is a commonly consumed street food and a popular Indian snack (Sakhale et al., 2011). Samosa's nutritional profile varies from region to region based on the type of filling and raw materials used in preparations. Characteristics of samosa to be measured are its sensory attributes, micro- and macroelements, nutritional composition, peroxide value, physical, structural, and functional properties, and proximate content (Diksha and Rajni, 2020a). A recent modification was made to extend the shelf life of samosa by storing par-fried samosa in frozen conditions to improve sensory attributes (Raj et al., 2017). The samosa was kept at a temperature from –18°C to –22°C to extend its shelf life from 3–4 to 90 days while maintaining its nutritional characteristics. The oil distribution in the crust is high and lipid oxidation was minimal during storage, with a less PV value, thiobarbituric acid value, and free fatty acid value of 3.97 ± 0.06 meqO_2/kg sample, 0.91 ± 0.52 mg MAL/kg sample, and $0.72\% \pm 0.04\%$, respectively (Raj et al., 2017).

2.4.2.2 Mathri

Mathri is a well-recognized and popular deep-fried Indian snack traditionally prepared from refined wheat flour and carom seeds with some seasonings (Arya et al., 1979). Mathri can be modified with dried green leafy vegetables for imparting better organoleptic properties (Verma and Jain, 2012) and adding micronutrients such as vitamin A, iron (Mehra and Singh, 2017), and β-carotene (De Benoist et al., 2006). This could be a possible way to deliver health benefits to consumers by adding nutritious elements (Negi and Roy, 2003). Mathri was fortified by iron using the cyanmethaemoglobin method by incorporating 17.45 mg of iron in 80 g of mathri, which helps maintain the hemoglobin level in the blood (Shazia et al., 2011). Mathri was enriched with phytonutrients, antioxidants, minerals, and vitamins that were extracted from Tandulaja (*Amaranthus spinosus*) powder (mixing 7 g of powder for preparing 100 g mathri) (Kadbhane et al., 2019). Mathri was prepared with some modifications in wheat flour and semolina by replacing it with bajra flour in different ratios that produced fried mathri with better sensory acceptance and average color properties (Mehra and Singh, 2017). The mathri formulated by incorporation of horsegram flour with some basic ingredients resulted in a better iron, protein, fat, and fiber contents of 6.24 mg, 13.48 g, 21.87 g, and 3.50 g, respectively (Jain et al., 2012).

2.4.2.3 Poori

Poori is a famous traditional fried food product prepared with refined wheat flour, salt, fine semolina, moreover bedmi poori in which wheat flour is partially substituted with pulse flour and added with some seasoning and oil as shortening for maintaining appropriate dough structure. All the ingredients were mixed properly to prepare the dough of fixed consistency with water (Maheshwari et al., 2017b). The appearance, texture, and umami flavor and acceptability of mono sodium glutamate added poori were observed, and the prepared poori had acceptable flavor, color, and texture (Maheshwari et al., 2017b). The study was conducted for determining the efficacy of monosodium glutamate in reducing salt addition in the plain and spiced poories by improving sensory attributes. Umami flavor is widely incorporated in food due to its high palatability and better acceptance in major fried foods. The poori palatability was compared with the control poori prepared with different salt levels with spices like chili, cumin, pepper, and omum. The poories prepared with monosodium glutamate had better qualities (Maheshwari et al., 2017a).

2.4.2.4 Kabab

Kababs are one of the most popular ready-to-eat meat dishes prepared from lamb, beef, chicken, and fish. The most common kabab is seekh kabab which is popular in all the affordable sections of the society (Bhat et al., 2013). Kababs are prepared with well-minced and structured meat or raw material cooked at 170°C–190°C at varying time durations which results in color variations, cooking yield, shear force, juiciness, tenderness, and overall sensory acceptability (Liu et al., 2012). The preparation of kabab is modified day by day by replacing the high-fat content of meat with some vegetables for creating economically satisfactory products. The replacement of meat could be achieved using vegetable products that fit into the daily protein intake of the human diet by providing 65 net protein utilization as per ICMR (Indian Council of Medical Research) recommendation (Bhat et al., 2013). A study was conducted for preparing chicken meat kabab with aqueous extract of Shatavari roots and powder to observe the quality characteristics of kabab. The kabab was prepared with 1% Shatavari root powder (SRP) and 2% Shatavari root aqueous extract (SRAE). The nutritional qualities of kabab were enhanced by the addition of SRP and SRAE while maintaining the original textural and sensory attributes of chicken meat (Hussain et al., 2020). Various other methods were adopted by different researchers

Figure 2.3 Representation of various fried food products.

to replace the meat and to prepare healthier kabab like "green kabab". But there is requirement of more research in the field. The various fried foods are shown in Figure 2.3.

2.5 Significance of Fried Food Products

Frying is generally recognized as the oldest method of food processing due to its rapidness, cheapness, ease of operation, and retention of sensory characteristics of food products. The main mechanism of frying is the heat and mass transfer that takes place between food and oil causing efficient cooking due to the high temperature of the oil and rapid exchange of oil penetration and moisture removal from the food surface (Oke et al., 2018). There is constant change in the composition of fried food products and oil used due to the quick exchange of moisture and oil. It is generally monitored by the formation of the crust layer and color development in the crust layer, and it can be optimized by modifying the heating medium and frying process (Oke et al., 2018). Some crucial changes took place during frying such as alternation in oil like hydrolysis, oxidation, and polymerization that are capable enough in altering the fried food quality (Oke et al., 2018). The constant modification in the quality of the fried products is analyzed on the nutritional front as various factors affect the same such as thermal, industrial, and culinary factors. As fat penetrates the food, it modifies the food matrix depending on size, type, texture, shape, frying time, and atmospheric conditions (Varela, 1998). Some of the major changes occurred in the lipid fraction of food are oil portion retention and deep penetration in the food matrix. Variation in the protein content of the fried food increased in grass carp fillets (Zhang et al., 2013), reduction in available lysine content was found in fish fillets by 17% (Oluwaniyi et al., 2010). Modifications of minerals, vitamins, and antioxidants were also observed by Oke et al. (2018). Fried food products are characterized by the formation of a crisp golden-colored layer with distinctive flavors influenced by physicochemical changes in the main food component causing microstructural changes that occurred due to exchange of moisture and oil from the main food components (Ngadi et al., 2008). Due to damage to the outer cells during cutting or chopping of starch containing target fried foods, the microstructure develops a 250 μm thin external layer. . This layer serves as an intermediate layer of shrunken, dehydrated cells that contains gelatinized starch that extends to evaporation during crust formation. (Sahin, 2000).

2.6 Impact of Fried Food Products on Nutrition and Health

The debauched growth of fried food consumption led to major changes in diets and lifestyles, which increased peoples' susceptibility to non-communicable diseases, coronary artery diseases, diabetes, hypertension, and obesity (Gadiraju et al., 2015). The high oil content of fried food products have a serious impact on human health. Major factors that pose threat to human health are use of unsaturated vegetable oils, hydrogenated fats, fats rich in trans-fatty acids (TFA), and the presence of 40%–45% saturated fatty acids (SFA) in palm oil (Sayon-Orea et al., 2013). Production of fried food products generates some harmful chemical compounds like acrylamide, hydroxy methyl furfural, heterocyclic amine (2-2-amino-1-methyl-6-phenylimidazo-pyridine), nitrosamines, and polyaromatic hydrocarbons like benzo a pyrene and chloro propanols (3-monochloro propane-1-diol) (Stadler, 2012). Production of these chemical compounds alters cellular mechanisms of human body and causes DNA damage by forming carcinogen-DNA adducts that form covalent bonds with DNA nucleotide and lead to alternations in body functions (Jägerstad and Skog, 2005). In fried foods, some harmful compounds are formed like acrylamide from free asparagine as precursor due to Maillard reaction. Sometimes caramelization takes place due to thermal treatments of foods by direct dehydration of sugars under acidic conditions (Kroh, 1994). The relation between the heating of lipid/oil and the generation of deleterious health effects is due to the oxidative and thermal degradation that causes the generation of oxidized and polymerized compounds rich in polarity (Gadiraju et al., 2015). The excess consumption of fried food products leads to grievous health impacts such as cardiovascular problems, heart failure, hypertension, type 2 diabetes, and obesity (Gadiraju et al., 2015) as shown in Figure 2.4.

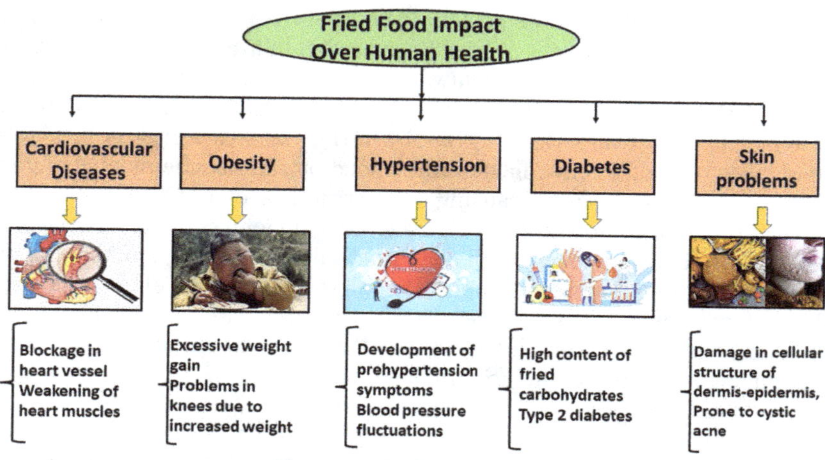

Figure 2.4 Impact of fried foods on human health.

2.7 Frying Issues and Recent Modifications

Recent innovations are always upbringing the techniques that care for human health and run the economy on smooth wheels. The most used technique is deep fat frying. The oil quality and frying temperature always plays a crucial role. As almost half of the food vendors exceeded the maximum recommended temperature of 175°C for frying which results to the formation of toxic compounds in the fried food products (Garayoa et al., 2021). The upcoming frying advancements include techniques like electromagnetic field, acoustic field, vacuum frying, pulsed electric frying, pressure frying, high-pressure processing pre-treatment in frying, electromagnetic-based frying, microwave frying, radiofrequency pre-treatment and infrared frying, ultrasonic-assisted frying, microwave-assisted frying, ultrasonic microwave-assisted vacuum frying (Su et al., 2022), radiant frying, air frying, and spray frying (Devi et al., 2021). These techniques can be used in combination or separately to improve frying efficiency and process parameters to obtain optimized fried food products with less oil uptake and better quality. This aids in the modification of frying by analyzing heat and mass transfer equations and the porous structure of the fried food products (Su et al., 2022). The recent techniques help reduce oil uptake along with alternations in the crispy crust of porous microstructure formation that is impacted by air flow and oil temperature that cause different changes in the upper surface of fried products (Ran et al., 2019). The formation of smooth surfaces, rough surfaces, and rigid surfaces is due to exposure of foods' surface to harsh stresses and higher moisture loss in vacuum frying, ultrasound, microwave frying, and air frying (Islam et al., 2019).

2.8 Conclusion

The concept of frying is discussed in this chapter along with its various aspects. Frying is the most used method to produce fried foods that increases the sensory attributes of vegetable and meat food products. The technique used in the frying process can alter the microstructure of the fried product due to absorption of oil and removal of moisture from the fried foods' surface, which can have harmful effects on human health. Fried foods always raise a concern regarding human health due to the high uptake of oils and lipids during frying. Modifications to the frying process can minimize the oil uptake such as air frying, pan frying, and vacuum frying instead of deep fat frying. Classification of various types of fried foods also depends on their raw materials and the methods adopted for frying. There are a variety of fried foods consumed in different regions of the globe like French fries, nuggets, kababs, poories, mathri, fritters, pancakes, and many more.

References

Al Faruq, A., Khatun, M. H. A., Azam, S. R., Sarker, M. S. H., Mahomud, M. S., & Jin, X. (2022). Recent advances in frying processes for plant-based foods. *Food Chemistry Advances*, 1, 100086, 1–13.

Al Faruq, A., Zhang, M., & Adhikari, B. (2019). A novel vacuum frying technology of apple slices combined with ultrasound and microwave. *Ultrasonics Sonochemistry*, *52*, 522–529. https://doi.org/10.1016/j.ultsonch.2018.12.033.

Alvis, A., Villada, H. S., & Villada, D. C. (2008). Efecto de la temperatura y tiempo de fritura sobre las características sensoriales del ñame (*Dioscorea alata*). *Información Tecnológica*, *19*(5), 19–26.

Amalia, U., Darmanto, Y. S., & Rianingsih, L. (2016). Chemical characteristics of fish nugget with mangrove fruit flour substitution. *Aquatic Procedia*, *7*, 265–270.

Ananey-Obiri, D., Matthews, L., & Tahergorabi, R. (2020). Chicken processing by-product: A source of protein for fat uptake reduction in deep-fried chicken. *Food Hydrocolloids*, *101*, 105500. https://doi.org/10.1016/j.foodhyd.2019.105500.

Ananthanarayan, L., Gat, Y., Kumar, V., Panghal, A., & Kaur, N. (2018). Extruded black gram flour: Partial substitute for improving quality characteristics of Indian traditional snack. *Journal of Ethnic Foods*, *5*(1), 54–59.

Andrés, A., Arguelles, Á., Castelló, M. L., & Heredia, A. (2013). Mass transfer and volume changes in French fries during air frying. *Food and Bioprocess Technology*, *6*(8), 1917–1924.

Arya, S. S., Natesan, V., Parihar, D. B., & Vijayaraghavan, P. K. (1979). Stability of β-carotene in isolated systems. *International Journal of Food Science and Technology*, *14*(6), 571–578.

Bai, S., Wang, Y., Luo, R., Shen, F., Bai, H., & Ding, D. (2022). Formation of flavor volatile compounds at different processing stages of household stir-frying mutton sao zi in the northwest of China. *LWT-Food Science and Technology*, *139*, 110735.

Behrends, J. M., Goodson, K. J., Koohmaraie, M., Shackelford, S. D., Wheeler, T. L., Morgan, W. W., & Savell, J. W. (2005). Beef customer satisfaction: USDA quality grade and marination effects on consumer evaluations of top round steaks. *Journal of Animal Science*, *83*(3), 662–670.

Belkova, B., Hradecky, J., Hurkova, K., Forstova, V., Vaclavik, L., & Hajslova, J. (2018). Impact of vacuum frying on quality of potato crisps and frying oil. *Food Chemistry*, *241*, 51–59. https://doi.org/10.1016/j.foodchem.2017.08.062.

Bhat, Z. F., Pathak, V., & Fayaz, H. (2013). Effect of refrigerated storage on the quality characteristics of microwave cooked chicken seekh kababs extended with different non-meat proteins. *Journal of Food Science and Technology*, *50*(5), 926–933.

Bhati, D., & Raghuvanshi, R. S. (2020). Nutritional quality evaluation of popular Indian fried snacks based on cowpea. *International Journal of Current Microbiology and Applied Sciences*, *9*(9), 650–655.

Bhosale, S. S., Biswas, A. K., Sahoo, J., Chatli, M. K., Sharma, D. K., & Sikka, S. S. (2011). Quality evaluation of functional chicken nuggets incorporated with ground carrot and mashed sweet potato. *Food Science and Technology International*, *17*(3), 233–239.

Bouchon, P. (2009). Understanding oil absorption during deep-fat frying. *Advances in Food and Nutrition Research*, *57*, 209–234.

Brannan, R. G., & Pettit, K. (2015). Reducing the oil content in coated and deep-fried chicken using whey protein. *Lipid Technology*, *27*(6), 131–133.

Calkins, C. R., & Hodgen, J. M. (2007). A fresh look at meat flavor. *Meat Science*, *77*(1), 63–80.

Cao, H., Wang, Z., Meng, J., Du, M., Pan, Y., Zhao, Y., & Liu, H. (2022). Determination of Arsenic in Chinese mitten crabs (Eriocheir sinensis): Effects of cooking and gastrointestinal digestion on food safety. *Food Chemistry*, *393*, 133345, 1–8.

Chauhan, D. S., Khare, A., Lal, A. B., & Bebartta, R. P. (2022). Utilising oleogel as a frying medium for deep fried Indian traditional product (Mathri) to reduce oil uptake. *Journal of the Indian Chemical Society*, *99*(3), 100378.

Chen, S. D., Chen, H. H., Chao, Y. C., & Lin, R. S. (2009). Effect of batter formula on qualities of deep-fat and microwave fried fish nuggets. *Journal of Food Engineering*, *95*(2), 359–364.

Chiou, A., Kalogeropoulos, N., Salta, F. N., Efstathiou, P., & Andrikopoulos, N. K. (2009). Pan-frying of French fries in three different edible oils enriched with olive leaf extract: Oxidative stability and fate of microconstituents. *LWT-Food Science and Technology, 42*(6), 1090–1097.

Da Silva, P. F., & Moreira, R. G. (2008). Vacuum frying of high-quality fruit and vegetable-based snacks. *LWT-Food Science and Technology, 41*(10), 1758–1767.

Dana, D., & Saguy, I. S. (2006). Mechanism of oil uptake during deep-fat frying and the surfactant effect-theory and myth. *Advances in Colloid and Interface Science, 128*, 267–272.

De Benoist, B., Dary, O., & Hurrell, R. (2006). Iron, Vitamin A and Iodine. In *Guidelines on Food Fortification with Micronutrients*, Allen, L., Ed. (Vol. 126, pp. 41–54). Geneva: World Health Organization.

de Oliveira, V. S., Chávez, D. W. H., Paiva, P. R. F., Gamallo, O. D., Castro, R. N., Sawaya, A. C. H. F., & Saldanha, T. (2022). Parsley (*Petroselinum crispum Mill.*): A source of bioactive compounds as a domestic strategy to minimize cholesterol oxidation during the thermal preparation of omelets. *Food Research International, 156*, 111199.

Debnath, S., Rastogi, N. K., Krishna, A. G., & Lokesh, B. R. (2012). Effect of frying cycles on physical, chemical and heat transfer quality of rice bran oil during deep-fat frying of poori: An Indian traditional fried food. *Food and Bioproducts Processing, 90*(2), 249–256.

Dehghannya, J., & Ngadi, M. (2021). Recent advances in microstructure characterization of fried foods: Different frying techniques and process modeling. *Trends in Food Science and Technology, 116*, 786–801.

deShazo, R. D., Bigler, S., & Skipworth, L. B. (2013). The autopsy of chicken nuggets reads "chicken little". *The American Journal of Medicine, 126*(11), 1018–1019.

Devi, S., Zhang, M., Ju, R., & Bhandari, B. (2021). Recent development of innovative methods for efficient frying technology. *Critical Reviews in Food Science and Nutrition, 61*(22), 3709–3724.

Devi, S., Zhang, M., & Law, C. L. (2018). Effect of ultrasound and microwave assisted vacuum frying on mushroom (*Agaricus bisporus*) chips quality. *Food Bioscience, 25*, 111–117.

Diksha, & Modgil, R.., (2020a). Physico-chemical, functional, nutritional quality and antioxidant level of Indian street food: Samosa. *International Journal of Current Microbiology and Applied Sciences, 9*(5), 3522–3530.

Diksha, & Modgil, R., (2020b). Effect of value addition on the functional properties nutritional and sensory quality of Indian street food (Pakoda). *International Journal of Current Microbiology and Applied Sciences, 9*(9), 3106–3113. http;//doi.org/10.20546/ijcmas.2020.909.383.

Dourado, C., Pinto, C., Barba, F. J., Lorenzo, J. M., Delgadillo, I., & Saraiva, J. A. (2019). Innovative non-thermal technologies affecting potato tuber and fried potato quality. *Trends in Food Science and Technology, 88*, 274–289. https://doi.org/10.1016/j.tifs.2019.03.015.

EFSA. (2015). Scientific opinion on acrylamide in food. EFSA Panel on contaminants in the food chain (CONTAM). *EFSA Journal, 13*(16), 4104.

Eichenlaub, S., & Koh, C. (2015). Modeling of food-frying processes. In *Modeling Food Processing Operations*, Bakalis, S., Knoerzer, K., Fryer, & P. J., Eds. (pp. 163–184). Cambridge: Woodhead Publishing.

Encyclopedia Britannica, Editors of Encyclopaedia. "Frying." *Encyclopedia Britannica*, March 4, 2022. https://www.britannica.com/topic/frying.

Fan, K., Zhang, M., & Bhandari, B. (2019). Osmotic-ultrasound dehydration pretreatment improves moisture adsorption isotherms and water state of microwave-assisted vacuum fried-purple-fleshed sweet potato slices. *Food and Bioproducts Processing, 115*, 154–164.

Fang, M., Huang, G., & Sung, W. (2021). Mass transfer and texture characteristics of fish skin during deep-fat frying, electrostatic frying, air frying and vacuum frying. *LWT-Food Science and Technology, 137*, 110494. https://doi.org/10.1016/j.lwt.2020.110494.

Farkas, B. E., Singh, R. P., & Rumsey, T. R. (1996). Modeling heat and mass transfer in immersion frying. I, model development. *Journal of Food Engineering, 29*(2), 211–226.

Gadiraju, T. V., Patel, Y., Gaziano, J. M., & Djoussé, L. (2015). Fried food consumption and cardiovascular health: A review of current evidence. *Nutrients, 7*(10), 8424–8430.

Garayoa, R., Sanz-Serrano, J., Vettorazzi, A., López de Cerain, A., Azqueta, A., & Vitas, A. I. (2021). Practices of deep-frying processes among food handlers in social food services in Navarra, Spain. *International Journal of Gastronomy and Food Science, 26*, 100432. https://doi.org/10.1016/j.ijgfs.2021.100432.

Gouyo, T., Mestres, C., Maraval, I., Fontez, B., Hofleitner, C., & Bohuon, P. (2020). Assessment of acoustic-mechanical measurements for texture of French fries: Comparison of deep-fat frying and air frying. *Food Research International, 131*, 108947. https://doi.org/10.1016/j.foodres.2019.108947.

Gouyo, T., Rondet, É., Mestres, C., Hofleitner, C., & Bohuon, P. (2021). Microstructure analysis of crust during deep-fat or hot-air frying to understand French fry texture. *Journal of Food Engineering, 298*, 110484.

Guillermo, A., Armando, A., & Pedro, R. (2021). Determination of the diffusion coefficient through oil absorption and moisture loss, such as the porosity of pieces of yam (Dioscorea rotundata) during deep fat frying. *Heliyon, 7*(9), e08036.

Heredia, A., Castelló, M., Argüelles, A., & Andrés, A. (2014). Evolution of mechanical and optical properties of French fries obtained by hot air-frying. *LWT - Food Science and Technology, 57*(2), 755–760. https://doi.org/10.1016/j.lwt.2014.02.038.

Hien, D. T., & Nguyet, H. T. M. (2021). Effects of vacuum frying on the quality of king orange peel in manufacture of chocolate candy fillings. *Food Science and Applied Biotechnology, 4*(2), 156–165.

Hu, J., Zeng, H., Deng, C., Wang, P., Fan, L., Zheng, B., & Zhang, Y. (2019). Optimization of vacuum frying condition for producing silver carp surimi chips. *Food Science and Nutrition*, *7*(8), 2517–2526.

Huang, M. S., Zhang, M., & Bhandari, B. (2018). Synergistic effects of ultrasound and microwave on the pumpkin slices qualities during ultrasound-assisted microwave vacuum frying. *Journal of Food Process Engineering*, *41*(6), e12835.1–21.

Hussain, S., Malik, A., Sharma, P., & Yadav, S. (2020). Study on development of chicken meat kabab by using shatavari roots powder and its aqueous extract. *Haryana Veterinarian*, *59*(1), 63–66.

Ikediala, J. N., Correia, L. R., Fenton, G. A., & Ben-Abdallah, N. (1996). Finite element modeling of heat transfer in meat patties during single-sided pan-frying. *Journal of Food Science*, *61*(4), 796–802.

Islam, M., Zhang, M., & Fan, D. (2019). Ultrasonically enhanced low-temperature microwave-assisted vacuum frying of edamame: Effects on dehydration kinetics and improved quality attributes. *Drying Technology*, *37*(16), 2087–2104.

Jägerstad, M., & Skog, K. (2005). Genotoxicity of heat-processed foods. *Mutation Research/Fundamental and Molecular Mechanisms of Mutagenesis*, *574*(1–2), 156–172.

Jain, S., Singh, V., & Chelawat, S. (2012). Chemical and physicochemical properties, of horse gram (Macrotyloma uniflorum) and its product formulation. *Journal of Dairying Foods and Home Sciences*, *31*(3), 184–190.

Jayasinghe, C. V. L., Silva, S. S. G., & Jayasinghe, J. M. J. K. (2013). Quality improvement of tilapia fish nuggets by addition of legume flour as extenders. *Journal of Food and Agriculture*, *6*(1–2), 32–44.

Ji, K., Liang, H., Ge, X., Ren, M., Pan, L., & Huang, D. (2022). Optimal methionine supplementation improved the growth, hepatic protein synthesis and lipolysis of grass carp fry (*Ctenopharyngodon idella*). *Aquaculture*, *554*, 738125.

Jumras, B., Inprasit, C., & Suwannapum, N. (2020). Effect of microwave-assisted vacuum frying on the quality of banana chips. *Songklanakarin Journal of Science & Technology*, *42*(1), 203–212.

Kadbhane, V. S., Shelke, G. N., & Giram, K. K. (2019). Preparation of Tandulaja powder and its fortified Mathari. *Journal of Pharmacognosy and Phytochemistry*, *8*(3), 4553–4557.

Kang, H., & Chen, H. (2015). Improving the crispness of microwave-reheated fish nuggets by adding chitosan-silica hybrid microcapsules to the batter. *LWT - Food Science and Technology*, *62*(1), 740–745. https://doi.org/10.1016/j.lwt.2014.04.029.

Kita, A. (2014). The effect of frying on fat uptake and texture of fried potato products. *European Journal of Lipid Science and Technology*, *116*(6), 735–740.

Kroh, L. W. (1994). Caramelisation in food and beverages. *Food Chemistry, 51*(4), 373–379.

Li, H., Kong, B., Liu, Q., Chen, Q., Sun, F., Liu, H., & Xia, X. (2022). Ultrasound pretreatment for improving the quality and protein digestibility of stir-frying chicken gizzards. *Food Research International 161*, 111782, 1–8.

Li, P., Wu, G., Yang, D., Zhang, H., Qi, X., Jin, Q., & Wang, X. (2020). Applying sensory and instrumental techniques to evaluate the texture of French fries from fast food restaurant. *Journal of Texture Studies, 51*(3), 521–531.

Liberty, J. T., Dehghannya, J., & Ngadi, M. O. (2019). Effective strategies for reduction of oil content in deep-fat fried foods: A review. *Trends in Food Science and Technology, 92*, 172–183. https://doi.org/10.1016/j.tifs.2019.07.050.

Liu, S. X., Xia, X. F., Kong, B. H., & Fu, Y. (2012). Influence of pre-fried time and temperature on the quality of microwave beef kebabs. *Advanced Materials Research, 554*, 1081–1085.

Maheshwari, H., Prabhavathi, S., Devisetti, R., & Prakash, J. (2017a). Determining efficacy of monosodium glutamate for salt reduction in plain and spiced 'poories' through sensory responses. *Journal of Experimental Food Chemistry, 3*(3), 1–9.

Maheshwari, H. M., Prabhavathi, S. N., & Prakash, J. (2017b). Exploring the flavour potentiating effect of monosodium glutamate on acceptability profile of spiced 'Poories'. *Indian Journal of Nutrition and Dietetics, 54*, 265–277.

Manzoor, S., Masoodi, F. A., Rashid, R., & Dar, M. M. (2022). Effect of apple pomace-based antioxidants on the stability of mustard oil during deep frying of French fries. *LWT-Food Science and Technology, 163*, 113576.

Mehra, A., & Singh, U. (2017). Development, organoleptic and nutritional evaluation of pearl millet based mathri. *International Journal of Recent Scientific Research, 8*(6), 17939–17942.

Moreira, R. G. (2014). Vacuum frying versus conventional frying–An overview. *European Journal of Lipid Science and Technology, 116*(6), 723–734.

Moreira, R. G., & Barrufet, M. A. (1998). A new approach to describe oil absorption in fried foods: A simulation study. *Journal of Food Engineering, 35*(1), 1–22.

Negi, P. S., & Roy, S. K. (2003). Changes in β-carotene and ascorbic acid content of fresh amaranth and fenugreek leaves during storage by low cost technique. *Plant Foods for Human Nutrition, 58*(3), 225–230.

Ngadi, M., Adedeji, A. A., & Kassama, L. (2008). Microstructural changes during frying of foods. In *Advances in Deep-Fat Frying of Foods*, Sahin, S., & Sumnu, S. G., Eds. (pp. 169–200). Boca Raton, FL: CRC Press.

Nguyen, K. H., Nielsen, R. H., Mohammadifar, M. A., & Granby, K. (2022). Formation and mitigation of acrylamide in oven baked vegetable fries. *Food Chemistry, 386*, 132764.

Nurlela, S., Hastuti, H., & Suparman, S. (2018). The quality of nugget of broiler chicken meat with addition of sago flour (Metroxylon Sp.). *Chalaza Journal of Animal Husbandry, 3*, 67–72.

Oke, E. K., Idowu, M. A., Sobukola, O. P., Adeyeye, S. A. O., & Akinsola, A. O. (2018). Frying of food: A critical review. *Journal of Culinary Science and Technology, 16*(2), 107–127.

Oluwaniyi, O. O., Dosumu, O. O., & Awolola, G. V. (2010). Effect of local processing methods (boiling, frying and roasting) on the amino acid composition of four marine fishes commonly consumed in Nigeria. *Food Chemistry, 123*(4), 1000–1006.

Oppong, D., Panpipat, W., Cheong, L. Z., & Chaijan, M. (2022). Rice flour-emulgel as a bifunctional ingredient, stabiliser-cryoprotactant, for formulation of healthier frozen fish nugget. *LWT-Food Science and Technology, 159*, 113241.

Pacheco, Y. E. G., Ramírez, J. R., Díaz, L. N., & Verbel-Vergara, J. (2020). Elaboración de un snack funcional tipo chips de ñame (*Dioscorea alata*) y batata (*Ipomonea batata*) fortificados con vitamina C. *Revista Gipama, 2*(1), 29–37.

Pedreschi, F. (2009). Fried and dehydrated potato products. In *Advances in Potato Chemistry and Technology*, Singh, J. & Kaur, L., Eds. (pp. 319–337). London: Academic Press.

Qin, R., Wu, R., Shi, H., Jia, C., Rong, J., & Liu, R. (2022). Formation of AGEs in fish cakes during air frying and other traditional heating methods. *Food Chemistry, 391*(15), 133213, 1–9.

Rahimi, D., Kashaninejad, M., Ziaiifar, A. M., & Mahoonak, A. S. (2018). Effect of infrared final cooking on some physico-chemical and engineering properties of partially fried chicken nugget. *Innovative Food Science and Emerging Technologies, 47*, 1–8. https://doi.org/10.1016/j.ifset.2018.01.004.

Rahimi, J. (2015). *Microstructure and Surface Characterization of Fried Batter Coatings*. Montreal: McGill University.

Raj, T., Kar, J. R., & Singhal, R. S. (2017). Development of par-fried frozen samosas and evaluation of its post-storage finish frying and sensory quality. *Journal of Food Processing and Preservation, 41*(4), e13049.

Ran, X. L., Zhang, M., Wang, Y., & Bhandari, B. (2019). Dielectric properties of carrots affected by ultrasound treatment in water and oil medium simulated systems. *Ultrasonics Sonochemistry, 56*, 150–159.

Rani, M., & Chauhan, G. S. (1995). Effect of intermittent frying and frying medium on the quality of potato chips. *Food Chemistry, 54*(4), 365–368.

Ren, A., Cao, Z., Tang, X., Duan, Z., Duan, X., & Meng, X. (2022). Reduction of oil uptake in vacuum fried *Pleurotus eryngii* chips via ultrasound assisted pretreatment. *Frontiers in Nutrition, 9*. https://doi.org/10.3389/fnut.2022.1037652.

Sabikun, N., Bakhsh, A., Ismail, I., Hwang, Y. H., Rahman, M. S., & Joo, S. T. (2019). Changes in physicochemical characteristics and oxidative stability of pre-and post-rigor frozen chicken muscles during cold storage. *Journal of Food Science and Technology, 56*(11), 4809–4816.

Sabikun, N., Bakhsh, A., Rahman, M. S., Hwang, Y. H., & Joo, S. T. (2021). Evaluation of chicken nugget properties using spent hen meat added with milk fat and potato mash at different levels. *Journal of Food Science and Technology, 58*(7), 2783–2791.

Sadeghi, R., Lin, Y., Price, W. J., Thornton, M. K., & Lin, A. H. M. (2021). Instrumental indicators of desirable texture attributes of French fries. *LWT-Food Science and Technology, 142*, 110968.

Sahin, S. (2000). Effects of frying parameters on the colour development of fried potatoes. *European Food Research and Technology, 211*(3), 165–168.

Sakhale, B. K., Badgujar, J. B., Pawar, V. D., & Sananse, S. L. (2011). Effect of hydrocolloids incorporation in casing of samosa on reduction of oil uptake. *Journal of Food Science and Technology, 48*(6), 769–772.

Salvador, A., Varela, P., Sanz, T., & Fiszman, S. M. (2009). Understanding potato chips crispy texture by simultaneous fracture and acoustic measurements, and sensory analysis. *LWT-Food Science and Technology, 42*(3), 763–767.

Sayon-Orea, C., Bes-Rastrollo, M., Basterra-Gortari, F. J., Beunza, J. J., Guallar-Castillon, P., De la Fuente-Arrillaga, C., & Martinez-Gonzalez, M. A. (2013). Consumption of fried foods and weight gain in a Mediterranean cohort: The SUN project. *Nutrition, Metabolism and Cardiovascular Diseases, 23*(2), 144–150.

Shazia, H., Swati, V., & Vibha, B. (2011). Efficacy of iron rich Mathri on hematological parameter of hostel girls in Udaipur city. *Food Science Research Journal, 2*(1), 69–72.

Shi, H., Zhang, M., & Yang, C. (2019). Effect of low-temperature vacuum frying assisted by microwave on the property of fish fillets (*Aristichthys nobilis*). *Journal of Food Process Engineering, 42*(4), e13050.

Shyu, Y. S., Hwang, J. Y., Shen, S. T., & Sung, W. C. (2021). The effect of different frying methods and the addition of potassium aluminum sulfate on sensory properties, acrylamide, and oil content of fried bread (Youtiao). *Applied Sciences, 11*(2), 549.

Singh, R. P., & Mermelstein, N. (1995). Heat and mass transfer in foods during deep-fat frying: Engineering aspects of deep-fat frying of foods. *Food Technology (Chicago), 49*(4), 134–137.

Stadler, R. H. (2012). Heat-generated toxicants in foods: Acrylamide, MCPD esters and furan. In *Chemical Contaminants and Residues in Food* Schrenk, D., & Cartus, A., Eds. (pp. 201–232). UK: Woodhead Publishing.

Su, Y., Gao, J., Tang, S., Feng, L., Azam, S. R., & Zheng, T. (2022). Recent advances in physical fields-based frying techniques for enhanced efficiency and quality attributes. *Critical Reviews in Food Science and Nutrition, 62*(19), 5183–5202.

Su, Y., Zhang, M., Zhang, W., Liu, C., & Adhikari, B. (2018). Ultrasonic microwave-assisted vacuum frying technique as a novel frying method for potato chips at low frying temperature. *Food and Bioproducts Processing, 108*, 95–104.

Teruel, M. R., García-Segovia, P., Martínez-Monzó, J., Linares, M. B., & Garrido, M. D. (2014). Use of vacuum-frying in chicken nugget processing. *Innovative Food Science and Emerging Technologies, 26*, 482–489. https://doi.org/10.1016/j.ifset.2014.06.005.

Van Dalen, G., Nootenboom, P., Van Vliet, L. J., Voortman, L., & Esveld, E. (2007). 3-D imaging, analysis and modelling of porous cereal products using X-ray microtomography. *Image Analysis and Stereology, 26*(3), 169–177.

Van Koerten, K. N., Schutyser, M. A. I., Somsen, D., & Boom, R. M. (2015). Crust morphology and crispness development during deep-fat frying of potato. *Food Research International, 78*, 336–342.

Varela, G. 1988. Current facts about the frying of food. In *Frying of Food: Principles, Changes, New Approaches*, Varela, G., Bender, A. E., & Morton, I. D. (pp. 9–25). Chichester: Ellis Horwood.

Verma, S., & Jain, S. (2012). Fortification of mathri with fresh and dehydrated vegetables and assessment of nutritional quality. *Rajasthan Journal of Extension Education, 20*, 155–158.

Waxman, A., Stark, J., Thornton, M. K., Olsen, N., Guenthner, J., & Novy, R. G. (2019). The effect of harvest timing on french fry textural quality of three processing potato varieties: Russet burbank, alpine russet, and clearwater russet. *American Journal of Potato Research, 96*(1), 33–47.

Wu, R., Jiang, Y., Qin, R., Shi, H., Jia, C., Rong, J., & Liu, R. (2022). Study of the formation of food hazard factors in fried fish nuggets. *Food Chemistry, 373*, 131562.

Yagua, C. V., & Moreira, R. G. (2011). Physical and thermal properties of potato chips during vacuum frying. *Journal of Food Engineering, 104*(2), 272–283.

Yogesh, K., Ahmad, T., Manpreet, G., Mangesh, K., & Das, P. (2013). Characteristics of chicken nuggets as affected by added fat and variable salt contents. *Journal of Food Science and Technology, 50*(1), 191–196.

Zhang, J., Wu, D., Liu, D., Fang, Z., Chen, J., Hu, Y., & Ye, X. (2013). Effect of cooking styles on the lipid oxidation and fatty acid composition of grass carp (*Ctenopharynyodon idellus*) fillet. *Journal of Food Biochemistry, 37*(2), 212–219.

Zhang, N., Sun, B., Mao, X., Chen, H., & Zhang, Y. (2019). Flavor formation in frying process of green onion (*Allium fistulosum* L.) deep-fried oil. *Food Research International, 121*, 296–306.

Zhu, Z., Fang, R., Yang, J., Khan, I. A., Huang, J., & Huang, M. (2021). Air frying combined with grape seed extract inhibits Nε-carboxymethyllysine and Nε-carboxyethyllysine by controlling oxidation and glycosylation. *Poultry Science, 100*(2), 1308–1318.

Ziaiifar, A. M., Courtois, F., & Trystram, G. (2010). Porosity development and its effect on oil uptake during frying process. *Journal of Food Process Engineering, 33*(2), 191–212.

Chapter 3

Fundamentals, Methodologies, and Developments in Food Frying

Anjali Thakur and Chandrakant Genu Dalbhagat
Indian Institute of Technology Kharagpur

DOI: 10.1201/9781003329244-3

3.1 Introduction

Frying is an ancient method of cooking wherein lipids are involved as a direct medium of heating at a temperature range from 160°C to 180°C or more according to the final product requirement (Banerjee et al., 2017). Another definition portrays frying as an expeditious result of both heat and mass transfers, commonly used in industrial production as well as household cooking (Andrés-Bello et al., 2011). Examples of such frying where the products come in direct contact with the heating medium are pan frying, stir frying, and deep fat frying, among which deep fat frying is the most popular.

Major phenomena occurring during the process of frying are as follows:

Cooking: It is the main phenomenon during frying that involves starch gelatinization, protein denaturation, Maillard reaction, caramelization, and other associated reactions responsible for change in organoleptic properties of food.

Dehydration: It is the process of abrupt removal of water by the high temperature (above 100°C) involved in frying operation.

Oil uptake: It is both a desirable or undesirable phenomenon due to absorbed frying media (fat or oil) in the product while frying.

Crust formation: This phenomenon is involved with the change in texture and structure of the fried products.

"The unit operation, frying is applied to modify and enhance the eating quality (texture and flavour) of a food. Moreover, it causes thermal degradation of various micro-organisms and enzymes, and reduces the food's water activity" (Fellows, 2009). The main theory behind the frying process involves the rapid increase in food surface temperature and vaporization of moisture into steam. As the temperature increases, moisture is removed on the surface of the food due to evaporation and formation of crust begins. The heat transfer rate is governed according to the product's coefficient of heat transfer and by difference between the temperatures of food material and hot oil. The thermal conductivity of a specific food has an impact on the rate of heat penetration. As there are different-sized capillaries present in the porous structure of the crust, the water and water vapor are removed while frying through the capillaries with larger size first. The surface moisture enters into the food across the lipids'/oils' boundary film, and the film thickness governs the rate of heat and mass transfer. This phenomenon can be compared with the process of hot-air dehydration as in both cases, the driving force for moisture loss is water vapor pressure gradient.

The time for the completion of frying of a particular food depends on:

- Food type
- Oil temperature
- Frying method
- Food thickness
- Desirable eating quality

Any food that retains a moist interior should be fried until the destruction of pathogenic microorganisms present in the thermal center and achievement of the desirable organoleptic properties. Temperature of frying is mostly set based on the final desirable quality and economic considerations. The time of processing can be brought down and the rate of food production can be raised at higher range of temperatures (180°C–200°C). However, temperatures at higher side bring about free fatty acids formation and degrade the oil quality in terms of color, viscosity, and flavor which consequently develop foaming. The temperature of frying is also decided by the final food quality concerns. High-temperature frying is used wherein a crust with a moist interior is desired for the product. On the other side, foods are fried at a lower temperature if the final requirement is to achieve a dry product. The evaporation front moves to higher depth of the food before crust formation at a lower temperature, hence forming dried foods.

3.2 Dietary Habits and Lifestyle

The amount of oil uptake may fluctuate between about 8% and 40% during frying operation depending on the dehydration level, which leads to several health effects such as obesity and related diseases (Chen et al., 2019). The fried foods mostly preferred by the people are potato chips, chicken nuggets, French fries, chicken fries, and fish fries (Devi et al., 2021). Frying, being a simple and quick cooking technique, occupies a large place in snack product industries.

3.3 Background and Trend of Frying

Currently advanced frying practices have been encouraged and subsidized in order to improve the quality of fried foods. The major issue comes to the mind during frying is the quantity of oil and fat absorbed in the process. Due to the increasing trend in healthy diet and health consciousness of people, the most difficult challenge for the fried product (snack) industry is to develop fried foods having the least amount of oil while still providing the desired organoleptic profile to the consumers. "The necessary efforts to decrease the fat content are coating the surface of the product using some cellulose and gums, modification of the frying medium and alteration of frying methods like pre- and post-frying treatments and frying conditions" (Devi et al., 2021).

3.4 Common Frying Methods

There are several types of food frying such as pan frying, shallow frying, stir frying, and deep frying. The two major methods of frying used commercially and differentiated by the heat transfer methods are shallow and deep fat frying (Fellows, 2009).

3.4.1 Shallow (or Contact) Frying

The shallow frying process is mostly used for the foods having a huge surface area to the volume ratio such as burgers, eggs, bacon slices, and patties. Here, the conduction heat transfer occurs to the food wherein heat transfer occurs directly from the hot pan surface to the thin oil layer and finally into the food material. The thickness of oil layer may vary due to the irregular shapes of the food. Food is turned often on the pan surface after bubbles formed which causes temperature fluctuations on food surface throughout the frying process, resulting in the distinctive asymmetrical browning in the foods fried by shallow frying.

3.4.2 Deep Fat Frying

In this case, the heat is transferred due to the combined conduction and convection process inside the oil and food product, respectively. Unlike shallow frying, uniform color and appearance of food is obtained as all food surfaces receive homogeneous heat treatment. This method is employed in the food industries for all shapes of foods except the foods having greater surface to mass ratio as it is evidenced that such foods tend to absorb higher amount of fat after being removed from the fryer (Selman, 1989). "Initially, before the onset of moisture evaporation from the surface the heat transfer coefficients are in the range of 250–300 W/m^{2}K which further increase to a range of 800–1,000 W/m^2 K because of escaping turbulent steam from the food" (Farinu, 2006).

3.4.2.1 Advantages of Deep Fat Frying

- Cooking speed;
- Economic and energy efficient;
- Maintains consistency of the cooked products;
- Enhance color, texture, and flavors;
- Retains micronutrients.

3.5 Fundamentals of Frying Process

High-temperature frying promotes the reactions between carbohydrates and proteins, enhancing crust dehydration and oil uptake. A thorough comprehension of the frying operation can aid in the process of optimization of cooking in terms of food quality, fat use life, and energy consumption (Gertz, 2014). The actual frying technology is said to have emerged near the Mediterranean region as a consequence of influence of olive oil (Moreira et al., 1999). Frying has gained popularity owing to its series of reactions forming better texture, flavor, and crust formation of the food. "To ensure the frying quality such as color, flavor, and crispy texture of the end product, it becomes necessary to monitor the critical points of process and conditions of frying" (Stevenson, 1984). The complexity of the process can be understood by the following phenomena.

3.6 Heat and Mass Transfer

Frying operation involves vaporization of water and formation of voids inside the crust surface determining the oil absorption volume. Deep fat frying involves both convective and conductive modes of heat transfer along with mass transfer due to diffusion. The responsible major frying variables are frying time and oil temperature, which control heat and mass transfer remarkably. Consequently, this process augments several reactions causing transformation in characteristics of food (Dash et al., 2022). "The more the difference in temperature between the food and heating oil, the more is the transfer of heat inside the food materials. Simultaneously, mass is also being transferred to the frying medium from the food and vice versa" (Gertz, 2014).

Both convection and conduction modes of heat transfer occur during deep fat or immersion frying (Sinha and Bhargav, 2015). The properties such as thermal diffusivity, specific heat, product and oil density, and thermal conductivity greatly influence the magnitude of heat transfer. In general, conduction heat transfer occurs deep inside the food, while convective heat transfer takes place between the frying oil and food (Asokapandian et al., 2020). Quick evaporation of moisture occurs in the starting stage of frying. The water escaped from the food arrives to the surface. Eventually the water at the food surface comes in contact with the frying medium, which results in bubbles formation that move vigorously, causing turbulence throughout the oil. Thus, the increased turbulence due to the surface bubbling water influences the heat transfer coefficient. Due to reduction of moisture in the product, the quantity of bubbles decreases with prolonged frying time.

Water evaporates quickly at first during frying. Because of the excessive vapor build-up caused by thick crust layer formation, pressure builds up inside the product, which starts the crust's fissure formation process. These fissures or cracks formed during deep frying act as the pathway for the oil to pass. The mechanism of heat and mass transfer during deep fat frying process can be observed in Figure 3.1.

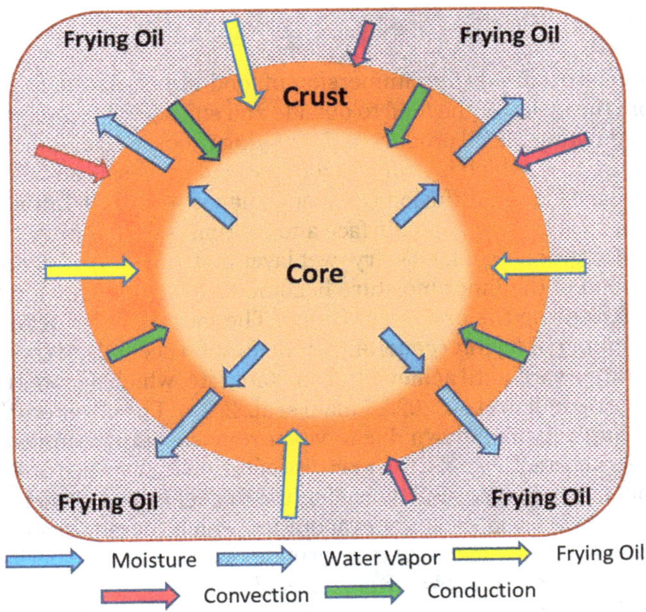

Figure 3.1 Heat and mass transfer mechanism during deep fat frying. (Adapted from Asokapandian et al., 2020.)

3.6.1 Heat Transfer in Deep Fat Frying Can Be Divided into Four Phases

Initial heating phase: During the first few seconds of frying, the rise of temperature to the boiling point of water occurs. It refers to the stage just before the commencement of vaporization of the product moisture. This phase is considered to be a very short phase, which causes negligible loss of water from the food product.

Surface boiling phase: The moisture at the surface starts to evaporate during this phase, forming a crust on the product surface. The surface heat transfer coefficient also appears to enhance in this phase.

Falling rate phase: This stage can be identified by bulk moisture evaporation from the core to outer region of food. Most of the physico-thermal changes, viz. crust thickening, denaturation of protein, and gelatinization of starch, occur during this phase. This is the longest phase during which the core of the product reaches the boiling temperature of water.

Bubble end point: This is the last and final stage, where the removal of moisture and release of water vapor bubbles from the food surface become less (Asokapandian et al., 2020).

3.6.2 Mass Transfer

Mass transfer occurs during immersion of food in hot oil due to transfer of water from the inside of the food to outside and surface oil replacement. This helps in regulating the oil product during frying process (Ziaiifar et al., 2010). The moisture loss, which is a diffusion-controlled process, can be elucidated by Fick's second law of diffusion (Asokapandian et al., 2020). Initial moisture transfer starts at the product's surface and continues at the dry-wet interface. The diffusion gradient of food's dry–wet layer and the pressure gradient due to evaporation of the inner moisture become the driving force for the moisture transfer during the process of frying. The moisture loss occurs rapidly at the beginning of frying (generally in the first 60 seconds of frying), but it decreases afterward until achieving a constant rate, which denotes the end of frying (Asokapandian et al., 2020; Lalam et al., 2013). The oil enters the empty spaces during cell breakdown due to water removal without infiltrating into the microstructure. However, the majority of the oil is absorbed after the food is taken out from the fryer. Furthermore, the thicker crust forms during deep frying because of rapid moisture evaporation which thereby restricts the oil migration inside while increasing absorption of oil during the period of cooling (Lumanlan et al., 2020; Van Koerten et al., 2017).

The following are three categorized periods of moisture loss during deep frying:

- Heating the product till it achieves the boiling point of water and evaporating the surface moisture content.
- Free as well as capillary water evaporation at a temperature near the boiling point of water.
- Finally, the temperature of product is increased to reach the oil temperature, while reducing the drying rates.

3.7 Factors Affecting the Oil Uptake during Frying of Foods

Factors influencing the oil uptake contribute to the fat content of the food and aids in low-fat food processing.

3.7.1 Product Parameters

3.7.1.1 Size, Shape, and Surface of Product

The quantity of oil uptake by a food while frying is heavily influenced by its shape and size. Majority of the frying oil remains on the surface due to oil uptake, which is a surface phenomenon, and it is inversely proportional to product thickness (Asokapandian et al., 2020). Guillaumin (1983) revealed a surge in uptake of oil when product thickness was decreased. For example, potato chips absorb more oil than French fries owing to its larger surface to volume ratio, which signifies that oil content and product surface area have a linear relationship. Oil penetration is generally limited to a depth of 1 mm (Saguy et al., 1997). The structural attributes also have an influence on uptake as the frying medium infiltrates into the food through the capillaries of the crust. Broken cells during cutting can promote greater oil absorption owing to its roughness and surface exposure to oil. Thus, use of sharp cutting blades can lower roughness of surface and subsequent uptake of oil.

3.7.1.2 Product Composition

The initial compositions like initial solid as well as moisture content influence the oil uptake during frying.

> The food product such as French fries and plantain cylinders exhibit intermediary moisture content, specifying that products having heavy initial water content lead to increased oil absorption in fried goods, which may be related to the water loss and oil uptake correlation.
>
> *(Yamsaengsung and Moreira, 2002)*

Yamsaengsung and Moreira (2002) also summarized that the distribution of pores during the frying process was the main reason of oil penetration while cooling. Thus, the quantity of oil absorption is greatly influenced by the structure and particle size of the fried products. Products having maximum dry matter yield a low-fat fried product (Asokapandian et al., 2020). It is also evidenced that frying tubers with more dry content produces a grainy texture while boiling tears it apart (Lulai and Orr, 1979; Lisinska and Leszczynski, 1989).

> Any leavening chemicals in the product affect how much oil is absorbed. This was noticed while frying squid with batter. The greater oil uptake may be attributed to the gas generation inside the product, which is retained by the oil during frying.
>
> *(Llorca et al., 2003)*

3.7.1.3 Porosity

The continual evaporation of water while frying causes structural changes on surface due to dehydration and the formation of porous structures. Conversion of moisture into steam occurs during deep frying, and the stream releases through microstructures, causing damages to the cells, expansion of the pores, and formation of capillary tunnels. Rapid water vapor release creates a barrier that may prevent oil from transferring into the amorphous crust and restrict the amount of fat that can be absorbed during frying. Nonetheless, after removing the food from the high-temperature frying medium, water vapor condensation occurs inside the crust which is porous due to which a 'vacuum effect' is formed absorbing additional oil (Lumanlan et al., 2020).

3.8 Frying Equipment

Batch fryers and continuous fryers are two categories of frying equipment. The smaller fryers mostly used in catering service are called batch fryers. Continuous fryers, which can process large quantities of frying oil and food, are typically used in industrial settings where there is a need for mass production. Fryers can be used in atmospheric pressure, low pressure, high pressure, and even in vacuum conditions (Oke et al., 2018; Mallikarjunan et al., 1997). However, the majority of industrial production takes place in an atmosphere. The equipment for shallow frying generally comprises a high-temperature metal surface, covered in a thin oil layer. The continuous type of deep fat fryers is most important for industrial uses. Batch operation is characterized by suspending a food to a hot oil bath and kept until the achievement of desirable degree of frying, generally identified by the changes in color of the food surface. Continuous deep fat frying, unlike static food frying in the case of batch-type fryers, consists of a movable mesh conveyor made up of stainless steel submerged inside an oil tank which is thermostatically controlled (Figure 3.2). In this case, the frying time is influenced by three major factors: frying oil temperature, speed of the conveyor, and the size of pieces. An inclined conveyor helps in removing the food and simultaneously drains the excess oil back into the tank. The operation is automatic, and the production rates are up to 15 tons per hour.

3.9 Pre-Frying Treatments

One of the most unavoidable quality factors is oil uptake; however, this is not up to the compatibility with the current trends of switching toward healthier choices and low-fat products. The following are some of the most crucial factors influencing oil uptake in fried foods: oil quality and composition, frying time and temperature, product shape, moisture content, and post- and pre-frying treatments. Before frying, some pre-treatments are necessary, including blanching, osmotic pre-treatment, freezing, and pre-drying. These pre-treatments have an impact on the final product's yield, fat content, distribution, and kinetics of moisture removal from the product (Zhang et al., 2020; Maity et al., 2018).

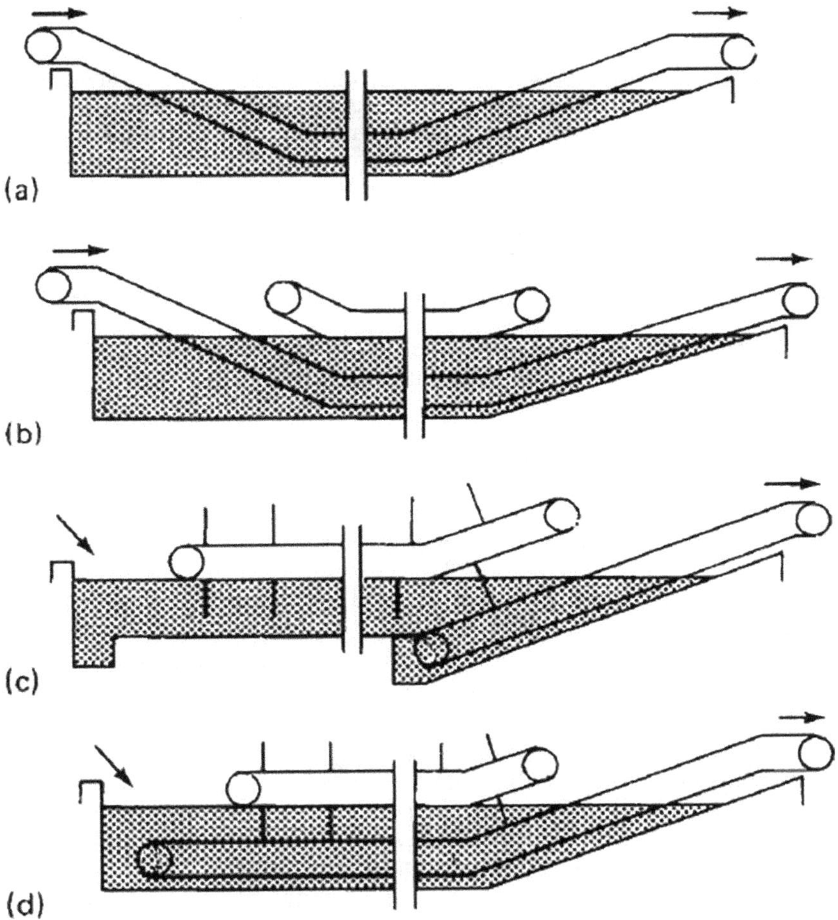

Figure 3.2 Arrangements of various conveyors: (a) delicate non-buoyant products (e.g., fish sticks), (b) breadcrumb-coated products, (c) dry buoyant bulk products (e.g., half-product snacks), and (d) dual purpose (e.g., nuts and snacks). (Courtesy of Coat and Fry Ltd., Fellows, 2009.)

3.9.1 Pre-Drying

The method for creating a dried and firm surface matrix surrounding the product is to dry it before frying it. This method reduces the product's overall water content and restricts oil absorption. The drying step can be done after blanching step which helps in increasing the crispiness of the product and reduction of the amount of oil absorption (Andrés-Bello et al., 2011).

> Pre-drying is a commonly used food pre-treatment in industries that increases evaporation of moisture and decreases the length of frying time. Hot air drying, vacuum oven drying, microwave heating, and baking are all common pre-drying processes in the food industry. Pre-drying is generally applied in the food frying industry to produce appealing colours and improved textures. This pre-treatment also aids in limiting the oil penetration into the food by reducing the open pores at the surface of the fried products. The temperature range during the pre-drying process commonly practised for plant foods is between 60°C and 70°C.
>
> *(Al Faruq et al., 2022)*

Pre-drying has long been used in order to reduce the oil quantity, and it obviously results in final fried product with lower moisture content. This creates lower interior vacuum pressure in the fried products after being removed from the frying equipment, consequently less oil penetration into the food (Zhang et al., 2020).

3.9.2 Superheated Steam Drying (SSD)

Superheated steam drying is another innovative pre-treatment procedure that differs from the norm (SSD). The idea behind SSD is that when it comes into contact with an object that has a lower boiling point than water, it rapidly condenses into water at atmospheric pressure. Therefore, in SSD, both drying and wetting occur simultaneously before the surface temperature of product reaches the boiling point of water. Zielinska et al. (2015) examined the benefits of employing SSD as a pre-treatment before deep fat frying to produce low-fat fried potato chips with enticing color and texture. The study also revealed that the oil uptake of the potato chips with SSD pre-treatment was 15%–17% lower than that of the chips without SSD pre-treatment. The researchers have also found that pre-treatment in the presence or absence of SSD gave rise to potato chips with comparable texture (hardness and crispiness) and color (lightness). Furthermore, some other researchers have demonstrated that SSD pre-treatment decreased both moisture and oil content while increasing the crispiness of fried snacks in the same frying process as untreated samples (Al Faruq et al., 2022).

3.9.3 Ultrasound-Assisted Osmotic Dehydration (UOD)

Prior to drying or frying, osmotic dehydration (OD) is used as a pre-treatment to reduce the water content in plant meals. The samples are immersed in a concentrated solution during this procedure, and water diffuses through a single semi-permeable membrane until the desired level is reached (Al Faruq et al., 2022). Two major countercurrent flows occur in this process: water movement from the inner part of product into the hypertonic solution, and solute moves to the product from the hypertonic solution (Muñiz-Becerá et al., 2017). Though OD has a poor mass transfer rate on its own, when combined with other non-thermal techniques like vacuum, ultrasound, and microwave, the mass transfer rate for drying plant foods increases. Micromechanical shocks come from cavities that form when ultrasound is applied to a liquid medium because they produce rarefaction and breakdown when compression occurs (Al Faruq et al., 2019). These shocks increase mass transfer and alter cellular structure, speeding up processes while using less energy and enhancing product quality. UOD has recently been successfully used as a pre-treatment for the frying process, having a considerable impact on the mass transfer rate (Al Faruq et al., 2022).

3.9.4 Pulsed Electric Field (PEF)

An electrical field chamber with a high voltage (20–400 kW) is used in the pulsed electrical field (PEF), a non-thermal pre-treatment, to improve permeability or break down biological cell membranes. The products are held at room temperature in a compartment between two electrodes. This mechanism results in electrical charging of the cell membrane when the opposing charges are mobilized due to the presence of electromagnetic field (Al Faruq et al., 2022). The trans-membrane electric charge produced by membrane polarization damages cell tissues, increasing heat and mass transfer throughout the frying process. Liu et al. (2018) observed that when PEF treatment and the traditional method of air drying were combined, the effects on potato frying were enhanced effective moisture diffusivity, reduced frying time, increased surface porosity, and softened cell walls.

3.9.5 Coating: Hydrocolloid Coating/ Coating/Edible Films and Types

Coating is a different pre-frying method that involves momentarily submerging a product in a coating suspension before frying it. This surface treatment creates a barrier against oil absorption while reducing surface porosity. Coating application is a potential approach because the surface characteristics of the food are most important for fat absorption (Andrés-Bello et al., 2011). Although the exact mechanism of action is ambiguous, its functionality is occasionally linked to a certain attribute. Low moisture permeability, low moisture content, and cross-linking or thermo-gelling are frequently stated coating characteristics in relation to fat uptake. Every feature is intended to reduce moisture loss and/or change the surface structure developed during frying. The most important attributes of good hydrocolloids are their capacity to form films, heat stability, fat and water transport properties, organoleptic properties, and nutritional values. Many food manufacturers use hydrocolloids as functional ingredients to give them a competitive edge. Furthermore, it enhances the distinctive benefits of flavor, texture, and moisture management while also improving product quality and stability. Due to their useful qualities, hydrocolloid coating has lately been used as a pre-treatment procedure before frying. For edible films and coatings, hydrocolloids are widely employed. These include gums like guar and xanthan, cellulose derivatives like methyl cellulose, hydroxypropyl cellulose, and carboxymethyl cellulose, and various polysaccharides and proteins (Al Faruq et al., 2022; Varela and Fiszman, 2011). Additionally, some of the most popular and efficient hydrocolloids are cellulose derivatives such as hydroxypropyl cellulose (HPC), methyl cellulose (MC), and hydroxypropyl methyl cellulose (HPMC) (Andrés-Bello et al., 2011; Albert et al., 2002).

3.9.6 Blanching in Water and Solutions

Food is "blanched" by submerging them in hot water or exposing them to steam (Arefi et al., 2022; Zhang et al., 2020). The decrease of acrylamide has been associated with the leaching of reducing sugars and/or asparagine during the hot-water blanching of potato slices (Maan et al., 2022; Schouten et al., 2020). Liyanage et al. (2021) blanched potato slices in distilled water mixed with ascorbic acid, citric acid, calcium chloride, and other chemical solutions at 65°C for 5 minutes, which resulted in the largest reduction in acrylamide concentration (up to 59%). By blanching in hot water for 4 minutes at 100°C, inhibitory effects on the production of acrylamide were repeated; in the case of French fries, this led to a 20% decrease (Zhang et al., 2021). When potato slices were soaked in water solutions containing 2% calcium chloride or 1% citric acid, acrylamide production was considerably inhibited because reducing sugars and asparagine decreased during soaking (Elbassiony, 2020). In terms of sensory aspects, blanching in water enhances the color and texture of meals before frying (Damto and Chala, 2019; Zhang et al., 2018; Zhang et al., 2020). According to panel studies on the crispness, color, and fragrance of potato chips, the best water-blanching temperature and timeframe are 85°C and 4 minutes (Asefa et al., 2016).

3.9.7 Radiofrequency Pre-Treatment

Radiofrequency (RF) in the range of 3kHz to 300MHz warms dielectric materials, forcing molecules to reverse directions and boosting the mobilization of individual molecules or atoms. This results in continual collisions and frictions between dipoles (Su et al., 2022a; Boreddy et al., 2019).

3.10 Post-Frying Treatments

When the food is taken out of the fryer, the cooling phase starts in which a complex process called oil absorption tends to occur. According to Andrés-Bello et al. (2011) and Garayo and Moreira (2002), the faster the rate of water loss, the higher the oil adhesion at the surface and the higher the rate of oil absorption. Moreover, from the moment fried foods are taken out of the fryer until the consumption, their quality is significantly altered over time. Therefore, post-frying procedures are crucial for satisfying consumer demands for keeping the best product quality as much as possible and affect the final oil content in the fried goods (Dehghannya and Michael, 2021). There are very few studies available regarding post-frying treatments in literature. The final oil content of the fried product can be decreased by holding the fried goods at high temperatures and wiping off the surface oil with a highly absorbent paper (Devi et al., 2021; Debnath et al. 2012). In the case of vacuum fryer, de-oiling is done by centrifugation, and it is evidenced that it can decrease the oil content in potato chips from 0.43 to 0.097 g/g (Moreira et al., 2009). Factors that have the greatest influence on acrylamide formation were discovered to be frying temperature and time, which is also affected by factors such as oil type, soaking, and type of commodities (Arefi et al., 2022; Maan et al., 2022). Although the post-frying treatments and oil absorption phenomena have been discussed in some

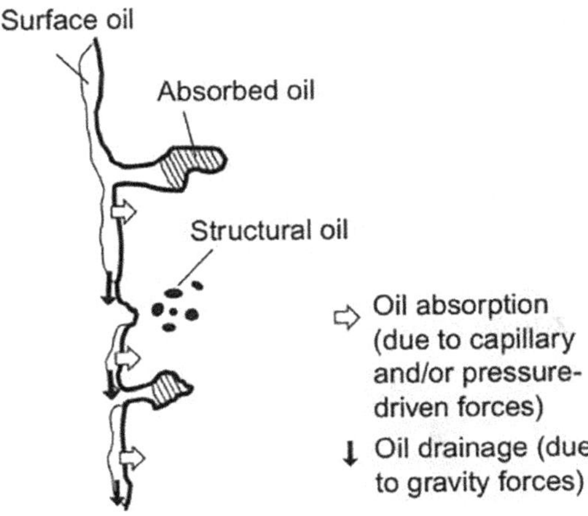

Figure 3.3 Three categories of oil in the product microstructure after the frying process, mechanism of drainage, and absorption post frying. (Adapted from Eichenlaub and Koh, 2015; Ouchon et al., 2003.)

literature, there is a research gap in the domain of gravity drainage modeling and simultaneous absorption. The procedure shows how the thickness of the oil coating gradually decreases over time as oil is either absorbed or lost (Eichenlaub and Koh, 2015) (see Figure 3.3).

3.11 Chemical Reactions of Oil during Frying

3.11.1 Hydrolysis of Oil

Steam forms when moisture condenses in heated oil while food is fried. As the frying process continues, the steam progressively dissipates as it evaporates as bubbles. Steam, oxygen, and water together begin the chemical reactions in the frying oil and food. Water, a weak nucleophile, attacks the triacylglycerol ester bonds, releasing glycerol, free fatty acids, and di- and monoacylglycerols. The amount of time a meal is fried directly relates to its free fatty acid concentration (Oke et al., 2018; Chung et al., 2004). The value of the free fatty acids is used to evaluate the frying oil's quality. Hydrolysis is preferred in this type of oil because short and unsaturated fatty acids are more water soluble than long and saturated fatty acids.

3.11.2 Oxidation of Oil

It is evidenced that oil interacts with oxygen during the process of frying (Oke et al., 2018). Thermal oxidation uses the same chemical mechanism as autoxidation. Although there are a dearth of specific information and thorough comparisons between them, the rate of thermal oxidation is significantly higher than that of autoxidation.

3.11.3 Polymerization of Oil

Volatile chemicals, which are present in oil at parts per million levels, are essential to the flavor attributes of frying oil and fried food products. Triacylglycerol dimers, polymers, and nonvolatile polar molecules are the major byproducts of frying oil's breakdown. Cyclic compounds are relatively rare compared to nonvolatile polar compounds, polymers, and dimers. Depending on the kind of reaction and the types of fatty acids present in the oil, dimers and polymers can be either acyclic or cyclic. Deep fat frying causes radical processes like dimerization and polymerization.

3.12 Effect of Frying on Quality Attributes

In recent years, the consumers' preference toward the fried products has been continuously changing and they are demanding the lower oil content products. Besides, consumers are afraid of the carcinogen compounds such as acrylamide which developed due to exposure of food to high temperature. Also, during frying deterioration of oil quality occurs in when comes in contact with oxygen and water leading to formation of toxic compounds. These compounds may change the physicochemical characteristics of a frying medium. Therefore, the effect of frying needs to be discussed on the different properties of oil.

3.12.1 Oil Uptake

Oil uptake is a complex process that involves simultaneous heat and mass transfer and changes the physical and chemical characteristics and transforms the structural attributes of the product (Lumanlan et al., 2020; Xie et al., 2022). Protein denaturation, starch gelatinization, oil hydrolysis and oxidation, and the creation of new molecules are some of the alterations. The oil uptake during frying can be divided into three categories: structural oil, which is oil absorbed into the microstructure; penetrated surface oil, which is oil that remains in the product after it is removed from the frying oil; and surface oil, which is oil that is present on the surface (Lumanlan et al., 2020). The mechanism of oil absorption is described by water–oil replacement, surfactant theory, and the cooling-phase effect (Xie et al., 2022). The various types of oil distribution are depicted in Figure 3.4.

The oil uptake in traditional fried foods can be up to 50% of the weight of fried foods. As high oil content is not preferred in fried products, efforts have been taken to develop improved frying techniques by

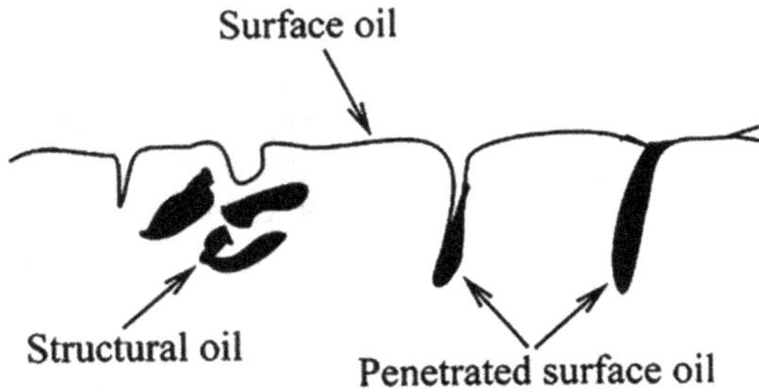

Figure 3.4 Types of oil distribution in deep drying. (Adapted from Lumanlan et al., 2020.)

changing the frying process operations, improving the properties of frying mediums, improving the coating treatments, etc. (Xie et al., 2022). The oil uptake in a product depends on its structure, type of frying system, frying conditions, and properties of the frying medium. In general, oil uptake is rapid at the initial stage of frying, and it slows down as frying proceeds; however, in longer frying, it increases (Al Faruq et al., 2019). Sometimes at higher frying temperatures, an outer crust is formed around the food, which hinders the penetration of the oil and thus reduces the oil content (Liu, Tian, Zhang, & Fan, 2021). A study conducted on normal maize starch revealed that an increase in the moisture content, frying temperature, and time increased the external oil content (Chen et al., 2019). Similarly, an increase in oil uptake was reported in potato French fries by Xu et al. (2022).

During frying, due to the high temperature of the oil, the temperature of the product rises, and eventually, the water inside the product converts into water vapor and escapes through the product, leading to the formation of internal pores, crevices, and outer crust (Liu, Tian, Hu, Yu, & Fan, 2021; Lumanlan et al., 2020). The pores and crevices provide the route for oil penetration, while the crust acts as a barrier. In microwave frying, it has been observed that oil uptake was significantly increased with microwave power and time. Moreover, the ultrasound-assisted microwave vacuum frying showed higher oil absorption. This may be due to the microscopic pores or channels created by ultrasound treatment. This, in turn, indicates that oil uptake is also dependent upon the ultrasound power (Al Faruq et al., 2019). Yang et al. (2021) compared electric field frying and conventional frying and found that electric field frying lowered oil uptake, retained higher water, and distributed oil evenly with lesser damage to the microstructure of the fried product.

3.12.2 Shrinkage

Shrinkage is the reduction in the volume of the product after frying as compared to its initial volume, i.e., before frying. It generally occurs in every fried product (Su et al., 2018b). It is dependent on the moisture loss of the product. Su et al. (2018a) reported that increased microwave and ultrasound power causes the moisture to evaporate quickly, and increased vapor pressure causes the product to expand, thus reducing shrinkage. Similar reports on the use of ultrasound are corroborated by Su et al. (2020). The more the puffing or expansion of fried foods, the less the shrinkage. Gouyo et al. (2021) reported that the pre-frozen French fries did not shrink during deep fat and hot air frying. This could be attributed to higher moisture loss (~92%–93%) from the French fries.

3.12.3 Texture

Texture is an important quality index that plays a major role in the acceptability of fried foods and is measured by the hardness and crispness of the crust (Gouyo et al., 2021). Textural changes are due to simultaneous heat and mass transfer through fried foods (Devi et al., 2018). According to Fang et al. (2021), fish skin's breaking force is a textural characteristic that indicates the fish skin's hardness and crispness. Moreover, Devi et al. (2018) discovered similar results when examining mushroom chips. They observed that the breaking force increased as the frying time increased up until the moisture content reached 10%, at which point it began to decrease. Sansano et al. (2018) stated that due to high water loss in microwave drying, it produced harder and crisper French fries. The breaking force was found to be decreased with an increase in the ultrasound power and frying temperature (Islam et al., 2019). This could be due to a higher drying rate, which is enhanced due to ultrasound that contributed mild damage to structure and caused less hardening. A similar effect was shown by ultrasound when coupled with a microwave, which can be attributed to the cumulative heat supply. In another way, the microwave produces a heat effect due to supplied microwave energy, while the ultrasound produces a sponge effect due to pore formations in the product (Su et al., 2018a). This indicates that the use of ultrasound provides a crisper food texture (Islam et al., 2019). A higher moisture removal rate in frying was also observed by Devi et al. (2018). Yu et al. (2019) reported the textural attributes of air-fried surimi in terms of hardness, springiness, gumminess, and chewiness, which were increased with an increase in the frying time and temperature.

3.12.4 Color

In fried foods, color is an important property that decides the consumer acceptability. If the products are over-fried, consumers reject them. Therefore, color attributes of fried products are very important. In general, the color is measured in Hunter color values, i.e., L^* (lightness to darkness), a^* (red to green), and b^* (yellow to blue) (Salehi, 2018). The L^* values are dependent on the temperature that promotes the non-enzymatic browning reaction, i.e., Maillard reaction (reaction between non-reducing sugar and amino acids) (Parikh and Takhar, 2016). The L^* values decrease as the temperature increases, irrespective of the heating source (e.g., microwave, infrared, or deep fat frying). The decrease in L^* is not desirable as it darkens the product due to intense browning reaction and crust formation. The higher values of L^* can be observed at lower frying temperature and in short frying. Lowering the boiling point (e.g., vacuum frying) of water can also improve L^*.

An increase in browning is exhibited by an increase in a* values due to non-enzymatic browning and phospholipid degradation. In general, a higher b* value is more desirable as it describes the yellowish color of the product. Salehi (2018) reported that a lower frying temperature gives lighter (L*) fried products with less a* and higher b* values. Sansano et al. (2018) reported microwave frying of French fries and found increments in a* and higher b* values at 430 and 600 W, respectively. In another study of microwave frying of potato, Parikh and Takhar (2016) stated that an increase in frying time and temperature increased the a* and b* values but decreased the L* value. The overall effect of L*, a*, and b* is reported as the total color score (ΔE). Su et al. (2020) stated that the products pretreated with ultrasound treatment in water showed a decrease in ΔE value. Islam et al. (2019) stated a decrease in ΔE of the ultrasound microwave vacuum fried samples than the non-ultrasound microwave vacuum samples.

3.12.5 Microstructural Changes

The microstructure of a fried product depends upon the moisture removal. As moisture is removed, the cells become compact, which compresses the holes or channels and develops the hard structure of the product (Devi et al., 2018). This can be attributed to the rapid heat transfer in the product due to the high temperature of frying oil (Su et al., 2018a). Gao et al. (2021) documented the deterioration of the cell structure in apple slices, along with the occurrence of curling, cracking, and deformation. The use of ultrasound develops channels through which moisture passes easily causing less curling and deformation. Furthermore, they tried samples with ethanol and ultrasound treatment which significantly kept the intact structure. This could be due to the formation of numerous micron-size pores on the surface of the product. Su et al. (2018b) applied microwave power in a vacuum and observed the formation of larger pores and cell fractures.

Acrylamide is a potentially harmful chemical that can form when certain foods are fried at high temperatures. This chemical is formed through a chemical reaction between amino acids and reducing sugars, known as the Maillard reaction. While acrylamide is present in many foods to some degree, it is particularly prevalent in fried foods such as French fries, potato chips, and other similar snacks. Research has shown that acrylamide is a carcinogen, meaning it has the potential to cause cancer in humans. As a result, food safety experts are concerned about the risks associated with acrylamide consumption and are working to reduce the levels of this chemical in foods through a variety of means, including changing cooking methods, using different ingredients, and developing new technologies to detect and measure acrylamide levels in food.

3.13 Alternative Trends in Frying Technologies

3.13.1 Microwave Frying

Nowadays, microwave heating is a widely used technique for various applications in the food industry, such as drying and dehydration, blanching, sterilization, pasteurization, baking, and roasting, due to its quicker processing time and lower processing temperature (Gharachorloo et al., 2010). Microwaves are electromagnetic radiations (waves) with their frequency varying between 0.3 and 300 GHz. For instance, the domestic microwave is designed at 2,450 MHz or 2.45 GHz, while industrial ones are designed either at 915 MHz or 2.45 GHz (Guo et al., 2017; Zhou et al., 2022). The mechanism of microwave heating involves volumetric heating in which the electromagnetic energy converts into thermal energy due to the presence of water molecules. This includes dipole rotation and ion conduction. In dipole rotation, the alteration in the direction of polar molecules caused by frequency leads to molecule collisions and heat generation. On the contrary, ionic conduction involves the movement of ions in response to an electric field. As the electric field direction shifts, ions move back and forth, creating friction and generating heat. In contrast to deep fat frying, thermal energy is transferred from within to outside of the product during microwave cooking. Internally, when the water boils, it enhances the water vapor pressure differential between the center and outer parts of food, thus removing the moisture (Al Faruq et al., 2022). This leads to more pores formation in the product as well as in crust, and thus, more volume is available for oil uptake (Zhou et al., 2022). Uneven heating is the problem of microwave heating, which needs to be addressed for quality fried products. Therefore, it is coupled with different techniques such as vacuum and ultrasound for desired outcomes.

3.13.2 Microwave Vacuum Frying (MVF)

Use of microwave heating along with vacuum frying technology has been proposed to enhance the frying process efficiency and the quality of the product. It is an integration of microwave and vacuum frying to improve the effectiveness of the process, enhance the drying time, and reduce the frying time and oil content (Zhang et al., 2020). The experimental setup for the microwave-assisted vacuum frying is presented in Figure 3.5 (Su et al., 2020). In this system, the microwave acts as a heating source, and the rest of the procedure is similar to the vacuum frying process.

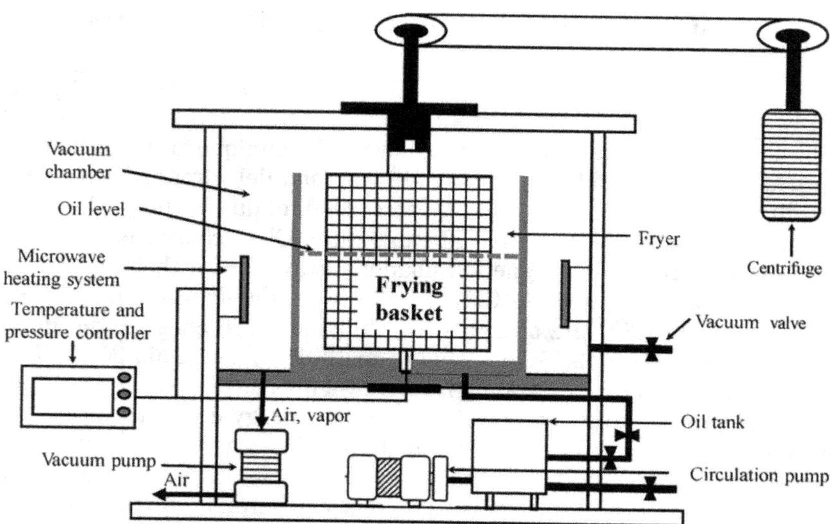

Figure 3.5 Microwave-assisted vacuum frying. (Adapted from Su et al., 2021.)

3.13.3 Ultrasound-Combined Microwave Vacuum Frying (UMVF) Technology

Ultrasound waves are a type of mechanical waves in the frequency range from 20kHz to 10MHz (Su et al., 2020). The low-frequency ultrasound is utilized for cell breakage, pore formations, heating, creating micro-vibrational effects, physicochemical alteration, and improvement in mass transfer. The ultrasound-assisted microwave fryer is a hybrid fryer in which ultrasound and microwave sources are used together. The microwave source acts as a heating source and has a frequency of 2,450MHz, and the output power can be varied between 0 and 1,200 W (Su et al., 2018b). The schematic of the ultrasound-assisted microwave vacuum fryer is shown in Figure 3.6.

The frying vessel in the fryer is generally fabricated using polytetrafluoroethylene (PTFE) material that does not absorb microwave and has good stability under the temperature up to 150°C. The ultrasound devices are installed alongside the frying chamber. The power can be up to 600 W, and the frequency is about 28kHz. Sometimes, the fryer is equipped with a centrifugal system that removes the excess oil from the product after frying.

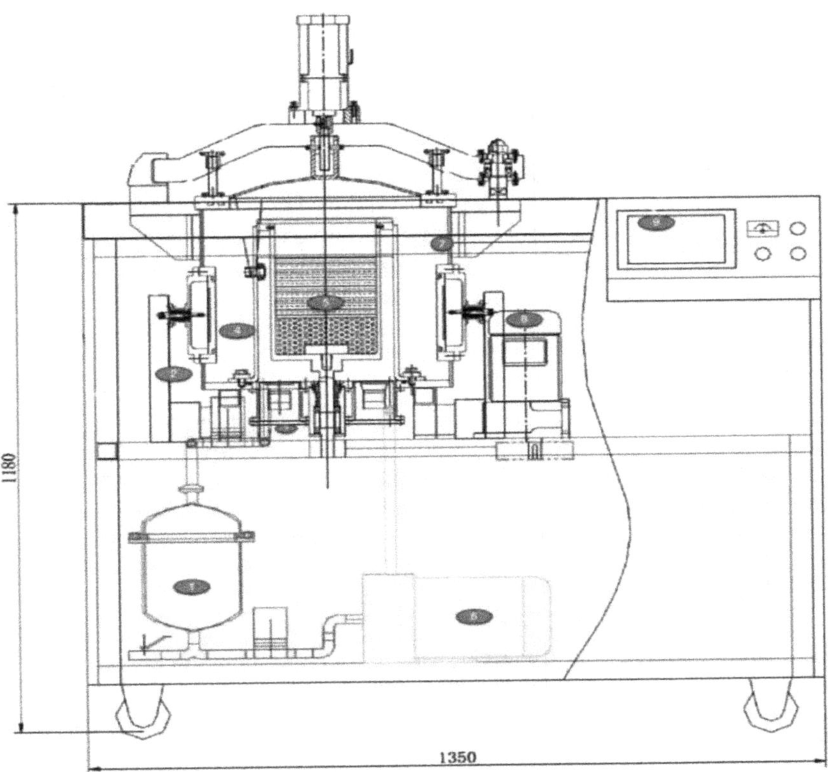

Figure 3.6 Ultrasound- and microwave-assisted vacuum fryer. (1) Oil tank, (2) microwave source and heating system, (3) ultrasound source and vacuum pressure balance system, (4) vacuum chamber, (5) frying chamber, (6) circulation pump, (7) electric cabin door system, (8) bending and centrifugation system, and (9) controller and operation panel. (Adapted from Islam et al., 2019.)

The evaporation rate is accelerated, and the frying time is decreased by combining ultrasonic and microwave (Su, et al., 2018a). Furthermore, if the ultrasound power increases, the rate of moisture evaporation also increases. The ultrasound causes an increase in the rate of evaporation because of increased turbulence at the soil-gas interface (Su et al., 2018a). Moreover, the ultrasound breaks the tissue structures and accelerates the moisture removal. Devi et al. (2018) also reported that the moisture removal was higher in ultrasound-assisted system as compared to vacuum and microwave vacuum frying.

When the water leaves through pores from the product during frying, oil occupies the porous area. It can be correlated that the oil uptake in the fried product is proportional to the moisture loss rate (Devi et al., 2018). Su et al. (2018b) reported a decrease in oil uptake with an increase in microwave and ultrasound power. In ultrasound-assisted frying, the ultrasound waves make the sample structure rigid, and due to this, the case hardening occurs, which hinders the oil transfer to the product (Devi et al., 2018).

3.13.4 Air Frying

As the name suggests, air frying technique is an alternative to the conventional frying process in which products are fried with little oil or no oil. In air frying, hot air, i.e., air superheated to high temperature (e.g. 180°C), circulates around the product to be fried (Abd Rahman et al., 2017; Ferreira et al., 2017). Air frying uses a fine mist of oil with frying air which surrounds the product during operation (Zaghi et al., 2019). The air fryer (Figure 3.7) is generally designed to circulate the air at a high temperature and mimic the air flow movement as that of the oil movement in the deep frying pan (Abd Rahman et al., 2017). In the air fryer, the heating elements are installed in the frying chamber. In addition, an exhaust fan is installed to circulate the air at a predefined operating condition. Due to the high and evenly distributed heat transfer, there is minimal variation in the product as compared to deep frying. The comparison between air frying and deep fat frying showed that the oil content in deep frying was ten times higher than that in the air frying (Zaghi et al., 2019). Air frying allows the preparation of low-fat food product with acceptable sensory attributes. But, the lipid oxidation and disruption of cell membranes are the disadvantages of air frying (Song et al., 2020). Yu et al. (2020) reported the use of air frying for surimi and found a crispy texture, flavor, and lower oil content.

3.13.5 Infrared Frying

Infrared (IR) technique has proved its potential in various food processing operations such as drying, pasteurization, baking, thawing, and roasting (Udomkun et al., 2019). In this method, the IR as electromagnetic radiation transmits and generates heat when it touches the food. As compared to conductive and convective frying methods, IR frying provides higher heat and mass transfer rate, higher thermal efficiency, and less degradation to food

Figure 3.7 Prestige air fryer (PAF-6). (Adapted from https://shop.ttkprestige.com/prestige-air-fryer-paf-6.html.)

components. The IR fryer (Figure 3.8) consists of a frying chamber, an IR heating system, a temperature sensor, and a control system. The details of the IR fryer are well described by Su et al. (2022a). IR fryer consists of an IR heater of 1–3 kW, operating at 220 V with 2–2.5 μm wavelength. The oil is heated to 160°C using the temperature controller and food samples are fried.

3.14 Packaging and Shelf Life of Fried Foods

Upon exposure to high temperature, frying oil undergoes various chemical reactions including oxidation, hydrolysis, and thermal decomposition. Also, when the product is packaged, it is likely to react with the gaseous composition around it, mostly oxygen, and moisture. Fat/oil oxidation is an

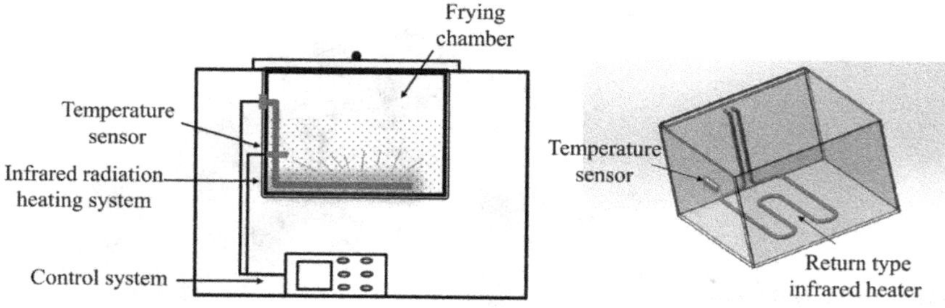

Figure 3.8 Schematic diagram of an IR fryer. (Adapted from Su et al., 2022a.)

important concern in fried foods that affects the shelf life of the product as well as creates safety issues (Juan-Polo et al., 2022). In addition, it develops the off-flavor, rancid taste, and discoloration of the product by forming the aldehyde and its derivatives (Oudjedi et al., 2019). It is well known that antioxidants can inhibit the oxidation process, and thus, the use of antioxidants in active packaging is now becoming popular. Choulitoudi et al. (2020) used a coating of *Satureja thymbra* extract, which is rich in phenolics and flavonoids, on the laminated film used in active packaging for fried potato chips. They found that the packaging film increased the protection and stabilized peroxide formation and oxygen consumption for up to 55 days. Similarly, the use of β-carotene antioxidants in multilayer low-density polyethylene-ethylene as an oxygen scavenger was used for the packaging of fried peanuts (Juan-Polo et al., 2022). Moula Ali et al. (2020) prepared the glycerol-plasticized gelatin films for the packaging of fried fish crackers (FFCs) and found lower hydroperoxides and secondary lipid oxidation after 30 days of storage, i.e., it improved the oxidative stability of the product. Ibadullah et al. (2019) packed the fried fish crackers in polyethylene terephthalate-polyethylene-aluminium-linear low-density polyethylene (PET-PE-ALU-LLDPE) and oriented polypropylene-polyethylene-metallized polyethylene terephthalate-linear low-density polyethylene (OPP-PE-MPET-LLDPE), and found that the product underwent lipid oxidations. Therefore, the packaging materials having better barrier properties to oxygen, moisture, and light have to be selected for extending the shelf life of packaged fried foods.

3.15 Summary

Currently, the lifestyle of people is changing due to their busy schedules, and they are preferring ready-to-eat foods instead of conventional foods as it is a tedious process to prepare food from raw materials. In this scenario, fried foods come into the picture. The process of frying is not as simple as we think about it. The contact between food material and hot oil causes physical, chemical, and organoleptic changes in the very complex cooking process of heat and mass transfer during frying. In order to make the world's food supply safer, the assessment of frying technologies may bring the attention of stakeholders, particularly that of decision-makers, to the need to evaluate the health hazards connected to the consumption of fried food products. Thus, it is crucial to verify the packaging materials used and the shelf stability of materials containing fried food. To protect the consumer's health from diseases, however, quality control, sufficient processes, and regulatory oversight of fried food products are required.

References

Abd Rahman, N. A., S. Z. Abdul Razak, L. A. Lokmanalhakim, F. S. Taip, and S. M. Mustapa Kamal. "Response surface optimization for hot air-frying technique and its effects on the quality of sweet potato snack." *Journal of Food Process Engineering* 40, no. 4 (2017): e12507. https://doi.org/10.1111/jfpe.12507.

Al Faruq, Abdulla, Min Zhang, and Benu Adhikari. "A novel vacuum frying technology of apple slices combined with ultrasound and microwave." *Ultrasonics Sonochemistry* 52 (2019): 522–529.

Al Faruq, Abdulla, Mst Husne Ara Khatun, S. M. Roknul Azam, Md Sazzat Hossain Sarker, Md Sultan Mahomud, and Xin Jin. "Recent advances in frying processes for plant-based foods." *Food Chemistry Advances* 1 (2022): 100086.

Albert, Susanne, and Gauri S. Mittal. "Comparative evaluation of edible coatings to reduce fat uptake in a deep-fried cereal product." *Food Research International* 35, no. 5 (2002): 445–458.

Ali, Ali Muhammed Moula, Koro de la Caba, Thummanoon Prodpran, and Soottawat Benjakul. "Quality characteristics of fried fish crackers packaged in gelatin bags: Effect of squalene and storage time." *Food Hydrocolloids* 99 (2020): 105378.

Andrés-Bello, A., Purificación García-Segovia, and Javier Martínez-Monzó. "Vacuum frying: An alternative to obtain high-quality dried products." *Food Engineering Reviews* 3, no. 2 (2011): 63–78.

Arefi, Arman, Oliver Hensel, and Barbara Sturm. "Intelligent potato frying: Time to say goodbye to the "good old" processing strategies." *Thermal Science and Engineering Progress* 34 (2022): 101389.

Asefa, B., D. Fikadu, and D. Moges. "Optimization of blanching time-temperature combination and pre-drying durations for production of high quality potato chips." *Optimization* 58 (2016): 54–62.

Asokapandian, Sangamithra, Gabriela John Swamy, and Haseena Hajjul. "Deep fat frying of foods: A critical review on process and product parameters." *Critical Reviews in Food Science and Nutrition* 60, no. 20 (2020): 3400–3413.

Banerjee, Soumitra, and Chandan Kumar Sahu. "A short review on vacuum frying-a promising technology for healthier and better fried foods." *International Journal of Nutrition and Health Sciences* 2 (2017): 68–71.

Boreddy, Sreenivasula Reddy, Devin J. Rose, and Jeyamkondan Subbiah. "Radiofrequency-assisted thermal processing of soft wheat flour." *Journal of Food Science* 84, no. 9 (2019): 2528–2536.

Chen, Long, Rongrong Ma, Zipei Zhang, David Julian McClements, Lizhong Qiu, Zhengyu Jin, and Yaoqi Tian. "Impact of frying conditions on hierarchical structures and oil absorption of normal maize starch." *Food Hydrocolloids* 97 (2019): 105231.

Choulitoudi, Evanthia, Aglaia Velliopoulou, Dimitrios Tsimogiannis, and Vassiliki Oreopoulou. "Effect of active packaging with Satureja thymbra extracts on the oxidative stability of fried potato chips." *Food Packaging and Shelf Life* 23 (2020): 100455.

Chung, J., J. Lee, and E. Choe. "Oxidative stability of soybean and sesame oil mixture during frying of flour dough." *Journal of Food Science* 69, no. 7 (2004): 574–578.

Damto, Teferi, and Geremew Chala. "Effect of blanching and frying time on the sensory quality of fried sweet potato chips." *Food and Nutrition Science-An International Journal* 3 (2019): 1–17.

Dash, Kshirod K., Maanas Sharma, and Ajita Tiwari. "Heat and mass transfer modeling and quality changes during deep fat frying: A comprehensive review." *Journal of Food Process Engineering* 45, no. 4 (2022): e13999.

Debnath, Sukumar, Navin K. Rastogi, A. G. Gopala Krishna, and B. R. Lokesh. "Effect of frying cycles on physical, chemical and heat transfer quality of rice bran oil during deep-fat frying of poori: An Indian traditional fried food." *Food and Bioproducts Processing* 90, no. 2 (2012): 249–256.

Dehghannya, Jalal, and Michael Ngadi. "Recent advances in microstructure characterization of fried foods: Different frying techniques and process modeling." *Trends in Food Science & Technology* 116 (2021): 786–801.

Devi, S., M. Zhang, and C. L. Law. "Effect of ultrasound and microwave assisted vacuum frying on mushroom (Agaricus bisporus) chips quality." *Food Bioscience* 25 (August 2018): 111–117. https://doi.org/10.1016/j.fbio.2018.08.004.

Devi, Shoma, Min Zhang, Ronghua Ju, and Bhesh Bhandari. "Recent development of innovative methods for efficient frying technology." *Critical Reviews in Food Science and Nutrition* 61, no. 22 (2021): 3709–3724.

Eichenlaub, S., and C. Koh. "Modeling of food-frying processes." In Serafim Bakalis, Kai Knoerzer and Peter J. Fryer (Eds.), *Modeling Food Processing Operations*, pp. 163–184. Woodhead Publishing, 2015.

Elbassiony, K. R. A. "Reduction of acrylamide formation in potato chips by gamma irradiation and some pretreatments processing." *Annals of Agricultural Science, Moshtohor* 58 (1) (2020) 45–52.

Fang, M., G. J. Huang, and W. C. Sung. "Mass transfer and texture characteristics of fish skin during deep-fat frying, electrostatic frying, air frying and vacuum frying." *LWT-Food Science and Technology* 137 (2021): 110494. https://doi.org/10.1016/j.lwt.2020.110494.

Farinu, Adefemi. "Heat and mass transfer analogy under turbulent conditions of frying." PhD diss., 2006.

Fellows, Peter J. *Food Processing Technology: Principles and Practice.* Elsevier, 2009.

Ferreira, F. S., G. R. Sampaio, L. M. Keller, A. C. H. F. Sawaya, D. W. H. Chávez, E. A. F. S. Torres, and T. Saldanha. "Impact of air frying on cholesterol and fatty acids oxidation in sardines: Protective effects of aromatic herbs." *Journal of Food Science* 82, no. 12 (2017): 2823–2831. https://doi.org/10.1111/1750-3841.13967.

Gao, J., Y. Su, C. Zhu, J. Li, T. Zheng, and B. Chitrakar. "Reduction of oil uptake in deep-fried apple slices by the combined ultrasonic and ethanol pre-treatment." *LWT-Food Science and Technology* 152 (August 2021). https://doi.org/10.1016/j.lwt.2021.112274.

Garayo, Jagoba, and Rosana Moreira. "Vacuum frying of potato chips." *Journal of Food Engineering* 55, no. 2 (2002): 181–191.

Gertz, Christian. "Fundamentals of the frying process." *European Journal of Lipid Science and Technology* 116, no. 6 (2014): 669–674.

Gharachorloo, M., M. Ghavami, M. Mahdiani, and R. Azizinezhad. "The effects of microwave frying on physicochemical properties of frying and sunflower oils." *JAOCS, Journal of the American Oil Chemists' Society* 87, no. 4 (2010): 355–360. https://doi.org/10.1007/s11746-009-1508-y.

Gouyo, T., É. Rondet, C. Mestres, C. Hofleitner, and P. Bohuon. "Microstructure analysis of crust during deep-fat or hot-air frying to understand French fry texture." *Journal of Food Engineering* 298 (2021). https://doi.org/10.1016/j.jfoodeng.2021.110484.

Guillaumin, R. "Animal fats in food-industry. 5. Desirable characteristics for fryings." *Revue Française des Corps Gras* 30, no. 9 (1983): 347–354.

Guo, Q., D. W. Sun, J. H. Cheng, and Z. Han. "Microwave processing techniques and their recent applications in the food industry." *Trends in Food Science and Technology* 67 (2017): 236–247. https://doi.org/10.1016/j.tifs.2017.07.007.

Ibadullah, Wan Zunairah Wan, Atiqah Aqilah Idris, Radhiah Shukri, Nor Afizah Mustapha, Nazamid Saari, and Nur Hanani Zainal Abedin. "Stability of fried fish crackers as influenced by packaging material and storage temperatures." *Current Research in Nutrition and Food Science Journal* 7, no. 2 (2019): 369–381.

Islam, M., M. Zhang, and D. Fan. "Ultrasonically enhanced low-temperature microwave-assisted vacuum frying of edamame: Effects on dehydration kinetics and improved quality attributes." *Drying Technology* 37, no. 16 (2019): 2087–2104. https://doi.org/10.1080/07373937.2018.1558234.

Juan-Polo, Adriana, Salvador E. Maestre Pérez, María Monedero Prieto, Carmen Sánchez Reig, Ana María Tone, Nuria Herranz Solana, and Ana Beltrán Sanahuja. "Oxygen Scavenger and Antioxidant LDPE/EVOH/PET-Based Films Containing β-Carotene Intended for Fried Peanuts (Arachis hypogaea L.) Packaging: Pilot Scale Processing and Validation Studies." *Polymers* 14, no. 17 (2022): 3550.

Lalam, Sravan, Jaspreet S. Sandhu, Pawan S. Takhar, Leslie D. Thompson, and Christine Alvarado. "Experimental study on transport mechanisms during deep fat frying of chicken nuggets." *LWT-Food Science and Technology* 50, no. 1 (2013): 110–119.

Lisinska, Grazyna, and Waclaw Leszczynski. *Potato Science and Technology.* Springer Science & Business Media, 1989.

Liu, Caiyun, Nabil Grimi, Nikolai Lebovka, and Eugene Vorobiev. "Effects of preliminary treatment by pulsed electric fields and convective air-drying on characteristics of fried potato." *Innovative Food Science & Emerging Technologies* 47 (2018): 454–460.

Liu, Y., J. Tian, B. Hu, P. Yu, and L. Fan. "Relationship between crust characteristics and oil uptake of potato strips with hot-air pre-drying during frying process." *Food Chemistry* 360 (April 2021a). https://doi.org/10.1016/j.foodchem.2021.130045.

Liu, Y., J. Tian, T. Zhang, and L. Fan. "Effects of frying temperature and pore profile on the oil absorption behavior of fried potato chips." *Food Chemistry* 345 (2021b). https://doi.org/10.1016/j.foodchem.2020.128832.

Liyanage, Dilumi W. K., Dmytro P. Yevtushenko, Michele Konschuh, Benoît Bizimungu, and Zhen-Xiang Lu. "Processing strategies to decrease acrylamide formation, reducing sugars and free asparagine content in potato chips from three commercial cultivars." *Food Control* 119 (2021): 107452.

Llorca, Empar, Isabel Hernando, Isabel Pérez-Munuera, Amparo Quiles, Susana M. Fiszman, and M. Lluch. "Effect of batter formulation on lipid uptake during frying and lipid fraction of frozen battered squid." *European Food Research and Technology* 216, no. 4 (2003): 297–302.

Lulai, Edward C., and Paul H. Orr. "Influence of potato specific gravity on yield and oil content of chips." *American Potato Journal* 56, no. 8 (1979): 379–390.

Lumanlan, Jane Cantre, Warnakulasuriya Mary Ann Dipika Binosha Fernando, and Vijay Jayasena. "Mechanisms of oil uptake during deep frying and applications of predrying and hydrocolloids in reducing fat content of chips." *International Journal of Food Science & Technology* 55, no. 4 (2020): 1661–1670.

Maan, Abid Aslam, Muhammad Adeel Anjum, Muhammad Kashif Iqbal Khan, Akmal Nazir, Farhan Saeed, Muhammad Afzaal, and Rana Muhammad Aadil. "Acrylamide formation and different mitigation strategies during food processing–A review." *Food Reviews International* 38, no. 1 (2022): 70–87.

Maity, Tanushree, A. S. Bawa, and P. S. Raju. "Effect of preconditioning on physicochemical, microstructural, and sensory quality of vacuum-fried jackfruit chips." *Drying Technology* 36, no. 1 (2018): 63–71.

Mallikarjunan, P., M. S. Chinnan, V. M. Balasubramaniam, and R. D. Phillips. "Edible coatings for deep-fat frying of starchy products." *LWT-Food Science and Technology* 30, no. 7 (1997): 709–714.

Mojaharul Islam, M., M. Zhang, B. Bhandari, and Z. Guo. "A hybrid vacuum frying process assisted by ultrasound and microwave to enhance the kinetics of moisture loss and quality of fried edamame." *Food and Bioproducts Processing* 118 (2019): 326–335. https://doi.org/10.1016/j.fbp.2019.10.004.

Moreira, Rosana G., M. Elena Castell-Perez, and Maria A. Barrufet. "Deep fat frying: Fundamentals and applications." Food Engineering Series, Edition 1 (1999): 1–350.

Moreira, Rosana G., Paulo F. Da Silva, and Carmen Gomes. "The effect of a de-oiling mechanism on the production of high quality vacuum fried potato chips." *Journal of Food Engineering* 92, no. 3 (2009): 297–304.

Muñiz-Becerá, Sahylin, Lilia L. Méndez-Lagunas, and Juan Rodríguez-Ramírez. "Solute transfer in osmotic dehydration of vegetable foods: a review." *Journal of Food Science* 82, no. 10 (2017): 2251–2259.

Oke, E. K., M. A. Idowu, O. P. Sobukola, S. A. O. Adeyeye, and A. O. Akinsola. "Frying of food: a critical review." *Journal of Culinary Science & Technology* 16, no. 2 (2018): 107–127.

Ouchon, P. B., J. M. Aguilera, and D. L. Pyle. "Structure oil-absorption relationships during deep-fat frying." *Journal of Food Science* 68, no. 9 (2003): 2711–2716.

Oudjedi, K., S. Manso, C. Nerin, N. Hassissen, and F. Zaidi. "New active antioxidant multilayer food packaging films containing Algerian Sage and Bay leaves extracts and their application for oxidative stability of fried potatoes." *Food Control* 98 (2019): 216–226.

Parikh, A., and P. S. Takhar. "Comparison of microwave and conventional frying on quality attributes and fat content of potatoes." *Journal of Food Science* 81, no. 11 (2016): E2743–E2755. https://doi.org/10.1111/1750-3841.13498.

Saguy, I. Sam, Eric Gremaud, Hugo Gloria, and Robert J. Turesky. "Distribution and quantification of oil uptake in French fries utilizing a radiolabeled 14C palmitic acid." *Journal of Agricultural and Food Chemistry* 45, no. 11 (1997): 4286–4289.

Salehi, F. (2018). "Color changes kinetics during deep fat frying of carrot slice." *Heat and Mass Transfer/Waerme- Und Stoffuebertragung*, 54(11), 3421–3426. https://doi.org/10.1007/s00231-018-2382-7.

Sansano, M., R. De los Reyes, A. Andrés, and A. Heredia. "Effect of microwave frying on acrylamide generation, mass transfer, color, and texture in French fries." *Food and Bioprocess Technology* 11, no. 10 (2018): 1934–1939. https://doi.org/10.1007/s11947-018-2144-z.

Schouten, M. A., J. Genovese, S. Tappi, A. Di Francesco, E. Baraldi, M. Cortese, G. Caprioli, S. Angeloni, S. Vittori, P. Rocculi, and S. Romani. Effect of innovative pre-treatments on the mitigation of acrylamide formation in potato chips. *Innovative Food Science & Emerging Technologies* 64 (2020): 102397.

Selman, J. D. *Factors Affecting Oil Uptake during the Production of Fried Potato Products*. Campden Food and Drink Research Association, 1989.

Sinha, Ankita, and Atul Bhargav. "Effect of food sample geometry on heat and mass transfer rate and cooking time during deep fat frying process." In *23rd National Heat and Mass Transfer Conference and 1st International ISHMT-ASTFE Heat and Mass Transfer Conference*, Thiruvananthapuram, India, 2015, pp. 17–20.

Song, G., L. Li, H. Wang, M. Zhang, X. Yu, J. Wang, J. Xue, and Q. Shen. "Real-time assessing the lipid oxidation of prawn (Litopenaeus vannamei) during air-frying by iKnife coupling rapid evaporative ionization mass spectrometry." *Food Control* 111 (2020): 107066. https://doi.org/10.1016/j.foodcont.2019.107066.

Stevenson, S. G., M. Vaisey-Genser, and N. A. M. Eskin. "Quality control in the use of deep frying oils." *Journal of the American Oil Chemists Society* 61 (1984): 1102–1108.

Su, Y., M. Zhang, B. Bhandari, and W. Zhang. "Enhancement of water removing and the quality of fried purple-fleshed sweet potato in the vacuum frying by combined power ultrasound and microwave technology." *Ultrasonics Sonochemistry* 44 (March 2018a): 368–379. https://doi.org/10.1016/j.ultsonch.2018.02.049.

Su, Y., M. Zhang, W. Zhang, C. Liu, and B. Adhikari. "Ultrasonic microwave-assisted vacuum frying technique as a novel frying method for potato chips at low frying temperature." *Food and Bioproducts Processing* 108 (2018b): 95–104. https://doi.org/10.1016/j.fbp.2018.02.001.

Su, Y., M. Zhang, B. Chitrakar, and W. Zhang. "Effects of low-frequency ultrasonic pre-treatment in water/oil medium simulated system on the improved processing efficiency and quality of microwave-assisted vacuum fried potato chips." *Ultrasonics Sonochemistry* 63 (2020). https://doi.org/10.1016/j.ultsonch.2020.104958.

Su, Y., M. Zhang, B. Chitrakar, and W. Zhang. "Reduction of oil uptake with osmotic dehydration and coating pre-treatment in microwave-assisted vacuum fried potato chips." *Food Bioscience* 39 (2021). https://doi.org/10.1016/j.fbio.2020.100825.

Su, Y., J. Gao, Y. Chen, B. Chitrakar, J. Li, and T. Zheng. "Evaluation of the infrared frying on the physicochemical properties of fried apple slices and the deterioration of oil." *Food Chemistry* 379 (2022a). https://doi.org/10.1016/j.foodchem.2022.132110.

Udomkun, P., P. Niruntasuk, and B. Innawong. "Impact of novel far-infrared frying technique on quality aspects of chicken nuggets and frying medium." *Journal of Food Processing and Preservation* 43, no. 5 (2019): e13931. https://doi.org/10.1111/jfpp.13931.

Van Koerten, K. N., D. Somsen, R. M. Boom, and M. A. I. Schutyser. "Modelling water evaporation during frying with an evaporation dependent heat transfer coefficient." *Journal of Food Engineering* 197 (2017): 60–67.

Varela, P., and S. M. Fiszman. "Hydrocolloids in fried foods. A review." *Food Hydrocolloids* 25, no. 8 (2011): 1801–1812.

Xie, D., D. Guo, Z. Guo, X. Hu, S. Luo, and C. Liu. "Reduction of oil uptake of fried food by coatings: A review." *International Journal of Food Science and Technology* 57, no. 6 (2022): 3268–3277. https://doi.org/10.1111/ijfs.15266.

Xu, L., X. Mei, J. Chang, G. Wu, H. Zhang, Q. Jin, and X. Wang. "Comparative characterization of key odorants of French fries and oils at the break-in, optimum, and degrading frying stages." *Food Chemistry* 368 (2022). https://doi.org/10.1016/j.foodchem.2021.130581.

Yamsaengsung, R., and R. G. Moreira. "Modeling the transport phenomena and structural changes during deep fat frying: Part I: Model development." *Journal of Food Engineering* 53, no. 1 (2002): 1–10.

Yang, D., G. Wu, Y. Lu, P. Li, X. Qi, H. Zhang, X. Wang, and Q. Jin. "Comparative analysis of the effects of novel electric field frying and conventional frying on the quality of frying oil and oil absorption of fried shrimps." *Food Control* 128 (April 2021). https://doi.org/10.1016/j.foodcont.2021.108195.

Yu, X., L. Li, J. Xue, J. Wang, G. Song, Y. Zhang, and Q. Shen. "Effect of air-frying conditions on the quality attributes and lipidomic characteristics of surimi during processing." *Innovative Food Science and Emerging Technologies* 60 (October 2019): 102305. https://doi.org/10.1016/j.ifset.2020.102305.

Zaghi, A. N., S. M. Barbalho, E. L. Guiguer, and A. M. Otoboni. "Frying process: From conventional to air frying technology." *Food Reviews International* 35, no. 8 (2019): 763–777. https://doi.org/10.1080/87559129.2019.1600541.

Zhang, Yachuan, Dieter H. W. Kahl, Benoit Bizimungu, and Zhen-Xiang Lu. "Effects of blanching treatments on acrylamide, asparagine, reducing sugars and colour in potato chips." *Journal of Food Science and Technology* 55, no. 10 (2018): 4028–4041.

Zhang, Xiaotian, Min Zhang, and Benu Adhikari. "Recent developments in frying technologies applied to fresh foods." *Trends in Food Science & Technology* 98 (2020): 68–81.

Zhang, Cheng, Xiaomei Lyu, Wei Zhao, Wenxu Yan, Mingming Wang, N. G. Kuan Rei, and Ruijin Yang. "Effects of combined pulsed electric field and blanching pretreatment on the physiochemical properties of French fries." *Innovative Food Science & Emerging Technologies* 67 (2021): 102561.

Zhou, X., S. Zhang, Z. Tang, J. Tang, and P. S. Takhar. "Microwave frying and post-frying of French fries." *Food Research International* 159 (May 2022): 111663. https://doi.org/10.1016/j.foodres.2022.111663.

Ziaiifar, Aman Mohammad, Francis Courtois, and Gilles Trystram. "Porosity development and its effect on oil uptake during frying process." *Journal of Food Process Engineering* 33, no. 2 (2010): 191–212.

Zielinska, Magdalena, Wioletta Błaszczak, and Sakamon Devahastin. "Effect of superheated steam prefrying treatment on the quality of potato chips." *International Journal of Food Science & Technology* 50, no. 1 (2015): 158–168.

Chapter 4

Quality Criteria for Selection of Suitable Oil for Frying

Frying Lipids (Oils/Fats): Quality Characteristics

Tejaswi Boyapati and Mohona Munshi

Vignan Foundation of Science, Technology and Research (Deemed to be University)

Koneru Lakshmaiah Education Foundation (KL – Deemed to be University)

Pradyuman Kumar

Sant Longowal Institute of Engineering and Technology (Deemed-to-be University)

DOI: 10.1201/9781003329244-4

4.1 Introduction

One of the earliest methods of preparation of food is frying (Chiou et al., 2012). As far back as 1600 BC, it existed. Over the years, many fried goods have been enjoyed by various civilizations. Foods that are deep-fried are getting more and more popular worldwide. Due to their distinctive flavor and texture, fried dishes have been a consumer favorite for decades. The type of oils and ingredients utilized throughout the procedure as well as the frying temperatures all affect the quality of the final items (Nayak et al., 2016). Frying is a common practice across the world, not just in industrial food production but also in restaurants and household kitchens. The goal of frying is to provide increased food quality (Dana & Saguy, 2001). During the frying process, food stuffs undergo a variety of biochemical, nutritional, and structural modifications (Figure 4.1). Consequently, different flavor elements are generated, protein gets denatured, certain micronutrients are degraded, carbohydrate component gets gelatinized, crusts are developed, pores are created, and distinctive microstructures are developed (Yu, 2011). It has been difficult to mimic the distinct texture and sensory qualities of fried dishes using any other method or another unit operation (Ngadi et al., 2008). Heat and mass are transferred at the same time while frying. Mass transfer is aided by the source of heating, which is oil (Vitrac et al., 2002). Convection and conduction mechanisms of heat transfer allows

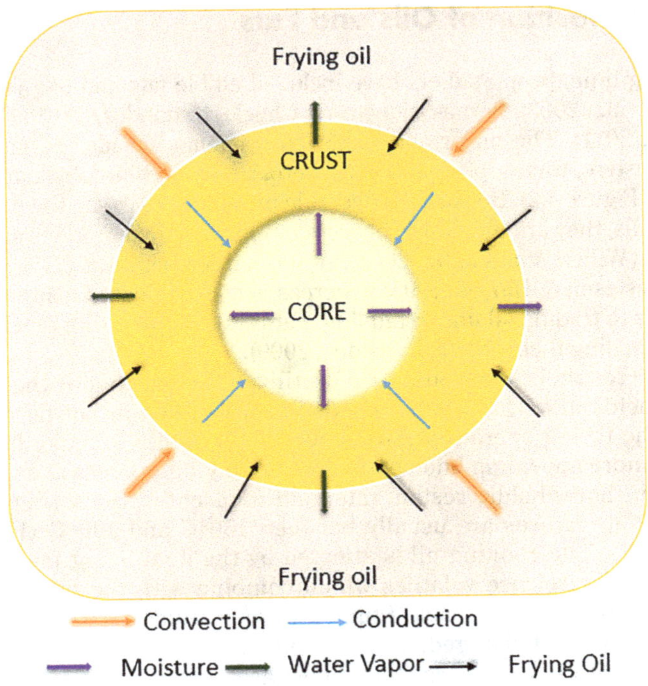

Figure 4.1 Schematic diagram of frying process.

heating from the surface of food material to its inside. The total temperature of the product rises, which causes the creation of water vapor (Farkas et al., 1996). Due to a difference in concentrations and pressure, the vapor then tunnels its way to the product's surface, leaving gaps that are crucial for the eventual absorption of oil. Heat additionally allows sugars to caramelize and triggers interactions between amino acids and reducing sugars, giving some fried meals their distinctive golden and or brown coloring and crunchy crust. Because it offers a nice fried taste that is not generated during other culinary processes, such as baking, frying is indeed a common way of preparing food. It's crucial to grasp how tastes form in oils throughout the cooking process in order to improve good flavors and suppress bad ones. In a nutshell, the flavors in cooking oil can come from one or more of the mentioned sources: flavors that are naturally present (such as those found in peanut and olive oils), flavors that are imparted during processing (including those from hydrogenation), and fatty acid degradation due to exposure to extreme temperatures while frying. Because of the impact of olive oil, it really is thought that the true science of frying began and was developed close to the Mediterranean area. Products manufactured from fried food generate hundreds of billions of USD.

4.2 Composition of Oils and Fats

For a long time, people's diets have included edible fats and oils (Table 4.1) (Sabir et al., 2003; Serna-Saldivar & Chuck-Hernandez, 2019; Varela & Fiszman, 2011). The major and minor components include the fatty acids, glycerol esters, waxes, phospholipids, sterols, tocols, chlorophyll, and hydrocarbons (Figure 4.2). Because triglyceride molecules make up the majority of fats and oils, they are thought to be less complicated than proteins and carbohydrates (Willett & Stampfer, 2003). At normal temperature (20°C), fats are solid (Westesen & Bunjes, 1995), whereas oils are liquids (Oommen, 2002), according to traditional understanding. Fats and oils have different compositions depending their origin (Gunstone, 2009).

As a result, how a fat or an oil is utilized is regulated by the quantity of fatty acids in it (Liu, 1994). Foods are processed in hot fats and oils during the frying operation (Stevenson et al., 1984). To make food components more appealing and appetizing, the frying process is extensively utilized in households, restaurants, and food enterprises (Figure 4.3). Frying temperatures are usually between 150°C and 190°C (Pedreschi et al., 2007). The cooking oil is affected by the item being fried as well. The meal may release volatiles which combine with the oil utilized for frying and turn the oil black. Studies were conducted to see if the oil's nutritive value had changed, as well as potential harmful effects. . There are issues with heat and mass transport as well as intricate interactions between the food material and the frying medium (Buenrostro-Gonzalez et al., 2004). The following are some of the most commonly used forms of frying (Figure 4.4):

Table 4.1 Fats vs. oils

	Fats	*Oils*
Similarities	One glycerol and three fatty acid molecules make up triglycerides	Triglycerides are substances made up of single glycerol molecule and three molecules of fatty acid
Differences	No C=C bonds/saturated	One or more C=C bonds/ unsaturated
	Saturated chains densely packed together	Unsaturated chains are packed more densely
	The van der Waals forces between molecules are stronger	The van der Waals forces between molecules are weaker

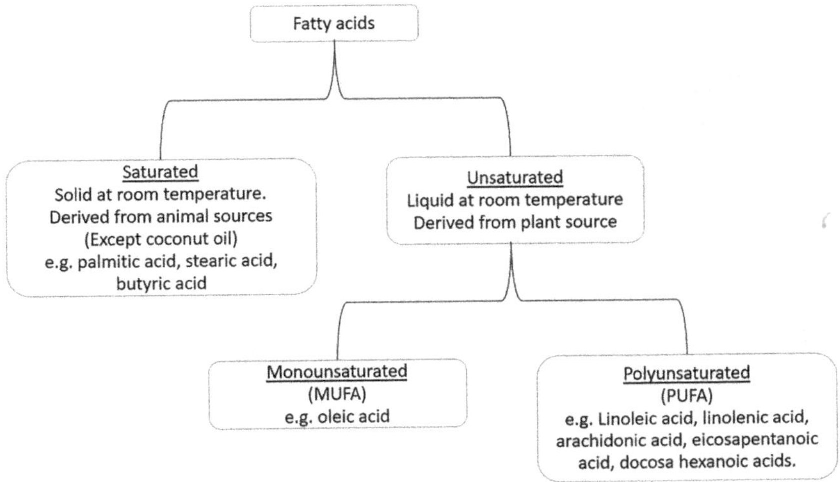

Figure 4.2 Classification of fatty acids.

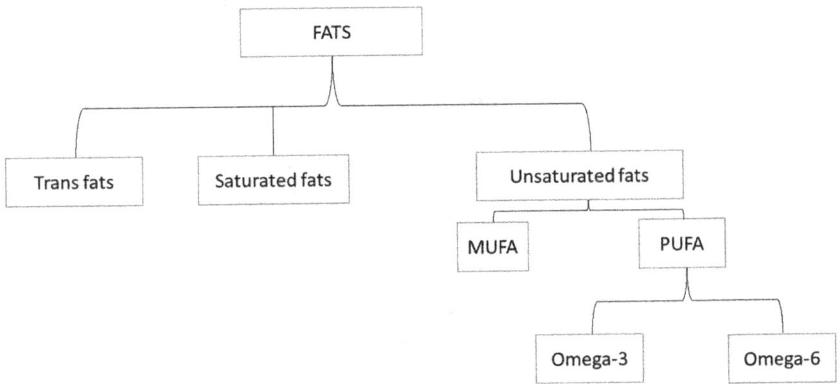

Figure 4.3 Classification of fats.

- Deep frying
- Pan frying
- Stir frying
- Shallow frying
- Deep fat frying

Deep frying is a demanding method that necessitates a lot of oil (Juáre et al., 2011). Deep frying imparts superb texture and a pleasing flavor to culinary items. As a result, worldwide acceptability and consumption of fried food

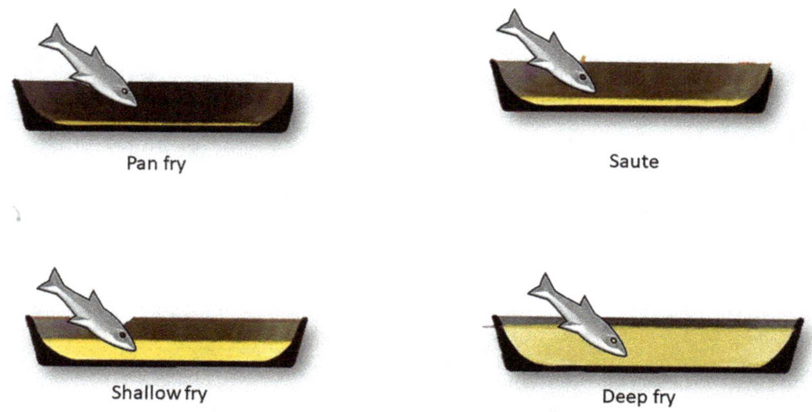

Figure 4.4 Forms of frying.

is increasing (Sánchez-Muniz & Bastida, 2006; Varela & Fiszman, 2011). Deep-fat frying is used by about 0.5 million commercial fryers in the United States alone (Serna-Saldivar & Chuck-Hernandez, 2019).

4.3 Degradation of Frying Oils While Frying Procedure

During frying, oils in turn undergo oxidation and hydrolysis, resulting in their degradation (Romero et al., 2000; Saguy & Dana, 2003). Both thermolytic and oxidative deteriorations are possible in oil while frying. At least on the surface, the oil in the frying pan is exposed to the atmosphere. Several substances, including hydroperoxides, epoxides, hydroxides, conjugated dienoic acids, and ketones, are produced when heated fat reacts with the oxygen in air. These components may fission into smaller fragments or stay in the triglyceride and produce cross-linking, resulting in dimeric or high polymeric triglycerides. The three main types of oxidation products are volatile, monomeric, and polymeric molecules. The steam that is produced when food is being fried removes certain volatile decomposition products (VDPs) from the oil. Additional reaction products that are non-volatile decomposition products (NVDP) might be found in oil. During deep frying, oxidation proceeds swiftly as temperature rises and oxidation progresses more quickly. Aside from temperature, frying fat quality, turnover rate, surface exposure to air, the presence and involvement of prooxidant metals such as iron or copper, the presence of elevated antioxidants, and the occurrence of silicone antifoams may all have an effect on how quickly frying fat oxidizes. It is difficult to fry at lower temperatures in a timely manner, as deterioration occurs at higher temperatures (Paul et al., 1997). Because of the presence of moisture in the meal and frying oil (Stevenson et al., 1984), hydrolysis happens quickly at higher temperatures resulting in the formation of FFAs, mono- and di-acylglycerols, and glycerol. Accordingly, the frying oil deteriorates because of oxidation and polymerization reactions (Senanayake, 2018). Polymerization occurs when frying fat is too oxidized. In the absence of oxygen, fats are able to generate new carbon–carbon bonds. When these bonds occur inside a single fatty acid, cyclic fatty acids are created. Dimeric acids and triglycerides are created via bonds between two fatty acids that are parts of the same triglyceride or between two distinct triglyceride molecules. Polymerization can lead to foaming. Foaming might happen when little bubbles start to develop on the fry kettle's walls. It is advisable to discard excessive foaming oil since it is a fire and safety hazard.

Additionally, it makes the oil gummy due to polymerization that causes the kettle, cookware, and frying basket edges start to gum up. The majority of the polymeric compounds that develop are quite bitter in addition to being a cleaning problem. Off-flavors in frying oils are caused by autoxidation (Velasco et al., 2010). Hydroperoxides are primary oxidation products, whereas aldehydes, hydrocarbons, ketones, alcohols, acids, and other compounds are secondary oxidation products (Ruberto & Baratta, 2000).

The amount of oxidation of oils which has been fried is determined using a variety of chemical parameters (Stevenson et al., 1984). They mostly depend on measurements of the by-products of primary or secondary oxidation (Capouet & Müller, 2006). A free-radical chain mechanism typically explains lipid oxidation in three phases (Recknagel et al., 2020):

- The earliest free radical's formation $RH \rightarrow R\cdot + H\cdot$
- Free-radical propagation and primary oxidation products are formed, such as hydroperoxide $R\cdot + O_2 \rightarrow ROO\cdot$, $ROO\cdot + RH \rightarrow ROOH + R\cdot$
- Termination stages and production of secondary oxidation products $R\cdot + R\cdot \rightarrow R\text{–}R$, $R\cdot + ROO\cdot \rightarrow ROOR$, $ROO\cdot\ ROO\cdot \rightarrow O_2 + ROOR$ or carbonyl compounds and alcohol.

Shelf life of fried foods is impacted by the increased rancidity in fried oils caused by frequent oil consumption (Syed, 2016). Thermal degradation happens when food is cooked at high temperatures, particularly where there is air, as a result of changes in its structural, biochemical, nutritive, and sensory properties (Bordin et al., 2013). The interaction between frying oil and water results in the formation of free fatty acids (FFAs) and partial glycerol esters. FFA formation is caused in part by oxidation and in part by hydrolysis.

It has been demonstrated that the presence of FFA in frying oil catalyzes further triglyceride hydrolysis. The FFA level is frequently unrelated to the quality of fried food. The quantity of steam that meal releases into the fat directly correlates to the amount of FFA that forms. FFA formation is accelerated when food with a significant amount of moisture is fried in higher amounts. Numerous oil degradation products are harmful to one's health because they deplete vitamins, stop enzymes from working, and irritate the stomach. Leukemia, insulin resistance, stroke, and cardiovascular disease are all well-known health problems (Paul et al., 1997). The fatty acid composition is an important component of frying oil that has a considerable impact on the flavor and durability of fried meals. Trans fatty acids are created when the most common frying technique modifies the frying oil's fatty acid composition (Gesteiro et al., 2019), raises the amount of saturated fatty acids, and decreases the level of unsaturated fatty acids (El-Hadad & Tikhomirova, 2018). As a result, chemical factors such FFAs, peroxide value (PV), and iodine value (IV) are altered (Ayyildiz et al., 2015). The time spent for frying determines the amount of total fatty acids formed, the temperature utilized, and used (Tsuzuki et al., 2010).

Total fatty acids have recently drawn more attention than ever before because of the strong link between total fatty acid consumption and the risk of cardiovascular disease (Hurt et al., 2010). In addition, it has been linked to chronic respiratory diseases, neural degenerative diseases, and cancer (Moro et al., 2016). The total fatty acids have also been shown to be associated with inflammatory markers, specifically the C-reactive protein (Baer et al., 2004). A more number of studies have concluded that consuming more trans fat might lead to endothelial dysfunction (Remig et al., 2010). Depending on the oils employed, heat treatments like frying have produced varying total fatty acid levels (Filip et al., 2011). However, it has been determined that the majority of total fatty acids in these items are derived from the oil used, rather than the process itself (Skeaff, 2009). Because of strict rules and regulations, as well as the use of the most up-to-date scientific industrial process to govern trans formation, food sources in technologically advanced nations are either trans free or to a restricted extent. Commercial fryers in poor countries do not use the correct oils for frying foods since there are no guidelines for trans fats and other levels of oxidation characteristics (Esfarjani et al., 2019). Because of more severe thermo-oxidative reactions than in heated oil containing food, there was a loss of C18:2 fatty acids and a fall in the iodine value of oil after heating. The iodine value decreases owing to the breakage of double bonds resulting from oxidation, scission, and polymerization. For frying, many oils can be utilized, including palm oil (PO), cottonseed oil (CSO), maize oil (CO), soy bean oil (SBO), sesame oil (SO), and sunflower seed oil (SFO), with the fatty acid composition and iodine value signifying fatty acid degradation (Rodriguez-Garcia et al., 2016). FFAs, among other things, are a measurement of the quantity of FAs hydrolyzed on the triacylglycerol backbone. They serve as a chemical gauge to assess how effective the frying process was (Stier, 2001). The oil rejection limit is set at 2% when fried oils are tested for FFAs to determine if they are suitable and safe for human consumption (Matthäus, 2006).

PV is also a technique for identifying preliminary lipid oxidation products such as hydroperoxides; PV is measured in $meqO_2$/kg of oil (Coupland et al., 1996). The PV technique is insufficient for identifying FOs because hydroperoxides in oil are unstable at frying temperatures, preventing their production. According to Dunford et al. (1997), secondary oxidation products with carbonyl groups are created from primary oxidation products (peroxides). The carbonyl value (CV) appears to be a useful measure for assessing frying oil oxidation levels since secondary compounds are more stable than primary ones. Secondary oxidation products can create rancidity and disagreeable tastes in fried meals, lowering their nutritional value (Endo et al., 2001). The production of high-molecular-weight molecules during the frying process has a significant impact on total polar compound (TPC) determination. A tried-and-true approach for choosing whether to reuse or discard frying oil is the TPC level (Al-Kahtani, 1991; Fritsch, 1981; Gertz et al., 2000).

In lipids containing methylene-intervallic dienes or polyenes, an oxidation process shifts the location of the double bond. CV, CD, and CT were also discovered to be excellent analytical indicators of changes caused by oxidation in fats and oils at higher temperatures (Farhoosh et al., 2008). As a result, UV absorptions at 420, 234, and 273 nm are increased.

4.4 Frying Process

Food is cooked in hot oil at 150°C–190°C during the frying process. One of the most popular and traditional methods of food cooking is deep-fat frying. It is frequently employed to create optimal textures and distinctive flavors in dishes that boost their palatability because of the simultaneous heat and mass transfer of oil, food, and air during the frying process (Figure 4.5) (Boskou et al., 2006; Durán et al., 2007). According to Cheng et al. (2006) and Paradis and Nawar (1981), oxygen absorption on the heated oil's surface, cooking temperature, length of frying, moistness, and fat content of the fried meal are all factors that affect how quickly food deteriorates.

Oil undergoes a number of chemical reactions while frying, including hydrolysis, oxidation, and polymerization. Significant changes take place while frying, when hot oil comes into contact with food being fried. The moisture present in the food may cause the oil to hydrolyze while frying. Hydrolysis is the main reaction in oil while frying. Adler-Nissen (1979)

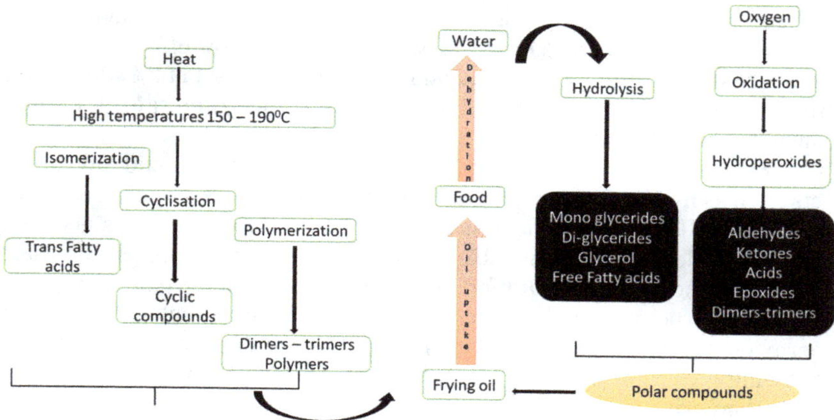

Figure 4.5 Illustration of the changes that occur in oil throughout the frying process.

states that a large amount of water immediately hydrolyzes as a result of chemical reactions that occur while frying food in oil. Lee et al. (2007) reported that water has weak nucleophile properties, which makes its triacylglycerol ester links susceptible to be attacked easily and produce oxidation products such as FFAs, mono and diacylglycerols, frying oil (FO) (formed due to oxidation of unsaturated fats), and a high level of unbalanced peroxides (Blumenthal & Stier, 1991; Fritsch, 1981; Stevenson et al., 1984). A new trend that has become more prevalent is the requirement for fried food products to be offered to customers or hotels/restaurants. These data suggest that fried food should be considered an important component of a person's diet. The selection of frying oil is difficult since it must take into consideration factors like price, nutritional value, and stability. According to Arroyo et al. (1995), oils with high saturated fatty acids (SFAs) offer unrivaled frying stability. However, in terms of nutritional and human health, these oils are the least desirable.

4.4.1 Color and Frying Process

After usage, frying oil's color darkens, which eventually has an impact on the appearance of the fried goods. There are color comparison kits on the market that enable rapid evaluation of oil color. Several foodservice businesses created color schemes that evaluate the appearance of fried food from acceptable to unsatisfactory (Krokida et al., 2001). A Lovibond tintometer allows for more precise color measurements (Wright, 1931). The acceptance of frying oils cannot be determined just by the color of the oil (Paul et al., 1997). The speed of darkening of oil varies based on the type of oil and the type of food that is fried.

4.4.2 Hydrolysis

When food is fried in boiling oil, the water from the foodstuff condenses and generates steam, which rises and finally evaporates as the food gets cooked. Moisture, steam, and O_2 trigger reactions in frying oils and food alike.

The ester bond of triacylglycerols is broken by water, a weak nucleophile, which causes the formation of di- and monoacylglycerols, glycerol, and FFAs. With each additional fry, the amount of FFAs rises in oil. The value of FFAs is used to measure the grade of frying oils. Oil with shorter, unsaturated fats is more susceptible to hydrolysis as they are highly soluble in water than longer, saturated fats.

4.5 Fatty Acid Formation and the Frying Process

Frying oils' fatty acid composition (FAC) has a big influence on the flavor and stability of fried dishes. Because frying in oil is a fundamental component of the diet, changes in quality during frying are significant. Researchers have found that variations in FAC have been regarded as reliable indicators of frying oils' quality according to modifications in the oil during frying.

For 40 hours at 180°C, C18:2 and C18:3 FAs were used (Miller & Ratti, 2009). It was revealed that after 40 hours of frying at 180°C, SFAs rose and PUFAs dropped. As a result, the amount of C18:3 acids (FFAs) reduced in comparison than C18:2 due to 10 times higher oxidation rates for (C18:1) oleic acid and 100× greater than those of (C18:0) stearic acid. It has also been reported that C18:3 fatty acid oxidizes 15× faster than C18:1 FAs (Rossell, 1989). In order for frying oil to meet consumer needs for good quality, a low level of C18:3 FA is required. 1.2 104, 5.3 104, 7.3 104, and 10.0 104 M/s are the reaction rates between O_2 and C18:0, C18:1, and C18:2 FAs, according to Alireza et al. (2010). C18:3 FA has a substantially faster relative reaction rate with oxygen than C18:2 and C18:1 FAs. As a result, the largest C18:3 FA decrease and the lowest C18:1 FA reduction are projected. The fundamental components of fats and oils are FAs, despite the fact that the human body cannot synthesize them and must instead obtain them through dietary intake (Tseng et al., 1996). As a result of greater frying temperatures, oxygen surrounding the food enters the oil and water is released from the food. It was discovered that frying causes thermal and oxidative changes. In addition to operation conditions, fryer design, and food type, oxidation products also depend on the operation conditions (Stevenson et al., 1984). The quantity of FAs created when frying reduced as a result of oxidative degradation, but the amount of SFAs rose (Chen et al., 1996). According to Kris-Etherton et al. (2002), elevated cholesterol levels, such as low-density lipoproteins (LDL), result from increased SFAs during frying, which may cause health problems. FA levels alter throughout the frying process as a result of repeated exposure of oils to higher temperatures, which leads to the exchange of fatty acids between fried foods and oils used for frying (Sanibal & Mancini-Filho, 2004). According to Jung and Min (1992), mono unsaturated fatty acids like C18:1 fatty acids and poly unsaturated fatty acids like C18:2 and C18:3 FAs undergo an auto-oxidation with a rate between 1:40 and 50:100 based on the level of oxygen uptake. It is crucial to have sufficient amounts of USFAs present while frying for oxidative deterioration of oil to occur. When USFAs are deep-fried at 180°C, they oxidize and break down into secondary oxidation products (Paradis & Nawar, 1981). The research conducted by Tyagi and Vasishtha (1996) reported that the heat degradation decreased (C18:2) linoleic and (C18:3) linolenic fatty acid concentrations in frying oils, e.g. SBO, by 79% and 60%, respectively. For FOs such as SFAs and UFAs, the FAC has been modified during frying to extend the frying process. According to this method, the first change that occurs during lipid oxidation is PUFA loss, and is monitored by this method (Sanibal & Mancini-Filho, 2004) examined how the profile of FA that a partly hydrogenated soybean fat (PHSF) and SBO changed when they were fried. When PHSF and SBO were fried for 50 hours, their PUFA content decreased by 11.8% and 59.9%, respectively. An analysis of FAC during the preparation of French fries over a 16-hour period in cottonseed oil, soybean oil, and potato oil was performed by Sulieman et al. (2006). The FAC levels in CSO, SFO, and PO were measured throughout frying, and the poly unsaturated fatty acids (C18:2 & C18:3) decreased from 52.87% to 44.26%, 65.68% to 63.61%, and 14.26% to 12.73%, respectively. However, the SFAs were elevated from 24.99% to 26.05%, 9.77% to 10.50%, and 41.97% to 44.18%, respectively.

4.6 Fatty Acid Ratio and the Frying Process

CSO was fried at 185°C for 12 hours by Houhoula et al. (2002). During frying, the ratio was found to decrease from 2.39 to 2.03. Aladedunye and Przybylski (2009a) looked at the ratios of C18:2/C16:0 and C18:3/C16:0 in canola oil (CLO) while French fries are being periodically fried at 185°C and 215°C and discovered that they decreased from 2.39 to 2.03. Temperatures ranging from 185°C and 215°C caused the ratio of C18:2/C16:0 to decrease from 4.73°C to 3.87°C and from 4.28°C to 3.23°C, respectively. At 215°C, the C18:3/C16:0 ratio decreased 1.9 times faster than at 185°C, since C18:3 FA rapidly oxidized during heating or frying. As a result of rapid oxidation in C18:3 FA, C18:3/C16:0 ratios decreased more rapidly than C18:2/C16:0 ratios. Ivanova et al. (2016) found that the C18:2/C16:0 ratio was an effective oxidation indicator for assessing vegetable oils at frying temperatures. They studied ground nut seed oil (GSO), coconut oil (CO), sunflower oil (SFO), and SBO. The C18:2/C16:0 ratios fell in the following order: SFO > GSO > SBO > CO. Cis FAs are transformed into trans FAs during oil frying, and trans FAs are swapped more often. A cis PUFA/SFA ratio of 0.45 is advised by the British Ministry of Health. According to Kris-Etherton et al. (2002), lower ratios are detrimental to health, particularly when it comes to cardiovascular disease.

4.6.1 Formation of Acrylamide during Frying

A hydrophilic substance known as acrylamide has indeed been identified as a possible carcinogenic substance to humans (Calleman, 1996). Many items that are fried produce acrylamide. These fried products include chips, French fries, oatmeal, ground coffee, snack foods, and bakery goods. After frying food, a common hazardous substance known as acrylamide is formed in the oil (Friedman & Levin, 2008). The acrylamide can be estimated by using the LC-MS (Figure 4.6).

Lard, sunflower oil, and SBO have all been utilized. Based on the findings, acrylamide wasn't really created while asparagine is being cooked without oil (control), but acrylamide formed when various amounts of asparagine were cooked using frying oil.

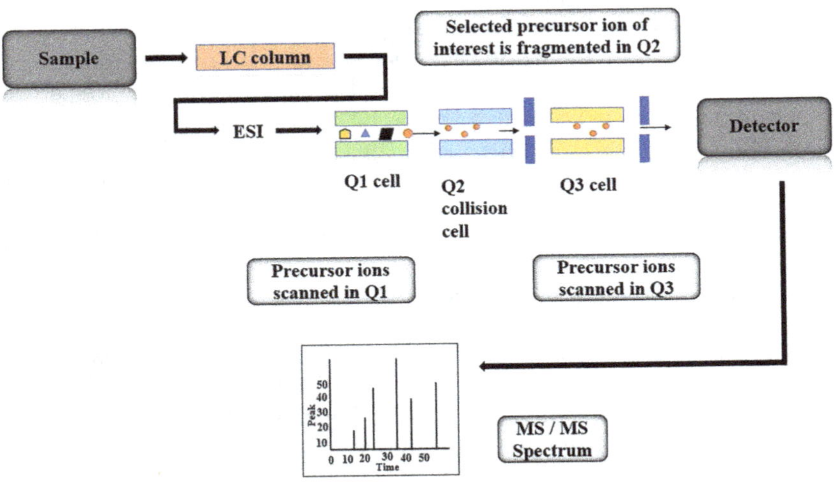

Figure 4.6 Schematic diagram for the analysis of acrylamide in foods.

4.7 Trans Fatty Acids and the Frying Process

Because of the link between trans fatty acid (TFA) consumption and various cancers, chronic respiratory diseases, neurological problems, and cardiovascular diseases (CVD), interest in TFAs has substantially increased (Wang et al., 2016). According to the research by Mozaffarian et al. (2004), C-reactive proteins, which are indicators of systemic inflammation, are linked to TFAs in females. The Lopez-Garcia et al. (2005) study suggests that a higher trans-fat diet can lead to endothelial dysfunction. Bharti et al. (2020) and Tsuzuki et al. (2008) found that heat techniques, such as frying, produce various levels of TFAs which are subjected to the oils employed. As described by Tsuzuki et al. (2020), however, the majority of TFAs in these items are believed to be due to the oil used, rather than the technique. Per Paiva-Martins and Kiritsakis (2017), the frying process increased the levels of (C18:1t) trans oleic acid and (C18:2t) trans linoleic acid in SFA, leading to the thermal oxidation. Bansal et al. (2009) determined that 1.10% of trans isomers can be produced at 200°C for 40 minutes, while 11.45% can be produced at 300°C for the same time period. Gertz et al. (2000) described how fast FO degrades when fried ingredients are hydrolyzed by water. Hydrolysis may also play a role in isomerizing double bonds in UFAs in the case of TFAs in FO. In this study, Suleman et al. (2020) were able to demonstrate that frying at a low temperature of 160°C could be advantageous in terms of consuming fewer TFAs as well as consuming fewer deteriorating edible oils. The generation of trans isomers occurs during food frying as discovered by Bansal et al. (2009). C18:1 TFA concentrations in FO are greater than those in oil being heated and hydrogenated. According to Aladedunye and Przybylski (2009b) and Prado et al. (2017), TFAs released during frying are higher in pre-fried frozen meals. This resulted in 2.1% and 14.3%, respectively, of trans MUFA. When FO was fried for 50 hours, 59.9% of the PUFAs were reduced to 32.6%. When partly hydrogenated soybean fat (PHSF) was used, trans MUFAs formed in FO and hydrogenated oils (HOs) increased from 20.2% to 28.5%. These findings clearly demonstrate that trans MUFAs are formed in oil being used for frying and hydrolyzed oil during the frying process. Since HO contains fewer UFAs than FO, trans isomers were created at a slower rate. Using gas liquid chromatography, Al Rehah and Anany (2012) investigated the synthesis of TFAs in various oil sample blends (SFO+CLO+PO) while frying at 180°C for 20 hours (GLC). TFA formation was seen to rise during the frying process, and low TFAs were generated by combining sunflower seed oil (SFO) with PO or canola oil (CLO). The impact of heated oils (i.e., HOs) on TFA intake was studied by Yang et al. (2014) through frying and heating procedures. The quantity of TFAs grew when the frying process was prolonged and edible oils were heated to high temperatures. At 160°C, 180°C, and 200°C, C18:1t FA was detected in FO and HO, but the oil content was below the permissible limit of 0.047%. A 0.02% increase in C18:2t and a 0.05% increase in C18:3t FAs were observed in FO. In HO, C18:2t FAs were slightly lower than in FO. The C18:1t FAs ranged from 0.05% to 0.14% in safflower oil (SAFO) and from 0.09% to 0.11% in rice bran oil (RBO). SO had FAs in concentrations ranging from 0.14% to 0.25% and 0.31% to 0.59%, respectively (Juárez et al., 2011). Trans-isomer contents of SBO, SFO (SO1 and SO2), and partially hydrogenated mixed vegetable oils (PHMVO1 and PHMVO2) ranged from 2.0% to 2.2%. Trans-isomers had been found in large quantities in PHMVO2, but they were only found in small quantities after the frying operation.

4.8 Free Fatty Acids and the Frying Process

The generation of monoacylglycerols, diacylglycerols, FFAs, and glycerol is another crucial oxidation factor when measuring FO degradation. As a result of hydrolysis, FFAs, glycerol, monoacylglycerols, and diacylglycerols are formed during the frying of foods. Latha and Nasirullah (2014) suggest that oxidative and thermal degradation of UFAs could be responsible for the synthesis of FFAs in oil used for frying during the heating and frying process. In accordance with Matthäus (2006), FFAs are widely used to determine whether FOs are suitable for human consumption. A limit of 2% is set for oil rejection. During the frying process, FFAs in SBO, canola oil (CLO), PO, and *Moringa oleifera* seed oil (MoO) were studied by Abdulkarim et al. (2007). In SBO, CLO, PO, and MoO, respectively, the amounts of FFAs were 0.22%, 0.10%, 0.19%, and 0.17%. A frying technique resulted in the lowest FFA levels (60%) in SBO, PO, MoO, and CLO after 5 days, and the highest FFA levels (71.4%) in CLO. A study conducted by Manral et al. (2008) examined the FFA concentration in SFO after 14 hours of frying at 180°C. After 14 hours, FFA concentrations were found to increase from 0.05% to 0.585% in SFO. Politi et al. (2017) studied the deterioration of seed oil and SFO during heating. During a 32-hour frying time at 147°C, 171°C, and 189°C, we examined cis-unsaturation and TFAs together with FFAs using Fourier transform infrared spectroscopy (FT-IR). FFAs are released into the oil during frying at 180°C (PHMVO1 and PHMVO2), but the cis-unsaturation is decreased. Juárez et al. (2011) demonstrated the presence of FFAs in SBO, SFO, and partially hydrogenated mixed vegetable oil. FFAs for oleic acid were increased in SBO, SO1, SO2, PHMVO1, and PHMVO2. Partially HOs showed the greatest changes, with 0.06%–0.36%, 0.05%–0.61%, 0.38%–2.02%, 0.11%–1.04%, and 0.03%–1.41%.

4.9 The Influence of Frying on Peroxide Value

Recent studies have demonstrated that PVs rise with frying time. PV is therefore employed to measure primary oxidation in oils and fats (Goburdhun & Jhaumeer-Laull, 2000). Che Man and Tan (1999) found that PV levels in oil increased during heating and frying. Man and Jaswir (2000) reported that during the process of deep frying potato chips, PV levels rose from 0.91 to 11.70 meqO_2/kg of oil. After 300 minutes of deep frying potato chips, their PV rose from 6.6 to 12.6 meqO_2/kg of oil (Goburdhun & Jhaumeer-Laull, 2001). According to Latha and Nasirullah (2014), RBO and refined SBO had a PV increase of 3.6–6.0 and 5.7–11.2 meqO_2/kg, respectively, during the process of frying. Bangash and Khattak (2006) analyzed the PV and other parameters that are indices for oxidation in standard mineral oil (SMO) and Sun flower seed oil (SFO) over the course of 5 days and 20 minutes at 180°C–190°C. With frying, PV was increased from 16.17 to 81.40 meqO_2/kg of oil, and SMO and SFO showed a PV enhancement of 18.60 to 85.50 meqO_2/kg of oil. Abdulkarim et al. (2007) investigated how PV increased during the frying process in *MoO*, PO, and SBO between 0 and 5 days. In CLO, SBO, PO, and MoO, PV changes each day are 0.72%, 0.69%, 0.67%, and 0.62%, respectively. A PV value of 5.08 was found in CLO, SBO, PO, and MoO (after 2 days), 4.42 was found after 2 days, 4.92 was found 4 days later, and 4.0 meqO_2/kg of oil after 4 days. As a result of unstable oxidative degradation in SBO and CLO when C18:3 FA was present, PV increased rapidly. Using C18:3 FA, CLO has been shown to be the least resistant to oxidation, followed by SBO, which has shown the least resistance to oxidation. Abdulkarim et al. (2007) evaluated the PV in refined SFO by frying fish at 180°C for 14 hours. As a result of the 12-hour frying, PV increased from 0.1 to 28.98 meqO_2/kg of oil. In addition, the PV decreased from 0.12 to 24.88 meqO_2/kg of oil after 14 hours of frying, which suggests that the drop in PV may be explained by the unpredictability of peroxides (Fritsch, 1981). The PV increased in all blended oils; however, PO:CLO showed the best oxidation stability. Siddique et al. (2015) measured PV in different 50:50 blends, such as PO:CLO, PO:SFO, and PO:SBO.

4.10 Iodine Value and the Frying Process

Goburdhun and Jhaumeer-Laull (2001) used FT-IR spectroscopy and other oxidation metrics during frying potato chips in SFO at 180°C for 600 minutes (Figure 4.7). There was an increase in FFAs and PVs, but a decline in IVs. In the presence of trans USFAs, the cis double bonds have been lost. The loss of unsaturation can be observed in FT-IR by a drop in IV. Moreover, Naz et al. (2008) found that in addition to a linear decrease in FO IV due to USFA loss and the lipid oxidation period, frying durations increased. Using commercially and laboratory available SFO with or without antioxidant substances, Ata-Ur-Rehman et al. (2006) fried potato chips and samosas at 190°C. At frying temperature, the IV was observed, but in laboratory frying, it was lower. Abdulkarim et al. (2007) found that laboratory frying is more efficient than industrial frying, and that adding antioxidants improves the outcome. They examined the effects of IV on SBO, CLO, PO, and MoO during the process of frying. During 5 days of frying, SBO and CLO showed

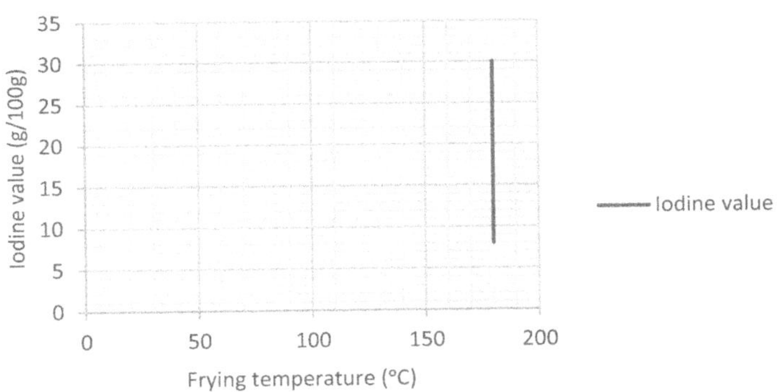

Figure 4.7 Trend for frying temperature vs. iodine value in fried foods (Naz et al., 2008).

the greatest changes at 116.9–111.8 and 109.9–103.0 g/100g, respectively. MoO and PO showed lower changes at 65.9–62.2 and 56.8 g/100g, respectively. The findings showed that PO and MoO were degradable to a lesser extent than CLO and SBO. Researchers used a commercial fryer at 190°C to measure the content of IV in SFO at frying temperature. The amount of IV in SFO was discovered to decrease during the frying or heating process.

4.11 Conjugated Diene and Triene and the Frying Process

According to literature, CLO conjugated diene (CD), conjugated trienes (CT), and CV were measured throughout the process of heating at 180°C (Farhoosh & Moosavi, 2009). CD and CT initially rose, but as the heating process progressed, they plateaued. Conjugated dienes between CD, CT, and polymer production. According to Farhoosh and Tavassoli-Kafrani (2011), CD was measured at 32 hours of frying. With longer frying times, CD rose from 10.6 to 61.6 mmol/L. The recommendation was made to discard the SFO as soon as CD reached 44 mmol/L. Using samples heated at 180°C for 16 hours (Farhoosh et al., 2012) in HO samples to determine the CD. From 7.0 to 29.3 mmol/L, the CD was observed to have risen.

4.12 Total Polar Compounds and Carbonyl Value Affected by Frying

The 1^0 and 2^0 oxidation products, respectively, are TPC and CV. As evidenced by Abdulkarim et al. (2007), one of the most trustworthy techniques for monitoring value changes over time in culinary oils and fats is to measure TPC and CV while the oil is being fried. TPC and CV, a worldwide standard for FOs and fats, have been established (Farhoosh & Tavassoli-Kafrani, 2011). The amount of TPC in potatoes was increased by 6.2%–18.7% after 15 days of frying in SFO according to Pokorný (1999). Sanibal and Mancini-Filho (2004) examined color index, IV, TPC, and FFAs after 50 hours at 180°C in SBO and partly hydrogenated soybean oil (PHSBO). During frying, color index, TPC, and FFAs rose in SBO and PHSBO, while IV fell. TPC concentrations in SBO and PHSBO, respectively, range from 8.6% to 40% and 6.6% to 23.0%, respectively, after 50 hours of frying. According to Babst (2012), changes occurred in SBO are more than PHSBO due to the increased amount of USFAs. Fresh oil samples contained less than 8% TPC, but after 80 hours of frying, spent oil samples still contained 30% TPC. Fresh FO samples' CV and TPC were larger than 2 mol/g, as reported by Farhoosh et al. (2008). During the frying process, they analyzed at how these two parameters were affected by frying. From 7.76 to 123.45 mol/g, CV and TPC rose linearly as frying time was increased. The CV and TPC were substantially correlated throughout the frying period, with an excellent coefficient correlation (R_2=0.9747). According to Jantan et al. (2008), bananas fried in edible oils like SFO, SBO, CLO, CO, and PO contained TPC. Over the same period, the TPC values for SFO, SBO, CLO, CO, and PO increased from 3.5% to 32.7%, 4.7% to 35.3%, 2.0% to 24.4%, 3.8% to 37.6%, and 3.3% to 20.5%, respectively. A comparison of TPC values of CO and SFO was conducted by Ersoy and Özeren (2009). When fish were fried for 10 and 14 hours at 180°C, the TPC grew from 3.58% to 26.9% and 35.2%, respectively. Aladedunye and Przybylski (2009b) investigated the TPC of French fries fried in CLO at 185°C and 215°C for 7 days/7 hours. During frying, the TPC rose linearly from 4.2% to 19.8% and 38%. The TPC after 4 days at 215°C was in the discard level (24%) because the value was higher than acceptable. RBD CLO, SO, PO, and their mixes (1:1, w/w) were examined by Alireza et al. (2010) for their frying stability at 180°C. From 0 to 5 days of frying, the TPC in all six FO samples varied as 8.50%, 6.94%, 9.94%, 8.18%, 6.00%, and 10.63%. All six FO samples contained a range of TPC values as 8.50%, 6.94%, 9.94%, 8.18%, 6.00%, and 10.63%. In all six FO samples, CLO contained the highest TPC concentration due to its high percentage of USFAs. Farhoosh and Tavassoli-Kafrani (2011) examined CV, TPC, and CD levels of SFO after 32 hours of frying. As the frying duration increased, CV, TPC, and CD ranges increased from 7.9 to 70.9 mol/g, 2.8% to 28.9%, and 61.6 mol/L. It was suggested to discard the SFO when the TPC and CDV values reached 20% and 44 mmol/L, respectively.

In İnanç Horuz and Maskan's (2015) study, 200 FO samples were collected from restaurants, cafeterias, and school canteens to assess TPC. The TPC was found to be 32.0% in 200 FO samples, 64 of which were unsatisfactory. Basuny et al. (2012) studied the chemical and physical characteristics of commercial chicken and samosa frying oils. Several physical and chemical characteristics were observed earlier and later phases of frying. Fresh and used oil samples had TPC values between 2.30% and 32.29%. Using inferior oil quality and performing a large number of frying cycles indicated higher TPC values. The TPC in SFO was investigated in a commercial fryer at 190°C by Guillén and Uriarte (2012). As the frying process progressed, the concentration of TPC increased. Rodríguez et al. (2021) determined the TPC concentrations in Oo, GSO, CO, SFO, and SBO. When the TPC was fried, it grew and took 41.7, 20.0, 37.2, 10.5 and 33.5 hours, respectively, to reach the recommended TPC value. In a study by Abdulkarim et al. (2007), potato chips were fried for 6 hours a day for 5 days. In SBO, CLO, PO, and MoO, TPC rose as 31.82%, 28.73%, 21.23%, and 20.78% after frying. CLO and SBO had the highest TPC compared to PO and MoO.

4.12.1 Changes in Mineral Content

Due to their hydrophilicity, minerals undergo significant modifications during culinary processes like boiling and frying. Their modifications while frying were essentially insignificant because they are hardly dissolved in cooking oil.

The mass of foods that are fried drops during the process of frying as a result of loss of water. Because the majority of mineral constituents are non-volatile, the amount of minerals should increase as wet weight does. Furthermore, an additional activity going on at the same time is the intake of cooking oil. If the metal percentage is represented on a dry-weight, then the amount of minerals would moderately drop as the mass of the fried medium increases.

4.12.2 Changes in Vitamin Content

Most vitamins are vulnerable to oxidation and rising temperatures. Although foods that are fried mainly reaches extreme temperatures in the outermost layers, their losses are unquestionably substantial. Core temperature, which typically ranges somewhere between 70°C and 90°C, has a major impact on losses in total. In this spectrum, interior temperature has a far greater impact on vitamin preservation than does the temperature of the cooking oil. Vitamins are thermosensitive, and the interior temperatures of the meal and the process of frying quickly oxidize them. Typically, ascorbic acid is the most thermosensitive. Vitamins B1, B12, B3, and B6 are indeed the B-complex vitamins that are mostly affected by the process of frying.

4.13 Techniques for Measuring the Characteristics of Frying Oil

In addition to analyzing edible oils and fats, a bunch of methods have been developed to measure FO quality (Figure 4.8). As explained by Ferreiro-Vera et al. (2013), the most popular methods for determining FAC, TFAs, and fatty acid ratios are GC and FT-IR spectroscopy. Mahesar et al. (2010), Saguy and Dana (2003), and Sherazi et al. (2010) conducted GC in margarines and biscuits. Aside from using harmful organic solvents and reagents, GC involves derivatization, which makes it more time-consuming and expensive than FT-IR. FT-IR, as described by Goburdhun and Jhaumeer-Laull (2001), is an intriguing analytical method since it's quick, easy, and doesn't need any sample preparation. Multivariate analysis is commonly used to investigate fats and oils in addition to FT-IR spectroscopy, including main fat groups, TFAs, FAC, and oxidation products including CD, CT, FFAs, PV, IV, and saponification number.

Additionally, unsaturation, carbonyl compounds, photovoltaics, FFAs, dielectric constant, density, and TPC can all be examined using FT-IR to identify the oxidative deterioration of edible oils (Goburdhun & Jhaumeer-Laull, 2001). In the previous study, the authors write that NMR (nuclear magnetic resonance) has been widely used to analyze oxidation products such as TFAs and FFAs (Sherazi & Mahesar, 2015). It is generally

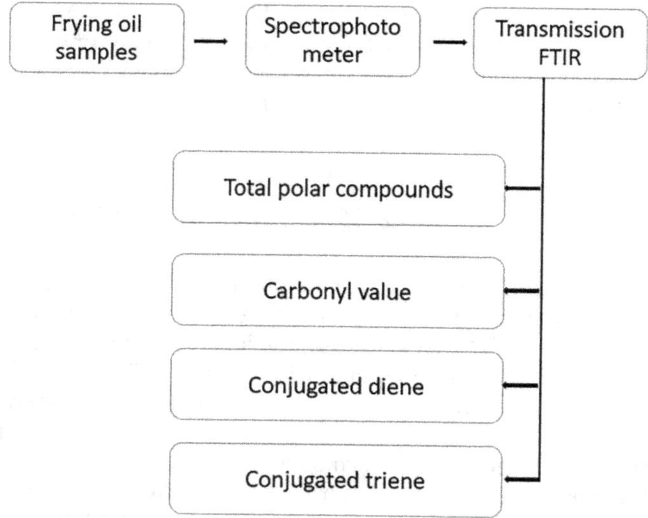

Figure 4.8 Spectrophotometric analysis of fried samples.

recommended to measure hydroperoxide levels using AOCS's iodometric method (Crowe & White, 2001), where meqO_2/kg of oil is used to represent PV. Titrimetric techniques are unsustainable because they take a long time and use a lot of chemicals and solvents in addition to being sensitive and reproducible. A number of other methods have been published in the literature to determine PV, including spectrophotometry and infrared spectroscopy. A conventional spectrophotometric approach is complex and involves a number of chemicals and reagents. Løvaas (1992) states that the method involves a number of chemicals and reagents. As a result of this research, we developed an FT-IR method for generating triphenylphosphine oxide in the presence of oil hydroperoxides, using triphenylphosphine (TPP) as a starting material.

4.14 Effects of Chicken Frying on Soybean, Sunflower, and Canola Oils

4.14.1 Frying Process

Fry operations were conducted using a West Point deep fryer (E-2016) after meticulously cleaning, drying, and washing the parts for 30 minutes. This fryer has a 3L capacity and can be set to a temperature between 0°C and 190°C. For two days in a row, 200g chicken pieces were fried at constant frying temperatures (190°C) for 15 minutes at a time (SFO, SBO and CLO). Each of the three oils (SFO, SBO, and CLO) was collected six times over the 12hours of frying (Figure 4.9). At the conclusion of each 2hours frying phase, a vial holding about 25mL of the FO was withdrawn within the fryer.

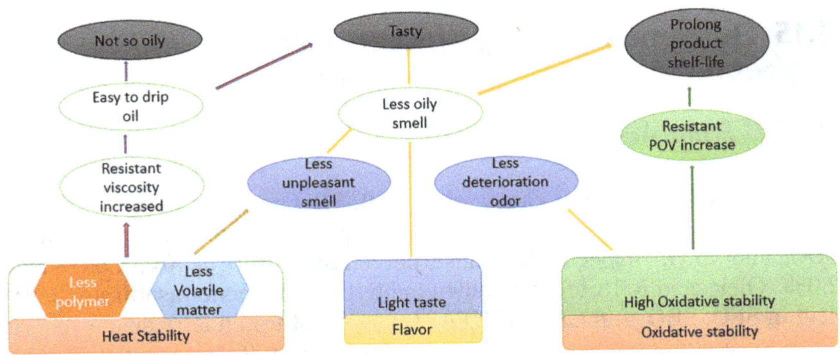

Figure 4.9 Oil stability and its effects on taste.

4.14.2 Fatty Acid Profile Determination and GC-MS Conditions

Mansor et al. (2012) determined the FA profile of fresh and used commercial oil samples using the IUPAC methodology. According to the standard procedure, 200 mg of oil samples were dissolved in 4 mL of sodium hydroxide solution and 5 mL of BF3 methanolic solution in 50 mL conical flasks. The extracted methyl esters were analyzed by GC-MS after 15 minutes of reflux.

Agilent Technologies gas chromatographs were outfitted with Agilent auto-sampler 7683-B injectors and inert XL Mass selective detectors for the detection of fatty acid methyl esters (FAMEs) in commercial FO samples using gas chromatography-mass spectrometry (MS) (MS-5975). FAMEs were separated using a capillary column that contained Rt-2560 biscyanopropylsiloxane capillary columns with a length of 100 m, a diameter of 0.25 mm, and a film thickness of 0.25 µm. The temperature was maintained at 140°C for 2 minutes and then increased by 4°C/min for 5 minutes to 230°C. The temperatures of the injection and detector are raised to 240°C and 260°C, respectively. 50:1 was the split ratio. The translating line was heated to 270°C, the quadruple to 150°C, and the ion source to 230°C. The mass spectrometer worked in the electron impact (EI) mode at 70 eV, with ion source temperatures of 230°C, 150°C, and 270°C for the translating line. Using an E_m voltage of 1,035 V, the mass scan identified FAs in the m/z range of 50–550. The peak FAs in both fresh and fried oil samples were identified using MS spectra and standard retention times (R_t). To validate the findings of the GC-MS libraries, the FAME standard of linolenic, linoleic, oleic, stearic, palmitic, myristic, elaidic, linolelaidic, and linolenelaidic acids was used. Only methyl esters whose mass spectra match the spectra of all other methyl esters at a proportion higher than 97% were included in the quantitative investigation.

4.15 Outcome of Frying Chicken with Soybean, Sunflower, and Canola Oils

4.15.1 Fatty Acid Composition

The fatty acid makeup of canola oil (CLO), SBO, and sunflower oil (SFO) is being studied. SFAs at 6.94%, 11.33%, and 4.78% were detected in SFO, SBO, and CLO, respectively. These percentages were 5.90%, 4.55%, and 2.03%. Oleic acid (C18:1c) was shown to be the main monounsaturated fatty acid in SFO, SBO, and CLO (MUFA) (monounsaturated fatty acids). There was also a significant concentration of elaidic acid (C18:1t) in SFO, SBO, and CLO, respectively, with 0.79%, 1.02%, and 1.29%. A high concentration of polyunsaturated fatty acids (PUFAs) was observed at SFO, SBO, and CLO, including linoleic (C18:2c), linoelaidic (C18:2t), linolenic (C18:3c), and linolenelaidic acids (C18:3t), with values of 64.87%, 54.67%, 24.33%. CLO produced the highest levels of saturated fatty acids during 12 hours of chicken at 190°C frying. CLO produced the highest levels of saturated fatty acids during 12 hours of frying (SFAs). CLO's C16:0 and C18:0 levels were 2.14% and 0.90%, respectively. In contrast to CLO, myristic acid (C14:0) was also raised between 0.10% and 0.56%, as well as 0.21% and 1.03% for both SFO and SBO, respectively. C18:1c and unsaturated fatty acids (USFAs) were reduced in the ranges of 19.44%–19.17%, 22.23%–22.12%, and 56.86%–56.63% in SFO, SBO, and CLO, respectively, according to the results of the tests. Approximately 1.69%, 0.09%, and 0.04% of the maximum C18:1t, C18:2t, and C18:3t were determined in CLO after 12 hours of frying with chicken, respectively.

When compared to their original FO samples, TFA levels in SFO, SBO, and CLO were considerably higher after 12 hours of frying chicken (Figures 4.10–4.12). The findings of Sanibal and Mancini-Filho (2004) supported the current findings by reporting that frequent usage of frying oils may raise TFAs content owing to the FAs interaction between the fried meal and oil as well as the high temperature and extended frying process. SFO, SBO, CLO, and CLO all showed a reduction in PUFA levels after frying chicken. Between the first and last frying cycles, all oils' percentages of C18:2 and C18:3 FAs declined. CLO showed a decrease in C18:2 and C18:3 FAs of 5.87% and 32.79%, respectively. The reduction in unsaturation is most likely due to the breaking of double bonds caused by oxidation and polymerization. It is shown that the amounts of C18:2 and C18:3 FAs decrease as the frying time increases. The amounts of C18:2 and C18:3 FAs for SFO, SBO, and CLO dramatically decrease from 0 to 12 hours of frying.

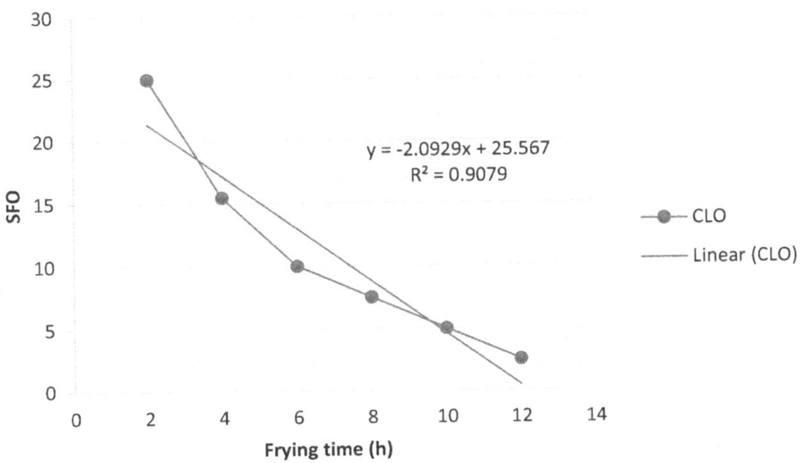

Figure 4.10 Trend of linoleic acid for SFO from 0 to 12 hours frying period (Talpur et al., 2009).

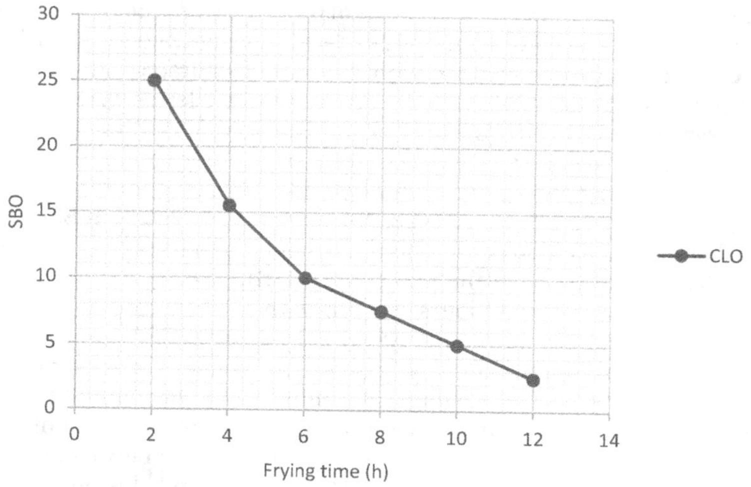

Figure 4.11 Trend of linoleic acid for SBO from 0 to 12 hours frying period (Talpur et al., 2009).

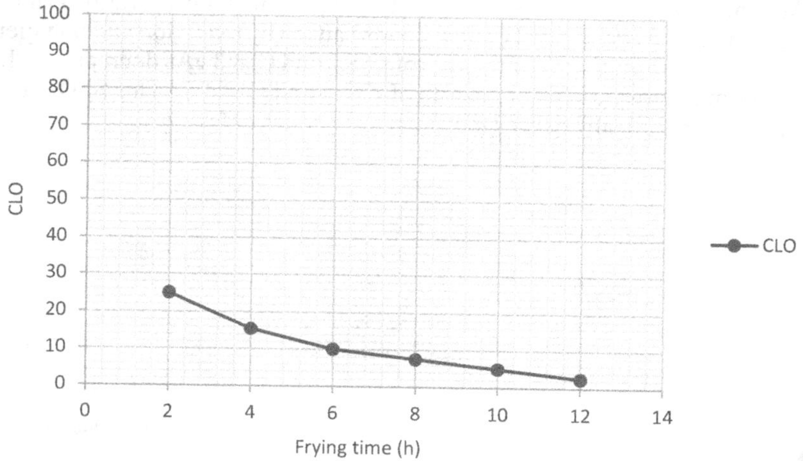

Figure 4.12 Trend of linoleic acid for CLO from 0 to 12 hours frying period (Talpur et al., 2009).

4.16 Conclusion

The FAC, macro- and micronutrient contents, color, purity, and other factors all have significant impacts on the quality of cooking oil. More than 90% of general public admit to eating fried food products often, and much more than

85% among these items have a high amount of fat (>10%). Given that it is clearly documented that this sort of food can incur from breakdown of its fat when exposed to elevated temperatures and the effect of this on the increased risk of experiencing cardiovascular illnesses, this condition has become a danger concern for the health of public. The research findings reported by many researchers on the evaluation of the lipid fraction of foods that have undergone deep frying or frying techniques are covered throughout this chapter. It has been amply proven that the fat contained inside them is both extensively oxidized and hydrolyzed, which are discussed in detail in this chapter. The results of the recorded research indicate that in addition to the seller's desire for profit, the oil should be discarded regularly for the health of customers.

References

Abdulkarim, S., Long, K., Lai, O. M., Muhammad, S., & Ghazali, H. (2007). Frying quality and stability of high-oleic Moringa oleifera seed oil in comparison with other vegetable oils. *Food Chemistry, 105*(4), 1382–1389.

Adler-Nissen, J. (1979). Determination of the degree of hydrolysis of food protein hydrolysates by trinitrobenzenesulfonic acid. *Journal of Agricultural and Food Chemistry, 27*(6), 1256–1262.

Al Rehah, F., & Anany, A. (2012). Physicochemical studies on sunflower oil blended with cold pressed tiger nut oil during the deep frying process; Estudios fisicoquimicos sobre mezclas de aceite de girasol con aceite de chufa prensado en frio durante el proceso de fritura. *Grasas y Aceites (Sevilla)*, 63, 455–465.

Al-Kahtani, H. A. (1991). Survey of quality of used frying oils from restaurants. *Journal of the American Oil Chemists Society, 68*(11), 857–862.

Aladedunye, F. A., & Przybylski, R. (2009a). Degradation and nutritional quality changes of oil during frying. *Journal of the American Oil Chemists' Society, 86*(2), 149–156.

Aladedunye, F. A., & Przybylski, R. (2009b). Protecting oil during frying: a comparative study. *European Journal of Lipid Science and Technology, 111*(9), 893–901.

Alireza, S., Tan, C. P., Hamed, M., & Che Man, Y. (2010). Effect of frying process on fatty acid composition and iodine value of selected vegetable oils and their blends. *International Food Research Journal, 17*(2), 295–302.

Arroyo, R., Cuesta, C., Sánchez-Montero, J., & Sánchez-Muniz, F. (1995). High performance size exclusion chromatography of palm olein used for frying. *Lipid/Fett, 97*(7–8), 292–296.

Ata-Ur-Rehman, F., Zahoor, T., & Tahira, R. (2006). Sensory evaluation of samosas and potato chips fried in different sunflower oils with and without butylated hydroxytoluene. *Pakistan Journal of Life and Social Sciences, 4*, 1–2.

Ayyildiz, H. F., Topkafa, M., Kara, H., & Sherazi, S. T. H. (2015). Evaluation of fatty acid composition, tocols profile, and oxidative stability of some fully refined edible oils. *International Journal of Food Properties, 18*(9), 2064–2076.

Babst, R. (2012). 13th European congress of trauma and emergency surgery. *European Journal of Trauma and Emergency Surgery, 38*(1), S1–S217.

Baer, D. J., Judd, J. T., Clevidence, B. A., & Tracy, R. P. (2004). Dietary fatty acids affect plasma markers of inflammation in healthy men fed controlled diets: a randomized crossover study. *The American Journal of Clinical Nutrition, 79*(6), 969–973.

Bangash, F., & Khattak, H. (2006). Effect of deep fat frying on physico-chemical properties of Silybum. *Journal of the Chemical Society of Pakistan, 28*(2), 121.

Bansal, G., Zhou, W., Tan, T.-W., Neo, F.-L., & Lo, H.-L. (2009). Analysis of trans fatty acids in deep frying oils by three different approaches. *Food Chemistry, 116*(2), 535–541.

Basuny, A. M., Arafat, S. M., & Ahmed, A. A. (2012). Vacuum frying: an alternative to obtain high quality potato chips and fried oil. *Banat's Journal of Biotechnology, 3*(5), 22–30.

Bharti, R., Guldhe, A., Kumar, D., & Singh, B. (2020). Solar irradiation assisted synthesis of biodiesel from waste cooking oil using calcium oxide derived from chicken eggshell. *Fuel, 273*, 117778.

Blumenthal, M. M., & Stier, R. F. (1991). Optimization of deep-fat frying operations. *Trends in Food Science & Technology, 2*, 144–148.

Bordin, K., Tomihe Kunitake, M., Kazue Aracava, K., & Silvia Favaro Trindade, C. (2013). Changes in food caused by deep fat frying-A review. *Archivos latinoamericanos de nutricion, 63*(1), 5–13.

Boskou, D., Blekas, G., & Tsimidou, M. (2006). Olive oil composition. In Dimitrios Boskou (Ed.), *Olive oil* (pp. 41–72). Elsevier.

Buenrostro-Gonzalez, E., Lira-Galeana, C., Gil-Villegas, A., & Wu, J. (2004). Asphaltene precipitation in crude oils: theory and experiments. *AIChE Journal, 50*(10), 2552–2570.

Calleman, C. J. (1996). The metabolism and pharmacokinetics of acrylamide: implications for mechanisms of toxicity and human risk estimation. *Drug Metabolism Reviews, 28*(4), 527–590.

Capouet, M., & Müller, J.-F. (2006). A group contribution method for estimating the vapour pressures of α-pinene oxidation products. *Atmospheric Chemistry and Physics, 6*(6), 1455–1467.

Che Man, Y., & Tan, C. (1999). Effects of natural and synthetic antioxidants on changes in refined, bleached, and deodorized palm olein during deep-fat frying of potato chips. *Journal of the American Oil Chemists' Society, 76*(3), 331–339.

Chen, J., Fu, J., Sheng, G., Liu, D., & Zhang, J. (1996). Diamondoid hydrocarbon ratios: novel maturity indices for highly mature crude oils. *Organic Geochemistry, 25*(3–4), 179–190.

Cheng, S.-S., Liu, J.-Y., Hsui, Y.-R., & Chang, S.-T. (2006). Chemical polymorphism and antifungal activity of essential oils from leaves of different provenances of indigenous cinnamon (Cinnamomum osmophloeum). *Bioresource Technology, 97*(2), 306–312.

Chiou, A., Kalogeropoulos, N., Boskou, G., & Salta, F. N. (2012). Migration of health promoting microconstituents from frying vegetable oils to French fries. *Food Chemistry, 133*(4), 1255–1263.

Coupland, J., Zhu, Z., Wan, H., McClements, D., Nawar, W., & Chinachoti, P. (1996). Droplet composition affects the rate of oxidation of emulsified ethyl linoleate. *Journal of the American Oil Chemists' Society, 73*(6), 795–901.

Crowe, T. D., & White, P. J. (2001). Adaptation of the AOCS official method for measuring hydroperoxides from small-scale oil samples. *Journal of the American Oil Chemists' Society, 78*(12), 1267–1269.

Dana, D., & Saguy, I. S. (2001). Frying of nutritious foods: obstacles and feasibility. *Food Science and Technology Research, 7*(4), 265–279.

Dunford, N. T., Temelli, F., & LeBlanc, E. (1997). Supercritical CO_2 extraction of oil and residual proteins from Atlantic mackerel (Scomber scombrus) as affected by moisture content. *Journal of Food Science, 62*(2), 289–294.

Durán, M., Pedreschi, F., Moyano, P., & Troncoso, E. (2007). Oil partition in pre-treated potato slices during frying and cooling. *Journal of Food Engineering, 81*(1), 257–265.

El-Hadad, S. S., & Tikhomirova, N. A. (2018). Physicochemical properties and oxidative stability of butter oil supplemented with corn oil and dihydroquercetin. *Journal of Food Processing and Preservation, 42*(10), e13765.

Endo, Y., Li, C. M., Tagiri-Endo, M., & Fujimoto, K. (2001). A modified method for the estimation of total carbonyl compounds in heated and frying oils using 2-propanol as a solvent. *Journal of the American Oil Chemists' Society, 78*(10), 1021–1024.

Ersoy, B., & Özeren, A. (2009). The effect of cooking methods on mineral and vitamin contents of African catfish. *Food Chemistry, 115*(2), 419–422.

Esfarjani, F., Khoshtinat, K., Zargaraan, A., Mohammadi-Nasrabadi, F., Salmani, Y., Saghafi, Z., … Bahmaei, M. (2019). Evaluating the rancidity and quality of discarded oils in fast food restaurants. *Food Science & Nutrition, 7*(7), 2302–2311.

Farhoosh, R., & Moosavi, S. (2009). Evaluating the performance of peroxide and conjugated diene values in monitoring quality of used frying oils. *Journal of Agricultural Science and Technology, 11*(2), 173–179.

Farhoosh, R., Niazmand, R., Rezaei, M., & Sarabi, M. (2008). Kinetic parameter determination of vegetable oil oxidation under Rancimat test conditions. *European Journal of Lipid Science and Technology, 110*(6), 587–592.

Farhoosh, R., Khodaparast, M. H. H., Sharif, A., & Rafiee, S. A. (2012). Olive oil oxidation: rejection points in terms of polar, conjugated diene, and carbonyl values. *Food Chemistry, 131*(4), 1385–1390.

Farhoosh, R., & Tavassoli-Kafrani, M. H. (2011). Simultaneous monitoring of the conventional qualitative indicators during frying of sunflower oil. *Food Chemistry, 125*(1), 209–213.

Farkas, B., Singh, R., & Rumsey, T. (1996). Modeling heat and mass transfer in immersion frying. I, model development. *Journal of Food Engineering, 29*(2), 211–226.

Ferreiro-Vera, C., Priego-Capote, F., Mata-Granados, J. M., & de Castro, M. D. L. (2013). Short-term comparative study of the influence of fried edible oils intake on the metabolism of essential fatty acids in obese individuals. *Food Chemistry, 136*(2), 576–584.

Filip, S., Hribar, J., & Vidrih, R. (2011). Influence of natural antioxidants on the formation of trans-fatty-acid isomers during heat treatment of sunflower oil. *European Journal of Lipid Science and Technology, 113*(2), 224–230.

Friedman, M., & Levin, C. E. (2008). Review of methods for the reduction of dietary content and toxicity of acrylamide. *Journal of Agricultural and Food Chemistry, 56*(15), 6113–6140.

Fritsch, C. (1981). Measurements of frying fat deterioration: a brief review. *Journal of the American Oil Chemists' Society, 58*(3), 272–274.

Gertz, C., Klostermann, S., & Kochhar, S. P. (2000). Testing and comparing oxidative stability of vegetable oils and fats at frying temperature. *European Journal of Lipid Science and Technology, 102*(8–9), 543–551.

Gesteiro, E., Guijarro, L., Sánchez-Muniz, F. J., Vidal-Carou, M. D. C., Troncoso, A., Venanci, L., … González-Gross, M. (2019). Palm oil on the edge. *Nutrients, 11*(9), 2008.

Goburdhun, D., & Jhaumeer-Laull, S. B. (2001). Evaluation of soybean oil quality during conventional frying by FTIR and some chemical indexes. *International Journal of Food Sciences and Nutrition, 52*(1), 31–42.

Goburdhun, D., Seebun, P., & Ruggoo, A. (2000). Effect of deep-fat frying of potato chips and chicken on the quality of soybean oil. *Journal of Consumer Studies & Home Economics, 24*(4), 223–233.

Guillén, M. D., & Uriarte, P. S. (2012). Aldehydes contained in edible oils of a very different nature after prolonged heating at frying temperature: presence of toxic oxygenated α, β unsaturated aldehydes. *Food Chemistry, 131*(3), 915–926.

Gunstone, F. (2009). *The chemistry of oils and fats: sources, composition, properties and uses*. John Wiley & Sons.

Houhoula, D. P., Oreopoulou, V., & Tzia, C. (2002). A kinetic study of oil deterioration during frying and a comparison with heating. *Journal of the American Oil Chemists' Society, 79*(2), 133–137.

Hurt, R. T., Kulisek, C., Buchanan, L. A., & McClave, S. A. (2010). The obesity epidemic: challenges, health initiatives, and implications for gastroenterologists. *Gastroenterology & Hepatology, 6*(12), 780.

İnanç Horuz, T., & Maskan, M. (2015). Effect of the phytochemicals curcumin, cinnamaldehyde, thymol and carvacrol on the oxidative stability of corn and palm oils at frying temperatures. *Journal of Food Science and Technology, 52*(12), 8041–8049.

Ivanova, S., Marinova, G., & Batchvarov, V. (2016). Comparison of fatty acid composition of various types of edible oils. *Bulgarian Journal of Agricultural Science, 22*(5), 849–856.

Jantan, I. b., Karim Moharam, B. A., Santhanam, J., & Jamal, J. A. (2008). Correlation between chemical composition and antifungal activity of the essential oils of eight cinnamomum. species. *Pharmaceutical Biology, 46*(6), 406–412.

Juárez, M. D., Osawa, C. C., Acuña, M. E., Sammán, N., & Gonçalves, L. A. G. (2011). Degradation in soybean oil, sunflower oil and partially hydrogenated fats after food frying, monitored by conventional and unconventional methods. *Food Control, 22*(12), 1920–1927.

Jung, M. Y., & Min, D. B. (1992). Effects of oxidized α-, γ-and δ-tocopherols on the oxidative stability of purified soybean oil. *Food Chemistry, 45*(3), 183–187.

Kris-Etherton, P. M., Harris, W. S., & Appel, L. J. (2002). Fish consumption, fish oil, omega-3 fatty acids, and cardiovascular disease. *Circulation, 106*(21), 2747–2757.

Krokida, M., Oreopoulou, V., Maroulis, Z., & Marinos-Kouris, D. (2001). Colour changes during deep fat frying. *Journal of Food Engineering, 48*(3), 219–225.

Latha, R. B., & Nasirullah, D. (2014). Physico-chemical changes in rice bran oil during heating at frying temperature. *Journal of Food Science and Technology, 51*(2), 335–340.

Lee, J., Lee, Y., & Choe, E. (2007). Temperature dependence of the autoxidation and antioxidants of soybean, sunflower, and olive oil. *European Food Research and Technology, 226*(1), 239–246.

Liu, K.-S. (1994). Preparation of fatty acid methyl esters for gas-chromatographic analysis of lipids in biological materials. *Journal of the American Oil Chemists' Society, 71*(11), 1179–1187.

Lopez-Garcia, E., Schulze, M. B., Meigs, J. B., Manson, J. E., Rifai, N., Stampfer, M. J.,... Hu, F. B. (2005). Consumption of trans fatty acids is related to plasma biomarkers of inflammation and endothelial dysfunction. *The Journal of Nutrition, 135*(3), 562–566.

Løvaas, E. (1992). A sensitive spectrophotometric method for lipid hydroperoxide determination. *Journal of the American Oil Chemists Society, 69*(8), 777–783.

Mahesar, S., Kandhro, A. A., Cerretani, L., Bendini, A., Sherazi, S., & Bhanger, M. (2010). Determination of total trans fat content in Pakistani cereal-based foods by SB-HATR FT-IR spectroscopy coupled with partial least square regression. *Food Chemistry, 123*(4), 1289–1293.

Man, Y. B. C., & Jaswir, I. (2000). Effect of rosemary and sage extracts on frying performance of refined, bleached and deodorized (RBD) palm olein during deep-fat frying. *Food Chemistry, 69*(3), 301–307.

Manral, M., Pandey, M., Jayathilakan, K., Radhakrishna, K., & Bawa, A. (2008). Effect of fish (Catla catla) frying on the quality characteristics of sunflower oil. *Food Chemistry, 106*(2), 634–639.

Mansor, T., Che Man, Y., Shuhaimi, M., Abdul Afiq, M., & Ku Nurul, F. (2012). Physicochemical properties of virgin coconut oil extracted from different processing methods. *International Food Research Journal, 19*(3), 837–845.

Matthäus, B. (2006). Utilization of high-oleic rapeseed oil for deep-fat frying of French fries compared to other commonly used edible oils. *European Journal of Lipid Science and Technology, 108*(3), 200–211.

Miller, J. I., & Ratti, R. A. (2009). Crude oil and stock markets: stability, instability, and bubbles. *Energy Economics, 31*(4), 559–568.

Moro, K., Nagahashi, M., Ramanathan, R., Takabe, K., & Wakai, T. (2016). Resolvins and omega three polyunsaturated fatty acids: clinical implications in inflammatory diseases and cancer. *World Journal of Clinical Cases, 4*(7), 155.

Mozaffarian, D., Pischon, T., Hankinson, S. E., Rifai, N., Joshipura, K., Willett, W. C., & Rimm, E. B. (2004). Dietary intake of trans fatty acids and systemic inflammation in women. *The American Journal of Clinical Nutrition, 79*(4), 606–612.

Nayak, P. K., Dash, U., Rayaguru, K., & Krishnan, K. R. (2016). Physio-chemical changes during repeated frying of cooked oil: a review. *Journal of Food Biochemistry, 40*(3), 371–390.

Naz, S., Siddiqi, R., & Asad Sayeed, S. (2008). Effect of flavonoids on the oxidative stability of corn oil during deep frying. *International Journal of Food Science & Technology, 43*(10), 1850–1854.

Ngadi, M., Adedeji, A., & Kassama, L. (2008). Microstructural changes during frying of foods. *Advances in Deep-Fat Frying of Foods*, 169–200.

Oommen, T. (2002). Vegetable oils for liquid-filled transformers. *IEEE Electrical Insulation Magazine, 18*(1), 6–11.

Paiva-Martins, F., & Kiritsakis, A. (2017). Olive fruit and olive oil composition and their functional compounds. In Fereidoon Shahidi & Apostolos Kiritsakis (Eds.), *Olives and Olive Oil as Functional Foods. Bioactivity, Chemistry and Processing* (pp. 81–116). Wiley.

Paradis, A., & Nawar, W. (1981). Evaluation of new methods for the assessment of used frying oils. *Journal of Food Science, 46*(2), 449–451.

Paul, S., Mittal, G., & Chinnan, M. (1997). Regulating the use of degraded oil/fat in deep-fat/oil food frying. *Critical Reviews in Food Science and Nutrition, 37*(7), 635–662.

Pedreschi, F., Kaack, K., Granby, K., & Troncoso, E. (2007). Acrylamide reduction under different pre-treatments in French fries. *Journal of Food Engineering, 79*(4), 1287–1294.

Pokorný, J. (1999). Changes of nutrients at frying temperatures. *Frying of Food, Oxidation, Nutrient and Non-Nutrient Antioxidants, Biologically Active Compounds and High Temperatures*. Technomic Publishing Co., Inc. Lancaster, Pennsylvania, 69–104.

Politi, F. A. S., Nascimento, J. D., da Silva, A. A., Moro, I. J., Garcia, M. L., Guido, R. V. C., ... Furlan, M. (2017). Insecticidal activity of an essential oil of Tagetes patula L. (Asteraceae) on common bed bug Cimex lectularius L. and molecular docking of major compounds at the catalytic site of ClAChE1. *Parasitology Research, 116*(1), 415–424.

Prado, T., Porto, B., & Oliveira, M. (2017). Method optimization for trans fatty acid determination by CZE-UV under direct detection with a simple sample preparation. *Analytical Methods, 9*(6), 958–965.

Recknagel, R. O., Glende, E. A., & Britton, R. S. (2020). Free radical damage and lipid peroxidation. In Robert G. Meeks & Steadman Harrison (Eds.), *Hepatotoxicology* (pp. 401–436). CRC Press.

Remig, V., Franklin, B., Margolis, S., Kostas, G., Nece, T., & Street, J. C. (2010). Trans fats in America: a review of their use, consumption, health implications, and regulation. *Journal of the American Dietetic Association, 110*(4), 585–592.

Rodríguez, G., Squeo, G., Estivi, L., Berru, S. Q., Buleje, D., Caponio, F.,... Hidalgo, A. (2021). Changes in stability, tocopherols, fatty acids and antioxidant capacity of sacha inchi (Plukenetia volubilis) oil during French fries deep-frying. *Food Chemistry, 340*, 127942.

Rodriguez-Garcia, I., Silva-Espinoza, B., Ortega-Ramirez, L., Leyva, J., Siddiqui, M., Cruz-Valenzuela, M. ... Ayala-Zavala, J. (2016). Oregano essential oil as an antimicrobial and antioxidant additive in food products. *Critical Reviews in Food Science and Nutrition, 56*(10), 1717–1727.

Romero, A., Cuesta, C., & Sánchez-Muniz, F. J. (2000). Trans fatty acid production in deep fat frying of frozen foods with different oils and frying modalities. *Nutrition Research, 20*(4), 599–608.

Rossell, J. (1989). Intermediate shelf life products as illustrated by fats and fatty foods. *Food Science and Technology Today, 3*, 235–240.

Ruberto, G., & Baratta, M. T. (2000). Antioxidant activity of selected essential oil components in two lipid model systems. *Food Chemistry, 69*(2), 167–174.

Sabir, S. M., Hayat, I., & Gardezi, S. D. A. (2003). Estimation of sterols in edible fats and oils. *Pakistan Journal of Nutrition, 2*(3), 178–181.

Saguy, I. S., & Dana, D. (2003). Integrated approach to deep fat frying: engineering, nutrition, health and consumer aspects. *Journal of Food Engineering, 56*(2–3), 143–152.

Sánchez-Muniz, F. J., & Bastida, S. (2006). Effect of frying and thermal oxidation on olive oil and food quality. *Olive Oil and Health*, 74–108.

Sanibal, E. A. A., & Mancini-Filho, J. (2004). Frying oil and fat quality measured by chemical, physical, and test kit analyses. *Journal of the American Oil Chemists' Society, 81*(9), 847–852.

Senanayake, N. (2018). Enhancing oxidative stability and shelf life of frying oils with antioxidants. *Inform Magazine, 29*, 6–13.

Serna-Saldivar, S. O., & Chuck-Hernandez, C. (2019). Food uses of lime-cooked corn with emphasis in tortillas and snacks. In Sergio O. Serna-Saldivar (Ed.), *Corn* (pp. 469–500). Elsevier.

Sherazi, S., Kandhro, A. A., Mahesar, S., Talpur, M. Y., & Latif, Y. (2010). Variation in fatty acids composition including trans fat in different brands of potato chips by GC-MS. *Pakistan Journal of Analytical & Environmental Chemistry, 11*(1), 6.

Sherazi, S. T. H., & Mahesar, S. A. (2015). Analysis of edible oils and fats by nuclear magnetic resonance (NMR) spectroscopy. In *Applications of NMR Spectroscopy* (pp. 57–92). Elsevier.

Siddique, B. M., Muhamad, I. I., Ahmad, A., Ayob, A., Ibrahim, M. H., & Ak, M. O. (2015). Effect of frying on the rheological and chemical properties of palm oil and its blends. *Journal of Food Science and Technology, 52*(3), 1444–1452.

Skeaff, C. (2009). Feasibility of recommending certain replacement or alternative fats. *European Journal of Clinical Nutrition, 63*(2), S34–S49.

Stevenson, S., Vaisey-Genser, M., & Eskin, N. (1984). Quality control in the use of deep frying oils. *Journal of the American Oil Chemists Society, 61*(6), 1102–1108.

Stier, R. (2001). The measurement of frying oil quality and authenticity. In *Frying* (pp. 165–193). Elsevier; Woodhead Publishing.

Suleman, R., Wang, Z., Aadil, R. M., Hui, T., Hopkins, D. L., & Zhang, D. (2020). Effect of cooking on the nutritive quality, sensory properties and safety of lamb meat: current challenges and future prospects. *Meat Science, 167*, 108172.

Sulieman, A. E. R. M., El-Makhzangy, A., & Ramadan, M. F. (2006). Antiradical performance and physicochemical characteristics of vegetable oils upon frying of French fries: a preliminary comparative study. *Journal of Food Lipids, 13*(3), 259–276.

Syed, A. (2016). Oxidative stability and shelf life of vegetable oils. In Min Hu & Charlotte Jacobsen (Eds.), *Oxidative Stability and Shelf Life of Foods Containing Oils and Fats* (pp. 187–207). Elsevier.

Talpur, M. Y., Sherazi, S., Mahesar, S., & Kandhro, A. A. (2009). Effects of chicken frying on soybean, sunflower and canola oils. *Pakistan Journal of Analytical and Environmental Chemistry, 10*(1–2), 59–66.

Tseng, Y. C., Moreira, R., & Sun, X. (1996). Total frying-use time effects on soybean-oil deterioration and on tortilla chip quality. *International Journal of Food Science & Technology, 31*(3), 287–294.

Tsuzuki, W., Komba, S., & Todoriki, S. (2020). Trans isomerization of unsaturated fatty acids in bovine liver sterilised by gamma-irradiation at low temperature. *Radiation Physics and Chemistry, 166*, 108458.

Tsuzuki, W., Matsuoka, A., & Ushida, K. (2010). Formation of trans fatty acids in edible oils during the frying and heating process. *Food Chemistry, 123*(4), 976–982.

Tsuzuki, W., Nagata, R., Yunoki, R., Nakajima, M., & Nagata, T. (2008). Cis/trans-isomerisation of triolein, trilinolein and trilinolenin induced by heat treatment. *Food Chemistry, 108*(1), 75–80.

Tyagi, V., & Vasishtha, A. (1996). Changes in the characteristics and composition of oils during deep-fat frying. *Journal of the American Oil Chemists' Society, 73*(4), 499–506.

Varela, P., & Fiszman, S. (2011). Hydrocolloids in fried foods. A review. *Food Hydrocolloids, 25*(8), 1801–1812.

Velasco, J., Dobarganes, C., & Márquez-Ruiz, G. (2010). Oxidative rancidity in foods and food quality. In *Chemical Deterioration and Physical Instability of Food and Beverages* (pp. 3–32). Elsevier.

Vitrac, O., Dufour, D., Trystram, G., & Raoult-Wack, A.-L. (2002). Characterization of heat and mass transfer during deep-fat frying and its effect on cassava chip quality. *Journal of Food Engineering, 53*(2), 161–176.

Wang, D. D., Li, Y., Chiuve, S. E., Stampfer, M. J., Manson, J. E., Rimm, E. B., … Hu, F. B. (2016). Association of specific dietary fats with total and cause-specific mortality. *JAMA Internal Medicine, 176*(8), 1134–1145.

Westesen, K., & Bunjes, H. (1995). Do nanoparticles prepared from lipids solid at room temperature always possess a solid lipid matrix? *International Journal of Pharmaceutics, 115*(1), 129–131.

Willett, W. C., & Stampfer, M. J. (2003). Rebuilding the food pyramid. *Scientific American, 288*(1), 64–71.

Wright, P. (1931). Calibration of the lovibond tintometer and of the actinometer for the Levy-West ultra-violet light pastilles. *The British Journal of Radiology, 4*(48), 715–721.

Yang, M., Yang, Y., Nie, S., Xie, M., Chen, F., & Luo, P. G. (2014). Formation of trans fatty acids during the frying of chicken fillet in corn oil. *International Journal of Food Sciences and Nutrition, 65*(3), 306–310.

Yu, L. (2011). *Extrusion Processing of Protein Rich Food Formulations*. McGill University.

Chapter 5

Insights into the Chemistry of Deep-Fat Frying Oils

Bharti Mittu, Anjali Chaturvedi, and Renu Sharma

National Institute of Pharmaceutical Education and Research (NIPER)

HNB Garhwal University

Akal Degree College Mastuana Sahib

5.1 Introduction

Food preparation methods like frying are among the most popular in the world. It represents a multibillion-dollar sector of global food trade that is valued for its special flavors and textures (Zeb, 2019). Oils are combustible organic liquids that are viscous and insoluble in water, although they are lighter in weight but soluble in alcohol, ether, and benzene. Since

DOI: 10.1201/9781003329244-5

cooking and frying oils are exposed to high temperatures, they need to be more resistant to oxidation. It is of prime importance for the oils used in frying to maintain their temperature stability. As a result, frying steps are the most critical stages for oil owing to its well-known instability terms under thermal stress and achieved degradation among its physicochemical makeup. Oil becomes viscous and dark after repeated frying due to oxidation, hydrolysis, polymerization, and free radical release. Frying oils generate several compounds that alter their flavor, stability, and quality (Dueik et al., 2010). An effective frying process is determined by choosing the correct oil as frying medium (Artz et al., 2005). Ideally, the food surface should be dehydrated and the frying medium should be well heated and thermally conducted. During deep frying, food is immersed in oil either partly or completely at atmospheric pressure. Frying oil's life span depends upon its composition, antioxidants used, and production processes. Deep frying decreases unsaturated fats and increases foam, viscosity, color, and, density. It is also possible for the food being fried to affect the frying oil negatively. There is a possibility that the food produces volatiles, interacts with the frying oil, and darkens the oil by its interaction. The frying temperature of the oil influences glyceride degradation and changes thermal and physical characteristics, color and viscosity, frying velocity, and efficiency. The temperatures at which frying is generally accomplished are 160°C and 190°C. Sunflower oil, canola oil, soybean oil, maize oil, peanut oil, and olive oil are among the most commonly used oils since they not only transmit heat but also provide flavor to the food.

5.2 Types of Frying Oils

Oils are of many types based on their smoke point. Some oils have a smoke point at particular temperatures. It is recommended to deep fry with oils having high smoke points because they have a lower risk to damage food (Katragadda et al., 2010). Avocado oil has the greatest smoke point (Table 5.1) and may be used for cooking. The oil adds flavor and sweetness to the dish. It has a smoke point of 520°F. This oil is healthy for the heart and allows the body to absorb other vitamins. Peanut oil when fried gives a nut-like flavor and has an extremely high smoke point of 450°F. This oil is much useful for making medicines and preventing heart diseases. Soybean oil offers a variety of frying applications. It has a smoke point of 450°F and is considered to be ideal for deep frying. Soybean oil supports the health of bones so it can be used for cooking for patients with bone-related problems. Canola oil has a smoke point of 400°F, making it one of the greatest frying oils. Canola oil includes phytosterols which inhibit the absorption of

Table 5.1 Smoke Points of Oils

S. No.	*Type of Oil*	*Approximate Smoke Point*
1	Avocado oil	520°F
2	Peanut oil, soybean oil, and safflower	450°F
3	Grape seed	445°F
4	Canola oil	400°F
5	Coconut oil	350°F
6	Olive oil	330°F
7	Flaxseed oil	230°F

cholesterol molecules (Guillaume et al., 2018). The food can be cooked at the temperature of 350°F–375°F which helps to not to absorb the oil content (Lee and Jeonghee, 2021). The smoke points of various oils are summarized in Table 5.1.

It is not recommended to use oil for deep frying with a low smoke point like olive oil (330°F), flaxseed oil (230°F), or coconut oil (350°F). In general, the hotter the oil gets, the more the breakdowns occur. Selection includes choosing cooking oil with a high smoke number. Oils that have higher smoke point numbers are the best to be used for frying. The best cooking oil has a temperature of 345°F–360°F. The best cooking oils for frying are those which have mostly unsaturated fats and a small percentage of polyunsaturated fats. By extending chain length, unsaturated fatty acids cause thermo oxidative degeneration (Totani et al., 2006). They increase the oil's oxidation rate. Hydration reduces fats. Filtering frying oil with esterified acyl glycerol as an adsorbent improves the quality of oil by reducing free fatty acids produced during hydrogenation (Hidalgo and Zamora, 2000).

5.3 Deep Frying

Deep frying corresponds to established heat and mass transfer principles. The driving factors that cause the chemical and other changes in both the oil and the food are radiative heat transfer from the heated oil and conductive heat transfer through the food. Dehydration (development of steam), distillation (volatile materials), extraction, and leaching from the food to the oil are all examples of mass transfer. The physical and chemical changes that occur during deep frying are discussed briefly below.

5.3.1 Physical Changes

Steam escapes from the surface of oil, carrying with it volatile compounds. As the volatiles increase, the smoke point gets decreased. Moisture produces a layer above the oil, decreasing the interaction of ambient oxygen with the meal. The formation of steam prevents oil from accessing the inside of the fried product. As a result, regardless of the temperature of the oil, the temperature of the meal is always 100°C as long as steam is created.

Polymer production causes an increase in viscosity. The numerous lipids that might seep into the frying oil modify its characteristics and performance. The darkening is caused by colored lipids that are dissolved in the oil. Emulsifiers are phospholipids. Liposoluble metal compound traces may serve as prooxidants. Antioxidants include liposoluble vitamins and phenolic compounds. Volatile chemicals (such as those found in fish or onions) lead to bad flavors. Oil penetration is regulated by the form of the food, its textural characteristics, the porosity and viscosity of the frying medium, and the temperature and duration of the frying. Increased viscosity leads to more oil being absorbed. Food with a high fat content does not absorb oil.

5.3.2 Chemical Changes

When food is deeply fried, various variables occur, such as oil absorption. The release of water droplets from raw food dipped in oil results in steam distillation. There are so many products that are prepared by deep frying at approximately 1,780°C–1,800°C. Oils undergo some of the changes at higher temperatures which give a negative impact on oil and food. Short frying has greater organoleptic quality. Edible oil or fat is a biological mixture of compounds that contains fatty acids and esters derived from glycerol. Oils are combustible organic liquids that are viscous and insoluble in water, although they are lighter by weight and soluble in alcohol, ether, and benzene. Since cooking and frying oils are exposed to high temperatures, they need to be more resistant to oxidation. Temperature stability is more important for oils to be fried. As a result, oil is the most critical part of the frying process due to its well-known instability under thermal stress and the formation of the degradation process and due to its physicochemical effects. Deep frying reduces unsaturated fats and increases viscosity, density, polar compounds, and polymerization (Capar and Yalcin, 2017). Fried products tend to have negative physicochemical and rheological characteristics because of the polarity of water–oil system caused by the reactivity of the oil used for frying (Venkata and Rajagopal, 2016). Furthermore, there is the possibility of reducing the nutritional value and interfering with enzymes during digestion altogether with undesirable biological effects (Rani et al., 2010).

A triglyceride is an ester created from three fatty acids linked together by glycerol, a tri-hydroxyl alcohol. Whenever the triglyceride glycerol contains all three OH groups, it is called a simple triglyceride (Figure 5.1). Despite their simplicity, triglycerides are rare in nature. They are synthesized in laboratories. As a complex mixture of triglycerides, natural triglycerides can't be represented by a single formula. An alternative would be to use a triglyceride derived from natural sources of fats. Oils contain two or three different fatty acid components. This is referred to as a mixed triglyceride. Fry oil is commonly infused with antioxidants and antifoams (Dash et al., 2022). The desired flavor and stability are achieved through the use of blends of source oils (Laguerre et al., 2015). A partial hydrogenation process is also available for inter-esterified oils derived from chemical synthesis (Jo and Jeonghee, 2020). Partially hydrogenated oils can't spoil or rot as regular oils do.

During deep-oil frying, the most common chemical reactions in frying oil include hydrolysis, oxidation, and polymerization of oil, and various volatile and nonvolatile components are also produced (Figure 5.2). Steam helps to evaporate most of the volatile components in the atmosphere, while the remaining volatile components in oil experience further chemical reactions or are absorbed in fried foodstuffs. The changes in the physical and chemical properties of oil and fried foods are mainly due to the existence of nonvolatile components in the oil. These components also affect quality, flavor, stability, and texture of fried food products. In spite of increasing color, viscosity, foaming, specific heat, density, polar substances, polymeric components, and free fatty acid content, deep-fat frying also reduces unsaturated fatty acids of oil (Figure 5.3).

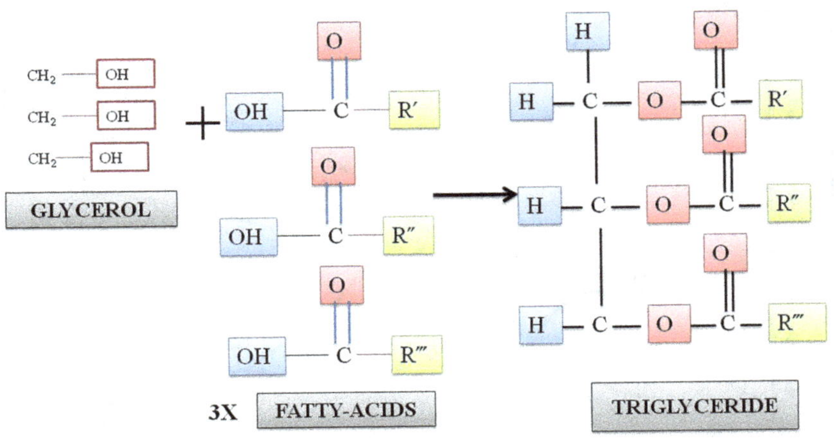

Figure 5.1 Formation of triglycerides from glycerol and fatty acids.

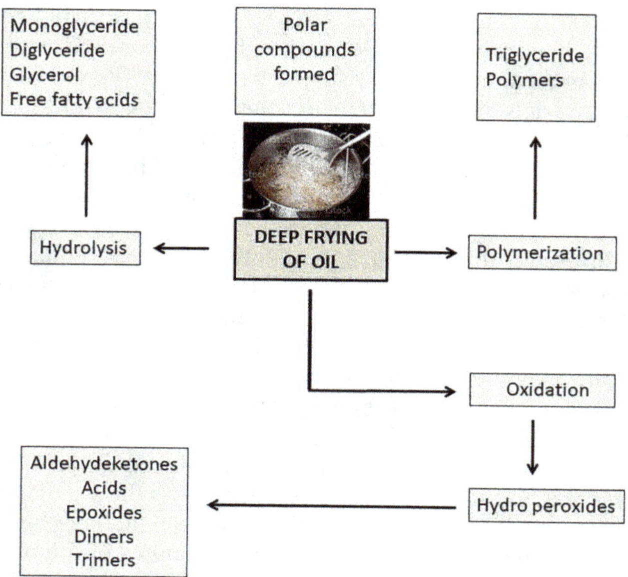

Figure 5.2 Chemical changes during deep frying of oil.

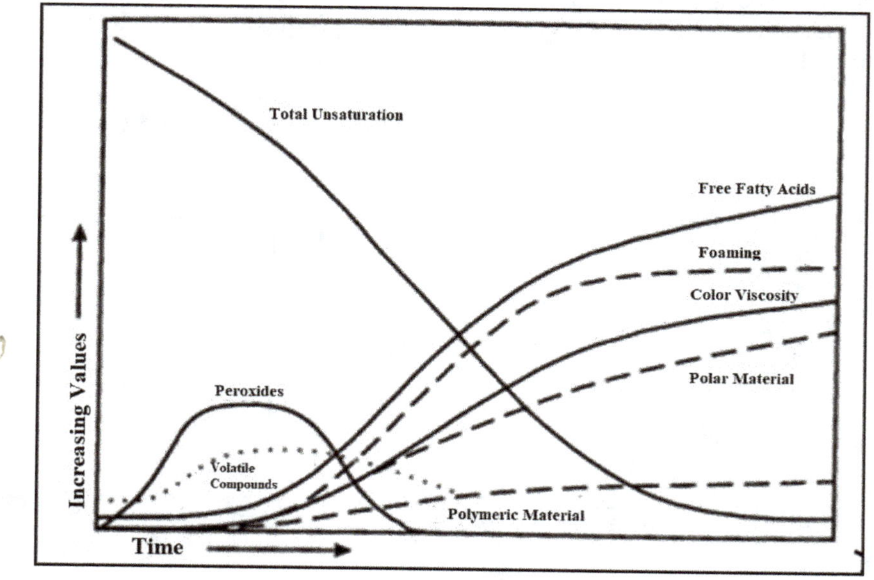

Figure 5.3 Physical and chemical changes of oil during deep-fat frying (Choe and Min, 2007).

The various parameters including frying oil and fryer type, frying time and temperature, and antioxidants affect the above-mentioned chemical reactions (hydrolysis, oxidation, and polymerization) of oil during frying.

5.3.2.1 Hydrolysis

Hydrolysis of oil takes place due to exposure to high temperatures and leads to breakdown of oil to produce free fatty acids, diglycerides, and monoglycerides. A larger proportion of free fatty acids lowers the oil's smoke point. Because mono- and diglycerides are polar in nature, they are more likely to froth. A maximum of 0.08%–1.00% free fatty acid content in oil is permissible (Thapar, 2019). The hydrolysis of frying oil can be reduced by frequent replacing of used oil with fresh oil. Alkalis like sodium hydroxide used for cleaning a fryer enhances the oil hydrolysis. The hydrolysis of oil is not affected by the time duration of frying process (Naz et al., 2005).

The heat from the oil evaporates moisture, which generates steam that does not persist long due to the bubbling movement; therefore, the steam subsides as the goods are fried. A chemical reaction starts to occur between the water, steam, and oxygen present in the frying oil and food. Water, being a weak nucleophile, attacks triacylglycerol ester bonds, yielding diacylglycerols, monoacylglycerols, glycerols, and free fatty acids. The amount of free fatty acids in the oil grows as the number of frying increases (Chung et al., 2004; Choe and Min, 2007) as shown in Figure 5.4. The free fatty acid content of frying oil is measured. Thermal hydrolysis occurs more often within the oil phase than at the water–oil contact. Hydrolysis is preferable for oils with fewer unsaturated fatty acids because short and unsaturated fatty acids are more soluble in water than long and saturated fatty acids (Choe and Min, 2007).

5.3.2.2 Oxidation

The oxygen in deep-fat frying reacts with oil. Both thermal oxidation and auto-oxidation reactions are based on same mechanisms (Houhoula et al., 2003). Thermal oxidation occurs at faster rate than the autoxidation, but comprehensive scientific information and comparison between oxidation rates of thermal oxidation and autoxidation is not available. The reaction mechanism of thermal oxidation involves the initiation, propagation, and termination as shown in Figure 5.4. Numerous reactions occur during autoxidation, leading to the production of volatile compounds as the food is ingested into the oil (Rodríguez et al., 2021). The oxidative processes occur because of the release of moisture, high temperatures are present, and the air is exposed to oil (Da silva et al., 2008).

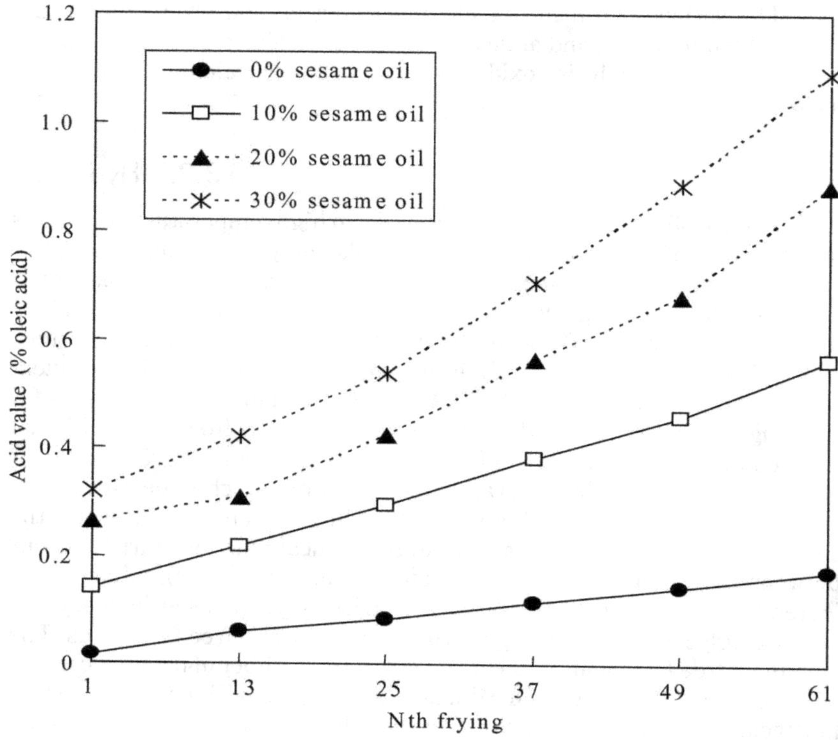

Figure 5.4 Free fatty acid formation in soybean and sesame oil mixture during consecutive frying of flour dough at 160°C (Chung et al., 2004).

Deep frying oxidation creates a wide range of chemicals and is classified as primary, secondary, and tertiary oxidations (Dana et al., 2003). The production of hydroperoxides connected to the double bond of unsaturated fatty acid results from primary oxidation. The oxidation process is always promoted by high temperatures. The breakdown of hydroperoxides increases during the secondary oxidation process, resulting in the creation of alcohols, carbonyls, and acids. Auto-oxidation occurs in unsaturated aldehydes. Polymerization of secondary oxidation products happens during the tertiary oxidation process, increasing oil viscosity and darkening the oil's color.

Oil in nonradical singlet state does not combine with diradical triplet state oxygen due to the different spin multiplicities. Ordinary air oxygen is a diradical compound. For the oxidation process, radical oxygen requires oil in radical state. In order to generate radical form of oil, the weakly bonded hydrogen attached with carbon of oil will be removed first. About

50 kcal/mol energy is required to break carbon–hydrogen bond in linoleic acid (Min and Jeffrey, 2002). The presence of double bonds reduces the carbon–hydrogen bond strength by withdrawing electrons. In oleic acid, the energy requirement for the breakage of carbon–hydrogen bond lying in α-position to the double bond is about 75 kcal/mole. Approximately 100 kcal/mol energy is required for radical formation in the compounds having carbon–hydrogen bond on the saturated carbon without any double bond next to it (Min and Jeffrey, 2002). The varying strengths of carbon–hydrogen bond of fatty acids result in the different rates of oxidation reactions in oleic, stearic, linoleic, and linoleic acids during thermal oxidation or autoxidation. The various factors including light, metal, heat, and reactive oxygen species facilitate the radical formation of oil.

The polyvalent metals like Cu^{2+} and Fe^{3+} facilitate the generation of alkyl radicals by extracting hydrogen from oil by oxidation–reduction mechanism even at low temperature. Both unsaturated and saturated fatty acids have different sites of radical formation. In saturated fatty acids, alkyl radical is formed at α-position of the carboxyl group having electron-withdrawing nature. In this oxidation reaction mechanism of oil, the step involving hydrogen abstraction from an oil molecule to generate alkyl radical is called the chain initiation step. Further reaction of the alkyl radical with other alkyl radicals, alkoxy radicals, and peroxy radicals results in the formation of dimers and polymers. Alkyl radicals having a reduction potential of 600 mV react rapidly with diradical oxygen which is in triplet state to produce peroxy radicals. The peroxy radicals with a reduction potential of 1,000 mV remove hydrogen from linolenic acid and oleic acid to generate hydroperoxide at a rate of 1×10 and 1×100/M/s, respectively (Choe and Min, 2007). From another molecule of oil, the peroxy radical abstracts a hydrogen atom and forms new hydroperoxide and alkyl radical. This chain reaction step is termed as propagation step.

The reaction of peroxy radicals with other radicals proceeds at the rate of about 1.1×10^6/M/s to form dimersor polymers (Choe and David, 2005). These chain reactions of peroxy radicals and free alkyl radicals accelerate the thermal oxidation of oil. The oxygen–oxygen bond in R–O–O–H is a relatively weak covalent bond. Hydroperoxides are not usually stable during the deep-fat frying and are decomposed to hydroxy radicals and alkoxy radicals by homolytic fission of the peroxide bond. The alkoxy radical either reacts with other alkoxy radicals or decomposed to nonradical products (Choe and Min, 2007). The generation of nonradical volatile and nonvolatile components at the end of oxidation reaction is termed as termination step as shown in Figure 5.5.

It is impossible to avoid oxidation. The rate of oxidation is affected by the oxidation of double bonds, the degree of unsaturation, the concentration of polyunsaturated acids, the concentration of oxygen, and the number of free radicals in the oil. The thermal oxidation involving initiation,

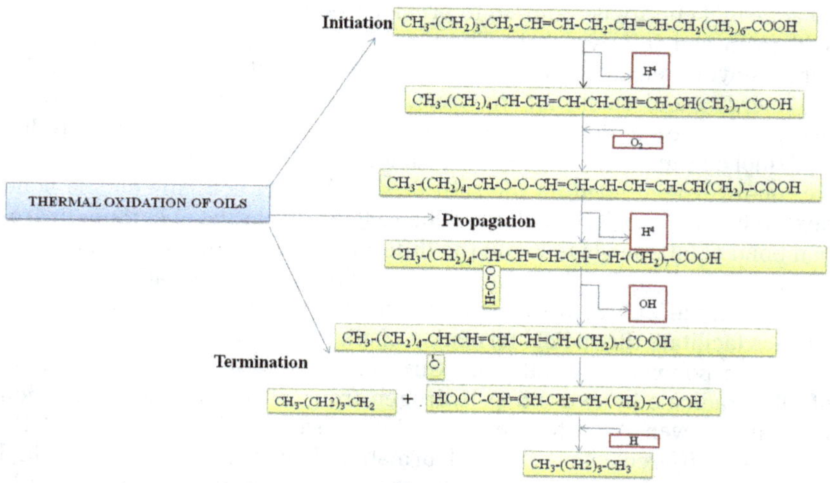

Figure 5.5 Thermal oxidation of oils.

propagation and termination of unsaturated fatty acids leads to the formation of conjugated hyperoxides, peroxides, and volatile aldehydes. The use of natural antioxidants like tocopherols slows down the oxidation of the oil.

5.3.2.3 Thermal Polymerization

Frying of oil results in disintegration of oil with the formation of products like cyclic monomers, dimers and non-polar polymers, and trans and position isomers (Bordin et al., 2013). Depending on the types of fatty acids composing an oil, polymers can be cyclic or acyclic compounds. When oleic acid like fatty acids is present in oil, the acyclic polymers are formed. Linoleic acid richer oil is more easily polymerized than oleic acid (Bordin et al., 2013). In comparison to nonvolatile polar compounds, dimers, and polymers, the amounts of cyclic compounds are relatively small. The volatile components contained in decomposition products of frying oil at the concentration of part per million levels are considered to be very significant to the flavor qualities of frying oil and fried foods (Dobarganes et al., 2000). Dimers and polymers formed by a combination of –C–C–, –C–O–C–, and –C–O–O–C– bonds are the macromolecules having a molecular weight range of 692–1,600 Daltons. Dimers and polymers possess epoxy, hydroperoxy, hydroxy, carbonyl groups, and –C–O–O–C– and –C–O–C– linkages. In deep frying, the dimer and polymer formation occur via free radical mechanism. Allyl radicals are mainly produced at methylene carbons α to the double bonds. The reactions among allyl radicals through C–C linkage produce dimers. The formation of acyclic

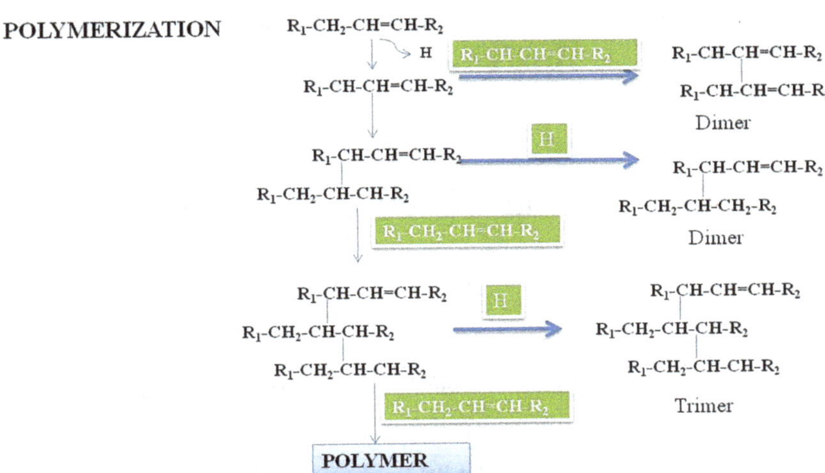

Figure 5.6 Formation of acyclic dimers and trimers during polymerization (modified from Choe and Min, 2007).

dimers and trimers from oleic acid is shown in Figure 5.6. Alkyl hydroperoxides (ROOH) or dialkyl peroxides (ROOR) are produced by reaction of triacylglycerols with oxygen. These peroxides are highly unstable and readily decomposed to alkoxy and peroxy radicals by RO–OH and ROO–R bond fission, respectively. Alkoxy radicals (RO·) can either abstract hydrogen from oil molecule to form hydroxy compounds, or combine with other alkyl radicals to form oxydimers. Peroxy radicals (ROO·) can combine with alkyl radicals to produce peroxy dimers (Figure 5.7).

The various factors including type of oil, frying temperature, and number of frying affect the formation of dimers and polymers. The amounts of polymers increased with increase in frying temperature and number of frying (Cuesta et al., 1993) as shown in Figure 5.8.

Cyclic polymers are produced by intermolecular or intramolecular reactions of triacylglycerols through free radical mechanism (Figure 5.9). In frying oil, the formation of cyclic compounds depends on the frying temperature and degree of unsaturation (Meltzer et al., 1981). The formation of cyclic monomers and polymers is reported to be directly linked with amounts of linolenic acid (Tompkins and Edward, 2000). Cyclic compounds are not formed to a noteworthy extent until the oil temperature reaches 200°C–300°C and the linolenic acid content exceeds about 20%.

The sole difference between thermal polymerization and tertiary oxidation is that the supply of oxygen is restricted. Thermal polymerization gives rise to cyclic monomers, dimmers, and polymers. The products

ETHER OR PEROXIDE POLYMER FORMATION DURING OIL FRYING

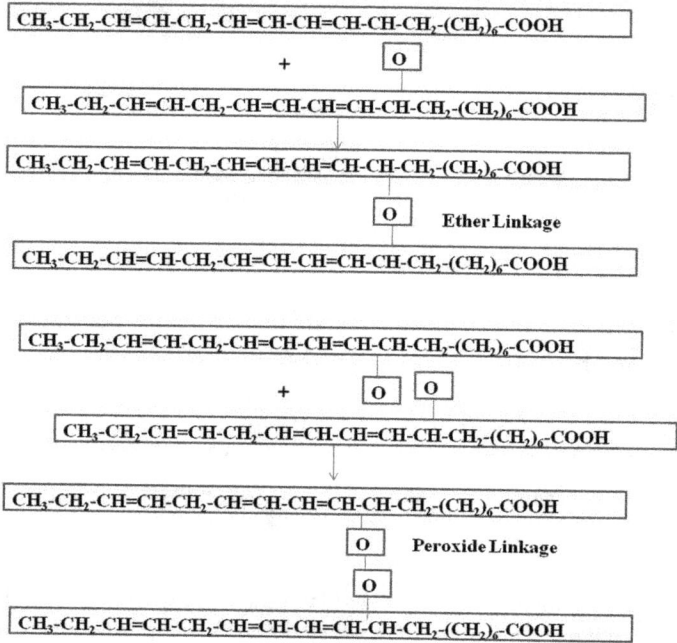

$CH_3\text{-}CH_2\text{-}CH{=}CH\text{-}CH_2\text{-}CH{=}CH\text{-}CH{=}CH\text{-}CH\text{-}CH_2\text{-}(CH_2)_6\text{-}COOH$

\+ O

$CH_3\text{-}CH_2\text{-}CH{=}CH\text{-}CH_2\text{-}CH{=}CH\text{-}CH{=}CH\text{-}CH\text{-}CH_2\text{-}(CH_2)_6\text{-}COOH$

↓

$CH_3\text{-}CH_2\text{-}CH{=}CH\text{-}CH_2\text{-}CH{=}CH\text{-}CH{=}CH\text{-}CH\text{-}CH_2\text{-}(CH_2)_6\text{-}COOH$

O **Ether Linkage**

$CH_3\text{-}CH_2\text{-}CH{=}CH\text{-}CH_2\text{-}CH{=}CH\text{-}CH{=}CH\text{-}CH\text{-}CH_2\text{-}(CH_2)_6\text{-}COOH$

$CH_3\text{-}CH_2\text{-}CH{=}CH\text{-}CH_2\text{-}CH{=}CH\text{-}CH{=}CH\text{-}CH\text{-}CH_2\text{-}(CH_2)_6\text{-}COOH$

\+ O O

$CH_3\text{-}CH_2\text{-}CH{=}CH\text{-}CH_2\text{-}CH{=}CH\text{-}CH{=}CH\text{-}CH\text{-}CH_2\text{-}(CH_2)_6\text{-}COOH$

↓

$CH_3\text{-}CH_2\text{-}CH{=}CH\text{-}CH_2\text{-}CH{=}CH\text{-}CH{=}CH\text{-}CH\text{-}CH_2\text{-}(CH_2)_6\text{-}COOH$

O **Peroxide Linkage**

O

$CH_3\text{-}CH_2\text{-}CH{=}CH\text{-}CH_2\text{-}CH{=}CH\text{-}CH{=}CH\text{-}CH\text{-}CH_2\text{-}(CH_2)_6\text{-}COOH$

Figure 5.7 Ether and peroxide formation during frying of oil.

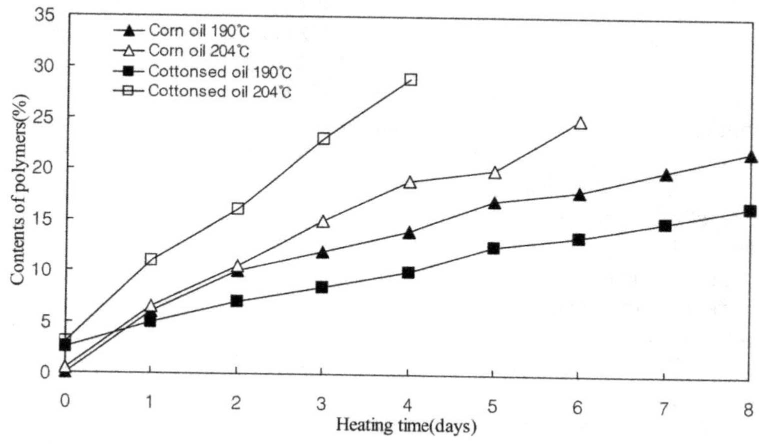

Figure 5.8 Polymer contents of cottonseed and corn oils heated at 190°C and 204°C (Takeoka et al., 1997).

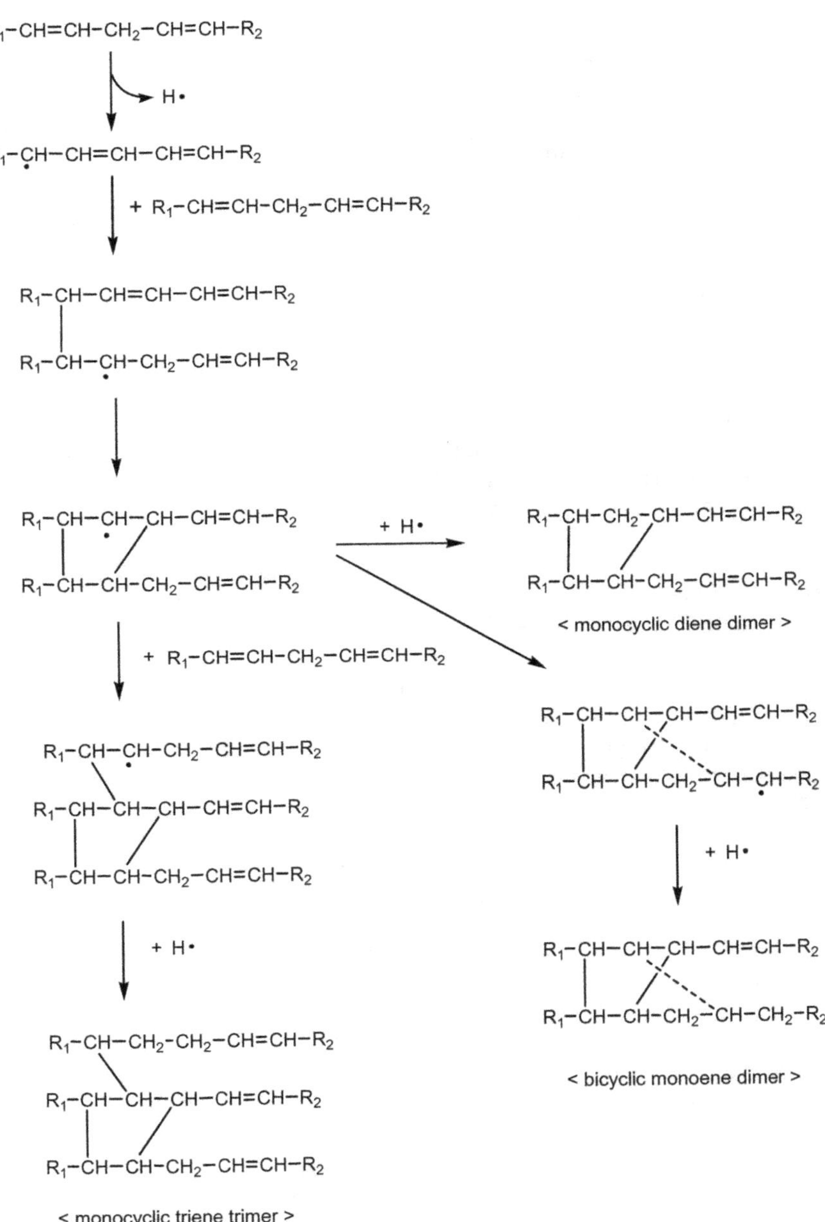

Figure 5.9 Formation of cyclic dimers and polymers from linoleic acids during deep-fat frying (Choe and Min, 2007).

formed during polymerization with varying polarity, stability, and molecular weight during termination reactions alkyl radicals are formed by initiation reaction, and alkoxyl radical is produced by hydrogen peroxide decomposition (Akil et al., 2015). Oxygen-rich polymers formed during deep-oil frying are reported to accelerate the oxidation of oil. The oil for frying can be made stable by relating its inherent stability to oxidation (Fedko et al., 2022). Oxidized oil's inherent stability is determined by the rate at which it reacts with oxygen (Zhang et al., 2012). Therefore, oils with low inherent stability are less susceptible to oxidation during frying (Erickson et al., 2022). Due to these reasons, high unsaturated fats and free fatty acids contribute to low frying quality in oils (Erickson, 2007).

Frying of the oil process accelerates the degradation of oil and viscosity (Trigo et al., 2022). Polymer quantity increases as the frying time and temperature increase (Vankar, 2011). A deep fryer creates more oxygen-rich polymers than an oleic acid producer. The rate of heat transmission is affected by thermal conductivity, thermal diffusivity, specific heat, and food density. These qualities alter as the oil and food are modified during frying. Interactions between dietary molecules also cause additional changes.

The processes of hydrolysis, oxidation, and thermal alteration are all interconnected. Hydrolysis and oxidation as well as cleavage of double bonds are the main methods of forming free fatty acids. A polymer can cause foaming in the oil, which holds steam bubbles for a long time and speeds up the process of hydrolysis. However, when the bubbles burst, oxygen is introduced into hot oil which causes it to oxidize. High frying temperatures hasten thermal oxidation and polymerization of oils in the first twenty times of frying, resulting in a decrease of polymers with peroxide and ether linkage. Oils deteriorate faster if they are heated and cooled continuously with the solubility of oxygen (Choe and Min, 2007).

The right temperatures for oil frying should be maintained on a regular basis to improve oil performance and extend the life of the food product (Liu et al., 2019). The amount of oil absorbed is determined by the moisture content and starch level of the food, the frying temperature, the time, and the kind of oil used in frying. The amount of oil absorbed is determined by the moisture content and starch level of the food, the frying temperature, the time, and the kind of oil used in frying (Selmi et al., 2010). The amount of oil absorbed by snacks tends to grow as the temperature of the oil drops and the frying time increases (Zhang et al., 2016).

5.4 Consequences of Refrying of Oil

During frying when oxygen is introduced, several complex changes occur. These changes consist of the main degradation of triacyl glycerols and some other minor components (Fedko et al., 2022). Lipid oxidation gives rise to liver problems and hypertension. Increased polymer content harms the liver by attacking the blood lipid profile. The viscosity and adsorption of oil increase calories in the body and also causes some cardiovascular diseases and obesity (Chao et al., 2007). It is critical to decreasing oil deterioration and the widespread usage of deep-fried oil. Carbonyl compounds or mono-epoxides, as well as certain aldehydes generated by linoleic acid, have been discovered as possibly carcinogenic chemicals. Some parameters including temperature, nature of frying oil, material type, food nature, heat transfer, presence of oxygen, capacity of frying, and the metal used during the frying of oil influence the quality (Jiang et al., 2019).

Cooking with a large amount of oil and reusing the remaining oil after it has been unbroken for an extended period without being used have many disadvantages that need to be considered; a rise in free carboxylic acid content occurs during oil reactions (Multari et al., 2019). Re-cooking of oil results in an increase in free carboxylic acid content which is insoluble in water, and so if the oil's free carboxylic acid content exceeds the acceptable limit, it can't be used (Maestre et al., 2011).

The polymerization of oil will rise as a result of refrying, hastening the breakdown of oil, increasing oil viscosity, lowering heat transmission, and froth while cooking, and causing food to acquire an unattractive hue owing to an increase in polymerization (Akil et al., 2015; Kim et al., 2018). Polymers lead to food absorbing a large amount of oil. Refrying produces cytotoxic compounds and trans fatty acids based on the texture of the oil being used (Kim et al., 2010). Trans fatty acids are responsible for cardiac diseases (Obidiegwu et al., 2020). When oil is fried intermittently at a lower temperature and a slower turnover rate, it oxidizes and polymerizes faster (Goswami et al., 2015). Oil is more stable when spinach extracts are added to slow down oxidation (Guillaume et al., 2018). Due to vitamin breakdown, gastrointestinal problems may arise as a result of vitamin loss. Intensified browning of food decreases essential amino acids in food (Koh et al., 2015). Various physical and chemical changes occur during the high-temperature deep frying of oil, culminating in the gelatinization of starch and protein denaturation (Oke et al., 2018). In addition to lose freshness in food, rancidity also affects the storage life of food and builds unpleasant flavors and aromas to it (Wang et al., 2022). Due to the rancidity of re-fried oil, cytotoxic compounds are bound such as acryl-organic compounds, which could be human waste (Granda et al., 2004).

Deep-fried food is quite popular in developing countries. Deep-fried foods are in even more demand. Although deep-fried food tastes sweeter but reusing the same oil for frying may pose several health risks to people. A little amount of fried food poses no health risk to an average person, but the true issue emerges when the same oil is cooked again. In addition of becoming highly viscous, edible oil also becomes contaminated with harmful products.

Hence oil should be heated as little as possible to avoid human hazards (Fan et al., 2005). Moreover, it is our responsibility to avoid this kind of food in large amounts. Also, it is to be noted that at home oil used once should always be discarded and not be used for refrying (Goh, 2020). The left oil can be given to the oil refineries for their re-use industrially.

5.5 Conclusion

The deep frying of oil gives rise to hydrolysis, oxidation, and polymerization. Hydrolysis gives rise to the development of free fatty acids which are finally enhanced through refrying if the oil has more free fatty acids than the limit oil becomes unstable to be used because of their insoluble nature in water and they get oxidized very easily.

Free fatty acids and oxidized compounds result in rancidity which helps to lose the freshness of food and enhances bad flavor and foul smell. The rancid oil produces toxic compounds like acryl amide which is carcinogenic.

Polymerization of oil is increased by refrying of oil, which results in the degradation of oil and dark colorizations of food with the high amount of absorption of oil. The formation of cyclic compounds increases cardiovascular diseases. Food browning results in the loss of essential amino acids and vitamins A, B, C, and E, which change the color of the oil and gives rise to health hazards like gastrointestinal irritations.

References

Akil, Emília, Vanessa Naciuk Castelo-Branco, André Mesquita Magalhães Costa, Ana Lúcia do Amaral Vendramini, Verônica Calado, and Alexandre Guedes Torres. "Oxidative stability and changes in chemical composition of extra virgin olive oils after short-term deep-frying of french fries." *Journal of the American Oil Chemists' Society* 92 (2015): 409–421.

Artz, William E., Patricia C. Osidacz, and Aline R. Coscione. "Iron accumulation in oil during the deep-fat frying of meat." *Journal of the American Oil Chemists' Society* 82, no. 4 (2005): 249.

Bordin, Keliani, Mariana Tomihe Kunitake, Keila Kazue Aracava, and Carmen Silvia Favaro Trindade. "Changes in food caused by deep fat frying-a review." *Archivoslatinoamericanos de nutricion* 63, no. 1 (2013): 5–13.

Capar, T. D., and H. Yalcin. "Effects of pre-drying on the quality of frying oil and potato slices." *Quality Assurance and Safety of Crops & Foods* 9, no. 3 (2017): 255–264.

Chao, Pei-Min, Hui-Ling Huang, Chun-Huei Liao, Shiau-Ting Huang, and Ching-jang Huang. "A high oxidised frying oil content diet is less adipogenic, but induces glucose intolerance in rodents." *British Journal of Nutrition* 98, no. 1 (2007): 63–71.

Choe, E., and D. B. Min. "Chemistry and reactions of reactive oxygen species in foods." *Journal of Food Science* 70, no. 9 (2005): R142–R159.

Choe, E., and D. B. Min. "Chemistry of deep-fat frying oils." *Journal of Food Science* 72, no. 5 (2007): R77–R86.

Chung, J., J. Lee, and E. Choe. "Oxidative stability of soybean and sesame oil mixture during frying of flour dough." *Journal of Food Science* 69, no. 7 (2004): 574–578.

Cuesta, C., F. J. Sánchez-Muniz, C. Garrido-Polonio, S. López-Varela, and R. Arroyo. "Thermoxidative and hydrolytic changes in sunflower oil used in fryings with a fast turnover of fresh oil." *Journal of the American Oil Chemists' Society* 70, no. 11 (1993): 1069–1073.

Da Silva, Paulo F., and Rosana G. Moreira. "Vacuum frying of high-quality fruit and vegetable-based snacks." *LWT-Food Science and Technology* 41, no. 10 (2008): 1758–1767.

Dana, Dina, Michael M. Blumenthal, and I. Sam Saguy. "The protective role of water injection on oil quality in deep fat frying conditions." *European Food Research and Technology* 217 (2003): 104–109.

Dash, Kshirod K., Maanas Sharma, and Ajita Tiwari. "Heat and mass transfer modeling and quality changes during deep fat frying: A comprehensive review." *Journal of Food Process Engineering* 45, no. 4 (2022): e13999.

Dobarganes, Carmen, Gloria Márquez-Ruiz, and Joaquín Velasco. "Interactions between fat and food during deep-frying." *European Journal of Lipid Science and Technology* 102, no. 8–9 (2000): 521–528.

Dueik, V., P. Robert, and P. Bouchon. "Vacuum frying reduces oil uptake and improves the quality parameters of carrot crisps." *Food Chemistry* 119, no. 3 (2010): 1143–1149.

Erickson, Michael D. *Deep Frying: Chemistry, Nutrition, and Practical Applications*. Elsevier, 2007. DOI: 10.1016/C2015-0-02457-1.

Erickson, Maxwell D., Dmytro P. Yevtushenko, and Zhen-Xiang Lu. "Oxidation and thermal degradation of oil during frying: a review of natural antioxidant use." *Food Reviews International* (2022): 1–32. https://doi.org/10.1080/87559129.2022.2039689

Fan, Liu-ping, Min Zhang, and Arun S. Mujumdar. "Vacuum frying of carrot chips." *Drying Technology* 23, no. 3 (2005): 645–656.

Fedko, Monika, Dominik Kmiecik, Aleksander Siger, and Małgorzata Majcher. "The stability of refined rapeseed oil fortified by cold-pressed and essential black cumin oils under a heating treatment." *Molecules* 27, no. 8 (2022): 2461.

Goh, Brandon Han Hoe, Cheng Tung Chong, Yuqi Ge, HwaiChyuan Ong, Jo-Han Ng, Bo Tian, Veeramuthu Ashokkumar, Steven Lim, Tine Seljak, and Viktor Józsa. "Progress in utilisation of waste cooking oil for sustainable biodiesel and biojet fuel production." *Energy Conversion and Management* 223 (2020): 113296.

Goswami, G., R. Bora, and M. S. Rathore. "Oxidation of cooking oils due to repeated frying and human health." *International Journal of Technology Management* 4, no. 1 (2015): 2–8.

Granda, C., R. G. Moreira, and S. E. Tichy. "Reduction of acrylamide formation in potato chips by low-temperature vacuum frying." *Journal of Food Science* 69, no. 8 (2004): E405–E411.

Guillaume, C., F. De Alzaa, and L. Ravetti. "Evaluation of chemical and physical changes in different commercial oils during heating." *Acta Scientific Nutritional Health* 2, no. 6 (2018): 2–11.

Hidalgo, Francisco J., and Rosario Zamora. "The role of lipids in nonenzymatic browning." *Grasas y Aceites* 51, no. 1–2 (2000): 35–49.

Houhoula, Dimitra P., Vassiliki Oreopoulou, and Constantina Tzia. "The effect of process time and temperature on the accumulation of polar compounds in cottonseed oil during deep-fat frying." *Journal of the Science of Food and Agriculture* 83, no. 4 (2003): 314–319.

Jiang, Tian, Ying Mao, Lushan Sui, Ning Yang, Shuyi Li, Zhenzhou Zhu, Chengtao Wang, Sheng Yin, Jingren He, and Yi He. "Degradation of anthocyanins and polymeric color formation during heat treatment of purple sweet potato extract at different pH." *Food Chemistry* 274(2019): 460–470.

Jo, Hyeri, and Jeonghee Surh. "Effects of curry powder addition and frying oil reuse on the oxidative stability of deep-fried oils used in croquette preparation." *Journal of the Korean Society of Food Science and Nutrition* 49, no. 5 (2020): 493–501.

Katragadda, Harinageswara Rao, Andrés Fullana, Sukh Sidhu, and Ángel A. Carbonell-Barrachina. "Emissions of volatile aldehydes from heated cooking oils." *Food Chemistry* 120, no. 1 (2010): 59–65.

Kim, Juyoung, Deok Nyun Kim, Sung Ho Lee, Sang-Ho Yoo, and Suyong Lee. "Correlation of fatty acid composition of vegetable oils with rheological behaviour and oil uptake." *Food Chemistry* 118, no. 2 (2010): 398–402.

Kim, Nuri, Ki Seon Yu, Jaecheol Kim, Taehwan Lim, and Keum Taek Hwang. "Chemical characteristics of potato chips fried in repeatedly used oils." *Journal of Food Measurement and Characterization* 12 (2018): 1863–1871.

Koh, Eunmi, Dayeon Ryu, and Katragadda Surh. "Ratio of malondialdehyde to hydroperoxides and color change as an index of thermal oxidation of linoleic acid and linolenic acid." *Journal of Food Processing and Preservation* 39, no. 3 (2015): 318–326.

Laguerre, Mickaël, Christelle Bayrasy, Atikorn Panya, Jochen Weiss, D. Julian McClements, Jérôme Lecomte, Eric A. Decker, and Pierre Villeneuve. "What makes good antioxidants in lipid-based systems? The next theories beyond the polar paradox." *Critical Reviews in Food Science and Nutrition* 55, no. 2 (2015): 183–201.

Lee, Jiyea, and Jeonghee Surh. "Effect of carrot powder coating on the oxidative stability of the oils used in deep-frying croquette." *Journal of the Korean Society of Food Science and Nutrition* 50, no. 7 (2021): 732–741.

Liu, Xiaofang, Shuo Wang, Eitaro Masui, Shigeru Tamogami, Jieyu Chen, and Han Zhang. "Analysis of the dynamic decomposition of unsaturated fatty acids and tocopherols in commercial oils during deep frying." *Analytical Letters* 52, no. 12 (2019): 1991–2005.

Maestre, Rodrigo, Manuel Pazos, and Isabel Medina. "Role of the raw composition of pelagic fish muscle on the development of lipid oxidation and rancidity during storage." *Journal of Agricultural and Food Chemistry* 59, no. 11 (2011): 6284–6291.

Meltzer, J. B., E. N. Frankel, T. R. Bessler, and E. G. Perkins. "Analysis of thermally abused soybean oils for cyclic monomers." *Journal of the American Oil Chemists' Society* 58, no. 7 (1981): A779–A784.

Min, David B., and Jeffrey M. Boff. "Lipid oxidation of edible oil." *Food Science and Technology-New York-Marcel Dekker* No.Ed.2 (2002): 335–364.

Multari, Salvatore, Alexis Marsol-Vall, Paulina Heponiemi, Jukka-Pekka Suomela, and Baoru Yang. "Changes in the volatile profile, fatty acid composition and other markers of lipid oxidation of six different vegetable oils during short-term deep-frying." *Food Research International* 122(2019): 318–329.

Naz, Shahina, Rahmanullah Siddiqi, Hina Sheikh, and Syed Asad Sayeed. "Deterioration of olive, corn and soybean oils due to air, light, heat and deep-frying." *Food Research International* 38, no. 2 (2005): 127–134.

Obidiegwu, Jude E., Jessica B. Lyons, and Cynthia A. Chilaka. "The Dioscorea Genus (Yam)-an appraisal of nutritional and therapeutic potentials." *Foods* 9, no. 9 (2020): 1304.

Oke, E. K., M. A. Idowu, O. P. Sobukola, S. A. O. Adeyeye, and A. O. Akinsola. "Frying of food: a critical review." *Journal of Culinary Science & Technology* 16, no. 2 (2018): 107–127.

Rani, Andrali K. Sandhya, Sunkireddy Yella Reddy, and Ramakrishna Chetana. "Quality changes in trans and trans free fats/oils and products during frying." *European Food Research and Technology* 230 (2010): 803–811.

Rodríguez, Alicia, Marcos Trigo, Santiago P. Aubourg, and Isabel Medina. "Optimisation of healthy-lipid content and oxidative stability during oil extraction from squid (*Illexargentinus*) viscera by green processing." *Marine Drugs* 19, no. 11 (2021): 616.

Selmi, Salah, Irineu Batista, Saloua Sadok, Narcisa M. Bandarra, and Maria L. Nunes. "Chemical composition changes and fat oxidation in sardine mince following sodium bicarbonate and sodium chloride washing." *Journal of Food Process Engineering* 33, no. 6 (2010): 1036–1051.

Takeoka, Gary R., Gerhard H. Full, and Lan T. Dao. "Effect of heating on the characteristics and chemical composition of selected frying oils and fats." *Journal of Agricultural and Food Chemistry* 45, no. 8 (1997): 3244–3249.

Thapar, Parul. "The chemistry in re-frying of foods." *Acta Scientific Pharmaceutical Sciences* 3, no. 6 (2019): 111–113.

Tompkins, Carol, and Edward G. Perkins. "Frying performance of low-linolenic acid soybean oil." *Journal of the American Oil Chemists' Society* 77, no. 3 (2000): 223–229.

Totani, Nagao, Tika Kuzume, Ayako Yamaguchi, Mitsunobu Takada, and Masafumi Moriya. "Amino acids brown oil during frying." *Journal of Oleo Science* 55, no. 9 (2006): 441–447.

Trigo, Marcos, Pedro Nozal, José M. Miranda, Santiago P. Aubourg, and Jorge Barros-Velázquez. "Antimicrobial and antioxidant effect of lyophilized Fucus spiralis addition on gelatin film during refrigerated storage of mackerel." *Food Control* 131(2022): 108416.

Vankar, Padma S. "Regeneration of used soybean frying oils with rampad adsorbent." *Electronic Journal of Environmental, Agricultural and Food Chemistry (EJEAFChe)* 10, no. 4 (2011): 2065.

Venkata, Rekhadevi Perumalla, and Rajagopal Subramanyam. "Evaluation of the deleterious health effects of consumption of repeatedly heated vegetable oil." *Toxicology Reports* 3(2016): 636–643.

Wang, Shi-Miao, Jian Li, Qi Zhao, Dan-Dan Lv, and Kanyasiri Rakariyatham. "The effect of frying process on lipids in small yellow croaker (Larimichthys polyactis) and frying oil." *Journal of Aquatic Food Product Technology* 31, no. 1 (2022): 83–95.

Zeb, Alam. *Food Frying: Chemistry, Biochemistry, and Safety.* John Wiley & Sons, 2019, pp. 175–205.

Zhang, Qing, Ahmed S. M. Saleh, Jing Chen, and Qun Shen. "Chemical alterations taken place during deep-fat frying based on certain reaction products: a review." *Chemistry and Physics of Lipids* 165, no. 6 (2012): 662–681.

Zhang, Qing, Ahmed S. M. Saleh, and Qun Shen. "Monitoring of changes in composition of soybean oil during deep-fat frying with different food types." *Journal of the American Oil Chemists' Society* 93, no. 1 (2016): 69–81.

Chapter 6

Flavor Constituents of Fried Foods

A Chemistry Perspective

Chirasmita Panigrahi
Indian Institute of Technology Madras

Sourav Misra
ICAR-National Institute of Natural Fibre Engineering and Technology

Siddharth Vishwakarma
Indian Institute of Technology Kharagpur

DOI: 10.1201/9781003329244-6

6.1 Introduction

Frying is a process that is being adopted in the food industry for decades due to its popularity, momentous market, and vast quantity of available products. The fried foods' palatability is associated to their exclusive organoleptic characteristics, including flavor, aroma, texture, and appearance. The distinctive flavor and crunchy crispy texture make them different from other products. So, flavor is an important factor contributing to the unique property of deep-fried foods (Chang et al. 2020). Frying is still a popular cooking method despite its high fat content and consumers' increased mindfulness of the interrelationship between food, health, and nutrition. However, despite having a high caloric value, fried dishes can be healthy and preferable over other cooking techniques like baking and boiling (Saguy and Dana 2003).

Flavor comprises mostly of taste and aroma, and is an inimitable characteristic of fried foods. Flavor is primarily contributed by the volatile compounds generated by the heat-induced complex reactions. The flavor of fried food is the outcome of complex physical and chemical interactions occurring between food components and the oil medium during the process of heating (Bansal et al. 2010). The simultaneous heat and mass transfer of oil (acts as a heat transfer medium), food, and air during deep-fat frying produces the desirable quality of fried foods (Choe and Min 2007). Oil/fat plays a crucial sensory role to derive, transmit, promote, and enhance oily, fatty, and nutty flavors (Shabbir et al. 2015). Frying produces physicochemical changes to the principal food components (fats, proteins, and polymeric carbohydrates). Elucidating the development mechanisms of deep-fried volatiles is highly important to examine perceived flavor, monitor frying process, and evaluate the desirability of fried food products. This chapter discusses the characteristics and formation mechanisms of deep-fried flavor, and describes the roles of frying media, frying types, frying conditions, and food components in the development of deep-fried volatiles.

6.2 Flavor Components Formed during Frying and the Mechanism Involved

Flavor and aroma components are majorly derived from either the Maillard reaction between proteins and sugars or thermal degradation of lipids. Numerous heterocyclic compounds are formed at high-temperature and low-moisture conditions that prevail during frying. The mechanism's pathways and important volatile compounds produced during frying are illustrated in Figure 6.1. Lipid oxidation and the Maillard reaction are the two most chief chemical reactions in the development of deep-fried flavor, which are determined by the complex volatiles. Both these oxidation and browning reactions follow similar reactions along with the formation of key intermediates (e.g., hydroperoxides in lipid oxidation, and Amadori rearrangement products in the Maillard reaction) which are then broken down, cyclized, degraded, and rearranged to produce volatile chemicals that give food its unique flavor (Figure 6.1).

Owing to the formation of chemical compounds with similar functional groups in these two reaction pathways, several studies have examined the interaction between lipid oxidation (intermediates/products) and

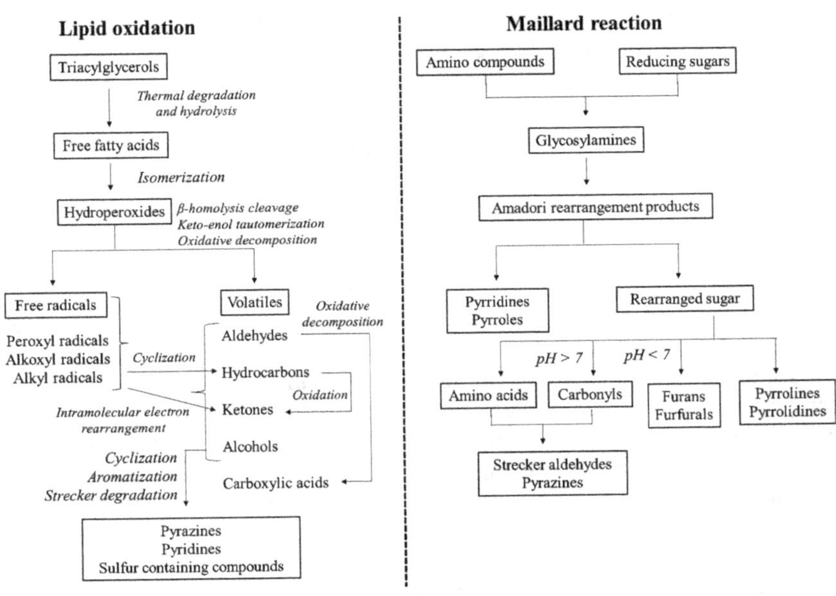

Figure 6.1 Predominant phenomena and common volatiles contributing to the development of flavor in frying process.

the Maillard reaction (intermediates/products). The focus is especially on the involvement of lipid oxidation products in the Maillard reaction to create volatiles (e.g., pyridines, pyrazines, and sulfur-containing compounds) accountable for deep-fried flavor. In addition, thermal degradation of amino acids and hydrolysis of triacylglycerols play minor but vital roles to enable or aid the formation of volatiles during frying. The furans, furanones, aldehydes, alcohols, and sulfur-containing compounds are produced in foods fried in oil heated to >140°C temperature. These were caused by the interactions between fats, proteins, and carbohydrates in the deep-fried system and thermo-degradations of sugars, fats, and amino acids. It was also noticed that the stage of frying with temperatures (140°C–165°C) was the most crucial period for the flavor formation (Zhang et al. 2019). Sulfur-containing compounds were recognized to contribute substantially to the aroma of frying oil of several foods, such as onion, garlic, meat, and tropical fruits. The most common substances found in deep-fried oil of onion were 11 sulfur-containing compounds (9 sulfides and 2 thiophenes), 3 saturated aldehydes (hexanal, heptanal, and nonanal), and 3 furans (dibenzofuran, furfural, and 2,3-dihydro-benzofuran). Other flavor compounds are lactones, alkanes, ketones, esters, heterocyclic pyrazines, hydroxy-nonenoic acids, decadienal, and furans (Hammouda et al. 2017). Numerous volatiles are produced in the crust of fried food, e.g., 2-octenal (nutty, tallowy flavor), 2-ethyl-3,6-dimethyl-pyrazine (earthy, roasted flavor), and dimethyltrisulfide (cabbage flavor). Sometimes, the compounds (e.g., 3,6-dimethylpyrazine for roasted flavor) identified in the raw 2-ethyl-samples were not found in the fried samples as those are either likely to be removed by evaporation or transformed to other substances with increase in oil temperature (Zhang et al. 2019).

6.2.1 Lipid Oxidation

It is well known that lipid oxidation takes place in three steps: initiation, propagation, and termination. In the initiation step, the abstraction of hydrogen from unsaturated fatty acids generates the alkyl radical, on which the free radical is delocalized on the carbon chain, and via molecular rearrangement, conjugated double bonds are formed (McClements and Decker 2007). The weakest carbon–hydrogen (C–H) bond is first attacked to release a free radical. The strength of C–H bond thus decides the oxidation rate of fatty acids. The order of fatty acids in terms of C–H bond strength is: stearic acid (100 kcal/mol) > oleic acid (75 kcal/mol) > linoleic acid (50 kcal/mol) > linolenic acid (25 kcal/mol) (Min and Boff 2002). This indicates that the hydrogen abstraction on linolenic acid will be faster than linoleic, oleic, and stearic acids. It is found that the oxidation rate of oleic acid into hydroperoxides is ten times slower than linoleic acid (Ho and Chen 1994).

Subsequently, the alkyl radical (reduction potential of 600mV) rapidly reacts with oxygen (at triplet state) to produce a peroxyl radical (reduction potential of 1,000mV), which has high energy to promote the hydrogen abstraction from another unsaturated fatty acids. The addition of hydrogen on the peroxyl radical produces hydroperoxides, and this process repeats causing the step of propagation. However, due to the instability of hydroperoxides, the homolytic cleavage of the –OOH group causes their decomposition into alkoxy radicals. Thereafter, β-scission reaction of –OOH group on the C–C bond occurs by which the volatiles, such as alkane, alkene, aldehydes, and alcohols, are produced (Whitfield and Mottram 1992). During termination, the peroxyl radical and alkoxyl radical combines, and the alkyl radicals combine among themselves to form non-radical volatile species (e.g., ketones, aldehydes, alcohols, hydrocarbons, acids, lactones, and esters) (Figure 6.1) (McClements and Decker 2007).

6.2.2 Maillard Reaction

The Maillard reaction, a non-enzymatic glycosidation of amino acids, is recognized as another most essential chemical reaction taking place during deep frying to determine the fried flavor. It is an extremely complex reaction starting with a free amino group (in proteins, peptides, or in amino acid) and a reducing sugar (e.g., fructose, glucose, and lactose) to yield an N-substituted glycosylamine, which then goes through an Amadori rearrangement to form Amadori rearrangement product (ARP). Further, ARP can be cyclized to produce nitrogen-containing heterocyclic compounds, like pyrroles and pyridines. In other case, it may either be broken into rearranged sugars, which are very critical for the volatile formation. These happen via four pathways (Zamora and Hidalgo 2005): (i) the rearranged sugars undergo keto-enol tautomerization (with 3-desoxy-1,2-diketones and/or 1-desoxy-2,3-diketones) to revert the original amino acid for subsequent Maillard reaction; (ii) the rearranged sugars suffer cyclization (in acid) to produce heterocyclic oxygen-containing compounds (e.g., furfurals and furans); (iii) cleavage of rearranged sugars into dicarbonyl fragments that further react with amine groups resulting from the Strecker degradation of amino acids [including oxidative decarboxylation and deamination of an α-amino acid (e.g., valine, serine, and leucine) in the presence of a dicarbonyl compound] to generate Strecker aldehydes (e.g., 3-methylbutanal, 2-methylpropanal, and 2-hydroxyethanal) and pyrazines (e.g., 2,5-dimethyl-3-ethylpyrazine, 2,5-dimethylpyrazine, and 2,3,5-trimethylpyrazine) in a base; and (iv) dicarbonyls on interacting with proline/hydroxyproline can obtain pyrrolines and pyrrolidines (Figure 6.1) (Mottram 1994; Jousse et al. 2002). Flavor compounds arising out of the Maillard reaction predominantly consist of aliphatic substances (e.g., fatty acids, ketones, and aldehydes) and heterocyclic derivatives (containing sulfur, oxygen, and/or nitrogen, e.g., pyrroles, pyrazines, pyridines, and furans). It is found that heterocyclic derivatives having lower odor thresholds are more efficient in carrying typical flavors compared to the aliphatic substances even at relatively higher concentrations (Pokorny 1980). Carbonyl compounds formed during frying can react with amines, amino acids, and proteins that in turn produce desirable nutty-flavored pyrazines (Negroni et al. 2001).

In some researches, the flavor formation in the fried potatoes is observed to be very complicated, which seems to develop not only from the chemical reactions within foods and frying media, but is also chiefly driven from complex interactions among reactive intermediates (Martin and Ames 2001). For example, unsaturated/saturated aldehydes (lipid oxidation products) may react with hydrogen sulfide (a Strecker degradation product) to generate roasted odorants [e.g., 2-ethyl-3,5-dimethylpyrazine, 3-(methylthio)propanal, and 2-nonenal] (Pokorny 1981; Whitfield and Mottram 1992).

6.3 Types of Volatiles Impacting Fried Flavor

6.3.1 Aldehydes

Aldehydes, including unsaturated (alkadienals and alkenals) and saturated (alkanals), are the most common volatiles developed through lipid oxidation in the frying process to exhibit green, oily, metallic, paint, and beany flavors. In general, hydrolysis and thermal degradation of triacylglycerols easily occur under frying conditions to generate free fatty acids, which then form hydroperoxides via keto-enol tautomerism, β-homolysis cleavage, or isomerization (Yin et al. 2011). Subsequently, each hydroperoxide can be degraded to volatiles of low molecular mass (e.g., aldehydes, ketones, hydrocarbons, carboxylic acids, and alcohols) (Zhang et al. 2015). So, the concentration of aldehydes is in association with the unsaturated fatty acids' oxidation and the oxidative intermediates' decomposition. The formation of alkanals has been positively correlated to the content of oleic acid, whereas alkenals are very well associated to the abundance of linolenic and linoleic acids (Perkins 2007). Moreover, aldehydes are considered a superior marker for detecting oxidative degradation to alternate common methods (e.g., anisidine value, total polar compounds, and peroxide value) (Molina-Garcia et al. 2017). However, there is a chance that the production of aldehydes might also be consumed by some complex reactions happened during frying. Zhang and coworkers (2015) claimed (E,E)-2,4-decadienal and (E)-2-hexenal were produced in oil while frying wheat dough, but were depressed by certain prevailing chemical reactions (e.g., Maillard reaction, degradation, and further oxidation) as chicken breast meat was further fried. In general, 2- and 3-methylbutanal (malty odor) and 3-(methylthio)propanal (cooked or fried potato-like smelling) contribute to the fried flavor in more than 50% of fried products, followed by hexanal (green grassy aroma), (E,E)-2,4-decadienal (oily odor), acetic acid (vinegar-like aroma), 1-octene-3-one (mushroom-like aroma), acetaldehyde (fresh fruity aroma), 2-acetyl-1-pyrroline (roasty odor), and (E)-2-nonenal (cardboard-like aroma) (Dunkel et al. 2014).

6.3.2 Hydrocarbons, Alcohols, Ketones, and Carboxylic Acids

In practice, hydrocarbons, alcohols, ketones, and carboxylic acids are formed in minor quantities during frying (Molina-Garcia et al. 2017). Saturated hydrocarbons (alkanes, including side-chain, straight-chain, and cycloalkanes) and unsaturated hydrocarbons (alkenes, including substituted monoenes, substituted dienes, straight-chain monoenes, straight-chain dienes, cyclic dienes, cyclic monoenes, and configurational isomers) are obtained in the volatile fraction via free radical reactions (e.g., combination and cyclization), oxidative degradation of unsaturated fatty acids, and thermal decarboxylation of saturated fatty acids (Chung et al. 1993; van der Klis et al. 2011). Similar to aldehydes, all hydrocarbons do not elevate in concentrations with the progress of frying process. For instance, the amounts of pentadecane and tetradecane were decreased with the prolongation of time, i.e., at the later stage of frying, owing to the complex reactions occurring (Zhang et al. 2018). A great extent of alkane evaporation (or boiling) might also result in volatile loss after frying due to their boiling points being much closer to the frying temperatures (Uriarte and Guillen 2010). Cyclic alkenes (e.g., cyclooctenes and cyclohexenes) were intermittently detected during the frying process since they are thermally instable (Nawar 1998). Alcohols (including unsaturated, saturated, and cyclic ones) are produced through positional double bond (C=C) isomerization and oxidation of unsaturated fatty acids (Zhang et al. 2018). In oxidative fatty acid decomposition, 1-octen-3-ol (a mushroom-like smelling) originating from linoleic acid-10-hydroperoxide decomposition is the most reported alcohol formed in frying (Asaaf et al. 1997). The thermal peroxidation of saturated fatty acids, further oxidation of hydrocarbons, keto-enol tautomerization of hydroperoxides, and breakdown and intramolecular electronic rearrangement of peroxides from unsaturated fatty acids are accredited to the creation of ketones (e.g., cyclic ketones and methyl ketones) (Chang et al. 1978; Yin et al. 2011). Moreover, carboxylic acids can be developed from lipid oxidation and even secondary decomposition of oxidative products (e.g., 2,4-decadienal and hexanal) (Molina-Garcia et al. 2017). The concentration of carboxylic acids remains almost constant during the frying phase (Dominguez et al. 2014). In a study done by ben Hammouda et al. (2017), most of carboxylic acids were consistent in concentrations, excluding 1-hexanoic acid (fatty odor) which is made out of the secondary degradation of 2,4-decadienal and hexanal (Molina-Garcia et al. 2017).

6.3.3 Furans, Pyridines, and Pyrazines

Furans, pyridines, and pyrazines are well-identified aroma sources typically found at trace quantities in the frying oil and fried foods. The appearances and amounts of these compounds are primarily related to time and temperature of frying as well as the nature and complexity of ingredients (especially reducing sugars and amino acids). Furans are made out of several routes, like thermal degradation of carbohydrates, Maillard reaction between amino acids and reducing sugars, and lipid oxidation (ben Hammouda et al. 2017; Zhang et al. 2018). They commonly produce the deep-fried foods with favorable aroma described as caramel-like, burnt, and sweet odors (Van Ba et al. 2012). 2-Pentylfuran is the most investigated volatile furan, which is formed by degradation of methyl linoleate hydroperoxide from linoleic acid oxidation (Vichi et al. 2003). Markedly, pyrazines are structure-odor active volatiles that accomplish great olfactory impact even at very minute concentrations.

Previous literatures about formation of several volatile fried flavor compounds are outlined in Table 6.1.

6.4 Factors Affecting Fried Flavor

The characteristic flavor of fried foods is developed by an amalgamation of reactions and compounds captivated by the frying oil. Generally, it can be declared that frying oils and food components indeed have huge impacts on the creation of fried volatiles. Lipid oxidation, hydrolysis, Maillard reaction, and amino acid degradation are primarily induced by the macronutrients available in frying oils (e.g., lipids) and food itself (e.g., proteins and carbohydrates) to form plentiful volatiles that govern the extent of fried flavor. In addition to oil and food constituents, a number of other relevant reactive parameters or conditions are there that extensively determine the generation of volatiles. This can be divided in two categories: (i) pH exhibits influences on the reaction pathways in regulating the nature of volatiles, and (ii) the frying temperature, time, moisture content, oxygen concentration, and frying type mainly determine the reaction kinetics, rather than the variety of volatiles.

Table 6.1 Various Volatile Flavor Compounds Developed during Frying of Foods

Food Sources	*Oil Used*	*Frying Condition*	*Chemical Reactions*	*Volatiles*	*References*
Wheat dough followed by chicken breast meat	Soybean oil	180°C for 5 minutes (dough) and 10 minutes (meat), 8 h/day for 7 days	Lipid oxidation, isomerization, hydrolysis, Maillard reaction, Strecker degradation	Hexanal, 2-pentenal, nonanal, 2-heptenal, 2,4-heptadienal, benzaldehyde, 2,4-decadienal, 3-methyl butanal, 3-nonen-2-one, 4-oxononanal, 2-propyl pyridine, 1-octen-3-ol, 2-pentylfuran, and 3-dodecene	Zhang et al. (2015, 2018)
Donut	Combinations of high-oleic rapeseed oil, high-oleic sunflower oil, palm mid fraction, and hydrogenated rapeseed oil	170°C for 1.5 minutes (each side) and the frying media was heated at 8 h/day for 10 days	Lipid oxidation, lipid degradation, saccharide degradation	Hexanal, octanal, nonanal, 2-octenal, decanal, heptanal, 2-decenal, 2-undecenal, and 2,4-decadienal	Merkle et al. (2016)
Pork loin	Corn oil	180°C for 5 minutes	Lipid oxidation, Maillard reaction, vitamin degradation	Hexanal, benzaldehyde, nonanal, 2-pentylfuran, 2,5-dimethylpyrazine, 2,6-dimethylpyrazine, trimethylpyrazine, 6-methyl-2-ethylpyrazine, and 2-methyl-3-propylpyrazine	Yang et al. (2017)

(Continued)

Table 6.1 (*Continued*) Various Volatile Flavor Compounds Developed during Frying of Foods

Food Sources	*Oil Used*	*Frying Condition*	*Chemical Reactions*	*Volatiles*	*References*
French fries	Rapeseed oil, sunflower oil, palm olein	Fried at 170°C initially for 4 minutes, and the oil temperature was maintained for 12 h over 3 days	Lipid oxidation	Hexanal, 2,4-heptadienal, nonanal, 1-octen-3-ol, 2,4-decadienal, 2-decanone, 2-decenal, 2-octenal, and 3-octanone	Petersen et al. (2013)
Shallot	Soybean oil, corn oil, lard	150°C–160°C for 5 minutes	Lipid oxidation, amino acid degradation	2-Hexenal, 2-heptenal, 2,4-heptadienal, 1-heptanol, 2-pentylfuran, nonanal, 2-octenal, 1-octen-3-ol, 2,4-decadienal, 2-decenal, and 2-undecenal	Chyau and Mau (2001)
Potato slices	Palm olein	180°C for 2.5 minutes	Lipoxygenase-mediated fat oxidation, Strecker degradation	Methylpropanal, penylacetaldehyde, 3-methylbutanal, dimethyl disulfide, 2-methylbutanal, methylpyrazine, 2-ethyl-5(3)-methylpyrazine, hexanal, 3-ethyl-2,5-dimethylpyrazine, and 2,4-decadienal	Martin and Ames (2001)

6.4.1 Frying Oils

The oil (or any kind of fat used for frying) is an essential flavor carrier. It makes food tasty by taking out the inherent flavors as well as developing new ones. The type of oil definitely has effect on kind and intensity of flavor owing to the variances in quality and quantity of their fatty acids. The fatty acid composition thus has a crucial influence on fried food flavor. For instance, higher concentration of linoleic acid in oil intensifies the fried food flavor by increasing the amount of volatile compounds like pentanal, hexanal, and 2,4-decadienal (Prevot et al. 1990). Flavor intensity is documented as the best indicator for determining overall flavor quality. The increasing oleic acid and decreasing linolenic acid have been found to produce more total polar compounds and in turn provide greater frying stability. Volatile compounds such as pentanal, hexanal, 2,4-decadienal, nonanal, and octanal were used to evaluate oxidation of oil during storage of fried potato chips. Previous experimentation showed that these volatiles are the decomposition compounds of thermally oxidized trilinolein and triolein. Pentanal, hexanal, and 2,4-decadienal are breakdown products of linoleic acid oxidative degradation, whereas nonanal and octanal are end products from oleic acid oxidation (Warner et al. 1994).

Since lipid oxidation is the most dominant in flavor formation, the frying oil comprised of triacylglycerols and minor components (e.g., phenols, phospholipids, free fatty acids, and sterols) enhances the sensory profile of fried products (Cao et al. 2017). Notably, fatty acid composition and oil type are supreme factors to affect the construction of volatiles by oxidative degradation of frying media (Whitfield and Mottram 1992). For example, oleic acid gets oxidized to yield higher quantities of unsaturated aldehydes (e.g., 2-undecenal and 2-decenal), followed by lower amounts of saturated aldehydes (e.g., octanal and nonanal) and then hydrocarbons (e.g., heptane and octane) (Nawar 1998). The oxidative decomposition of linoleic acid generates various volatile compounds (e.g., 2-heptenal, hexanal, 2,4-nonadienal, 2,4-decadienal, 2-octenal, and 2,4-octadienal) that are highly important in flavor creation (Buttery 1989). 2,4-Decadienal is the most characteristic flavor compound, which is derived from the decomposition of 9-diene hydroperoxide coming out of linoleic acid oxidation and is subsequently oxidized to give (E)-4,5-epoxy-(E)-2-decenal (having a crumb flavor) (Schieberle and Grosch 1991). In addition, typical volatile compounds arising from the linolenic acid oxidation are propanal, butenal, 2-pentenal, 2,4-hexadienal, 2,4-heptadienal, and acrolein (Katsuta et al. 2008).

According to a sensory analysis, a stronger fried flavor of potato chips was released when fried in the canola oil having 64% oleic acid, 24% linoleic acid, and 3% linolenic acid than the oil with 78% oleic acid, 9% linoleic acid, and 4% linolenic acid, but its justification was not properly addressed

(Warner et al. 1994). The same group of researchers further stated pentanal, hexanal, 2,4-decadienal, nonanal, and octanal to be the representative volatiles for monitoring the changes in flavors. The concentrations of pentanal, hexanal, and 2,4-decadienal (the most protuberant volatiles resulting from linoleic acid oxidation) were elevated as oleic acid amount was reduced and linoleic acid was raised. But nonanal and octanal (liberating from oleic acid oxidation) were not changed significantly (Warner et al. 1997). This phenomenon was also verified by Wagner and Grosch (1998), who reported that a higher number of decadienal isomers [(E,E)- and (E,Z)-2,4-decadienal] were released from French fries prepared in palm oil (10.5% linoleic acid) compared to those from coconut fat (1.9% linoleic acid). In another study by Peng et al. (2017), sunflower oil (containing more linoleic acid) liberated a higher amount of aldehydes (e.g., hexanal, butanal, 2,4-nonadienal, 2-heptenal, and 2,4-decedienal) than palm oil (rich in saturated fatty acids) and rapeseed oil (rich in oleic acid). As a result, the adequate proportion of oleic acid and linoleic acid in frying oil are 16%–42% and 37%–55%, respectively, which are suggested to offer valuable flavor quality and satisfactory flavor intensity (Warner et al. 1997).

Regarding linolenic acid, Frankel et al. (1985) appealed that the soybean oil (8.5% linolenic acid) was degraded to generate undesirable fishy, pungent, and paraffin-like odor during frying of bread at 190°C. The oxidation of linolenic acid was also depicted in the case of potatoes fried in linseed oil with 55% linolenic acid to produce extremely greater quantities (~5 times more) of toxicological compounds (e.g., glycidamide and acrylamide) than safflower and coconut oils (<1% linolenic acid) (Thurer and Granvogl 2016). In view of all these results, lowering linolenic acid (<3%) has been set as an objective to improve sensory quality, oxidative stability, and safety of frying oils and deep-fried food products (Liu and White 1992).

Fried flavor compounds are mostly the volatile compounds from linolenic or linoleic acid and are hydrocarbons, alkenals, dienal, and various cyclic compounds (Pokorny 1989). Some useful flavor compounds noticed in frying oil that are synthesized from oxidation of linoleic or linolenic acid are 4-hydroxy-3-nonenoic acid and its lactone, 4-hydroxy-2-nonenoic acid and its lactone, *trans*, *trans*-2,4-octadienal, *trans*-2-heptenal, *trans*-7-octenal, *trans*, *trans*-2,4-decadienal, *trans*, *trans*-2,4-nonadienal, nonenlactone, *trans*-2-octenal, and trienals and all these are produced from the oxidation of linolenic or linoleic acid (Buttery 1989).

In comparison with oleic acid, linoleic acid oxidation releases more flavor volatiles, whereas linolenic acid is more liable to oxidation that possibly can generate toxic chemicals. So, in the frying oil, it is recommended that linoleic acid content might be twice than that of oleic acid, but linolenic acid should be <3%. Depending on deamination reaction, pKa values, and nucleophilicity of amine groups, amino acids in foods act differently in releasing the fried flavor. While considering the frying condition, it can be specified that the formation of volatiles is more favorable in an alkaline environment (i.e., high pH) and at higher temperatures. Thus, both frying media and frying conditions are important factors to be judiciously selected for regulating the volatile development during deep frying. Apart from the above-mentioned two parameters, aging of oil has a huge effect. It is witnessed that the levels of volatile compounds in fried potato chips were greater during the earlier stage of the frying cycle than while the later phase of frying.

In reality, flavor perception is beyond than complex as it not only depends on the action of individual volatile compound, but also relies on the synergistic, additive, masking, and interaction actions between the volatiles (Nawar 1998). Thus, a mere simple study of volatile compositional analysis is scarce in understanding the complicated flavor pattern. Rather, repetitive inspections during various phases of frying process and investigation of volatile integrations are felt really necessary to reflect the incessant volatile changes and expansively elucidate the fried flavor acuity. Nevertheless, deep-fat frying has some major side effects, viz. the sugar degradation through caramelization and loss of macronutrients (sugars and amino acids) via Maillard reaction and Strecker degradation. According to Oluwaniyi et al. (2010), palm oil frying caused 2%–19% declination in total amino acid concentration of *Scomber scombrus*, whereby the available lysine was particularly damaged. They observed some notable phenomena responsible for this, such as the Maillard reaction between amine groups and carbonyl groups (of oxidized fats) or aldehyde groups (of reducing sugars), protein–protein interactions at frying temperatures, and various cross-linking reactions which made amino acids metabolically unavailable. Hence, regulating frying by-products and intermediate materials as well as balancing frying conditions to maintain ideal fried flavor along with simultaneously minimizing the production of toxicological substances ought to be carried out critically.

Furthermore, Ramirez et al. (2004) analyzed the volatile compositions of pork loin chips fried in conventional fats (e.g., pig lard and butter with 53% saturated fatty acids) and vegetable oils (e.g., sunflower oil and olive oil having 86% unsaturated fatty acids). As a result of lipid oxidation, more aliphatic aldehydes, like pentanal and hexanal, were detected when the degree of unsaturation elevated as in the case of vegetable oils. In contrast, the Strecker degradation of amino acids (e.g., phenylalanine and isoleucine) produced unique Strecker aldehydes like benzeacetaldehyde and 2-methylbutanal which were released from the conventional fats. However, fried volatiles are not found to be always dependent on the frying oil. The creation of pyrazines (e.g., 2-ethyl-3,5-dimethylpyrazine and 2,3-diethyl-5-methylpyrazine) was not very pronounced during frying in olive oil, safflower oil, coconut oil, and rapeseed oil in a study by Thurer and Granvogl (2016). This finding was also affirmed by Wagner and Grosch (1998), who reported the insignificant differences on alkylpyrazines in French fries fried in coconut oil and palm oil. In consequence, several other factors might be acted in determining deep-fried flavor in addition to the frying medium.

6.4.2 Food Components

Food components have a profound impact on flavor profile. The addition of food components (e.g., sugars and amino acids) conceivably affects the profile of volatiles by decreasing the amount of ketone compounds during frying. In scientific terms, the oxidative degradation of oils generates the products with peroxyl, hydroxyl, hydroperoxyl, and carbonyl groups that seem to react with amide- and amino-containing compounds inducing covalent-bonding complexes (Pokorny and Kolakowska 2011). Actually, the flavor of deep-fried food relies much on the food components than the frying oil (Dobarganes et al. 2000). For instance, Peng et al. (2017) concluded that butanal, acetaldehyde, hexanal, and 2,4-decadienal were significantly higher in the fried pork loins than those in fried potatoes. In another study by Zhang et al. (2018), nitrogen-containing volatiles (e.g., 2-propylpyridine, 2-pentylpyridine, and pyrrole-3-butyronitrile) were only spotted in fried chicken breast meat rather than the fried wheat dough. So, the complexity and characteristic of food components intensely affects the formation of fried flavor. Lu et al. (2018) observed that fried chicken meatballs had less heterocyclic amines than the deep fat–fried beef meatballs, since a high content of non-heme iron in beef reacted with peroxyl and hydroxyl radicals that accelerated lipid oxidation and Maillard reaction compared to chicken (Gibis 2016). Though carbonyl groups on lipid oxidation products were found to replace the reducing sugar in the Maillard reaction (Mandin et al. 1999; Zhang and Ho 1989), their result was interrogated by Adams et al. (2011), who identified glucose to be more competitive than 2-hexenal in reacting with lysine for the volatile production. This was further proved by Miyagi and Ogaki (2014a), who monitored that free sucrose (broken down to glucose and fructose) and amino acid contents were reduced by 25% and 89% during frying by caramelization and Maillard reaction, respectively. Therefore, reducing sugar is still necessary to provoke the release of aroma compounds during fat frying.

Moreover, the effect of amino acids on the flavor should not be disregarded by their noteworthy sensory attributes. For instance, methionine contributes to cooked potato flavor, alanine raises sweet and malt odors, tryptophan is responsible for the burnt meat aroma, and more lysine increases burnt and nutty smell (Pokorny 1980). Chun and Ho (1997) found a total of 29 pyridines and pyrazines in the mixtures of glutamine/asparagine fried in corn oil. They presented that more pyridines (particularly 2-pentylpyridine) and alkylpyrazines were released from the glutamine model system, because a stronger deamination reaction of glutamine occurred to generate ammonia under frying conditions. This specifies that the type of amino acid affects the production of volatiles. With regard to nucleophilicity, lysine produced less conversion (22%) of hexanal to 2-butyl-2-octenal (a ham-like flavoring compound) than glycine, which was more accessible to the Maillard reaction and caused 95% hexanal conversion (Adams et al. 2011). Additionally, the fried flavor is also distinguished between acidic and basic amino acids. In a research work (Kavousi et al. 2015), it was noticed that the acidic amino acids (e.g., aspartic acid and glutamic acid) lowered the pH that pushed the Amadori products degradation pathway toward formation of 5-hydroxymethylfurfural, whereas basic amino acids (e.g., lysine, histidine, and arginine) increased pH (>6) which substantially decelerated 5-hydroxymehylfurfural production. Hence, food components play an indispensable role in the fried flavor development.

6.4.3 Frying Conditions

A proper frying condition (e.g., oxygen concentration, frying time, frying temperature, and pH) is critical in developing a desirable fried flavor.

6.4.3.1 Oxygen Concentration

Oxygen concentration is a vital parameter that decides the intensity and quality of flavor. Quality deterioration due to browning and lipid oxidation often results in unpleasant flavor in case of a higher amount of oxygen. Therefore, it is supposed that the typical required flavor is produced only at the optimal oxygen concentration so as to encourage a proper lipid oxidation rate. Poor and weak flavor is obtained at low amounts of oxygen, while high levels of oxygen produce bad flavor due to rancidity (Pokorny 1989). Chyau and Mau (2001) witnessed a stronger flavor in corn and soybean oil in contrast to the lighter fried shallot flavor in medium-chain triglycerides being caused by little oxygen involvement.

6.4.3.2 Time

The concentration of volatile flavors varies as the frying time is prolonged, but it completely depends on the nature of compound. The aroma of deep-fried peanut increased with frying time (2–15 minutes), and the highest overall liking score was given to the peanuts fried for 4–9 minutes (Miyagi and Ogaki 2014b). Suppose 2-nonenal is taken as an example; it was slightly raised during wheat dough frying in soybean oil. But sometimes an unusual trend might be seen, like an initial increase followed by a reduction in 2-nonenal amount was observed when chicken breast meat was fried, due to an elimination action among volatiles (Zhang et al. 2015). In a Maillard model framed by lipid oxidation products (e.g., 2,4-decadienal and hexanal) and amino acids (e.g., glycine and lysine) with/without glucose, Adams and coworkers (2011) also verified that the volatile concentration was decreased by twice when the heating time was extended up to 2 hours.

6.4.3.3 Temperature

Frying temperature is another crucial parameter to determine the sensory appeal of fried foods. Higher temperature greatly facilitates complex chemical reactions (by opening structures of reactants) to stimulate the generation of volatiles. Thurer and Granvogl (2016) determined that as frying temperature of safflower oil was decreased from 180°C to 140°C, it resulted in a significant loss (around 86%) of prime odorants [e.g., 2,4-decadienal, 3-(methylthio)propanal, and 3-methylbutanal]. Mandin et al. (1999) claimed that lower reaction temperature (100°C–105°C) was accountable for the very light cooked flavor in a food system comprising linoleic acid, methionine, and glucose due to the absence of furans, pyrazines, oxazoles, and thiazoles. Irrespective of potent toxicologic complexes (e.g., acrylamide), relatively high temperature is authenticated to improve odor impression of fried foods.

6.4.3.4 Moisture

Moisture is an inevitable feature participating in a multiple series of reactions (e.g., hydrolysis, isomerization, oxidation, and Maillard reaction) to induce flavor formation in frying. In most cases, moisture is regarded as a negative factor that often gives rise to undesirable flavor by intensive deterioration reactions to rancidify frying oils (Bazina and He 2018). Further, Mandin et al. (1999) identified the absence of pyridines, pyrazines, and thiazoles under high-moisture conditions resulting from negative effects of the Maillard reaction. In addition, deep-fried banana chips having higher moisture content displayed worse odor acceptability to the sensory panelists (Ammawath et al. 2001). In contrast to the above-discussed effects, an innovative frying technique was developed by Ma and coworkers (Ma et al. 2016), who used a mixture of oil and water as an alternate to pure oil. In this system, the oil was at the upper layer maintained at the frying temperature, whereas the water was at the lower layer with temperature below 55°C. So, the food descended into water prior to carbonization at the high temperature of oil and was finally removed through plughole in the fryer. Hence, the rate of oil deterioration could be mitigated by the presence of water during the frying; meanwhile, desirable fried sensory characteristics were retained. As a result, the application of moisture will have a great influence on the development of fried flavor.

6.4.3.5 pH

pH is a factor having a pronounced effect on volatile production by influencing ARPs degradation to alternate kinetics of the Maillard reaction (Jousse et al. 2002). At acidic pH (<7), 1,2-enolization of ARPs corroborates to the formation of furfural (with pentoses) and hydroxymethylfurfural (with hexoses), whereas at basic pH (>7), 2,3-enolization produces 4-hydroxy-5-methyl-2,3-dihydrofuran-3-one and fission products (e.g., acetal, diacetyl, and pyruvaldehyde) (Martins et al. 2000). Yaylayan and HuyghuesDespointes (1996) mentioned that ARPs can be degraded to acetic acid and pyruvaldehyde with free amino acids under alkaline conditions, so the pH>7 is regarded as a key pathway for flavor development via the Maillard reaction. Moreover, pH also affects the hydrolysis of carbohydrates and the generation of unprotonated amino acid (a reactive species for the Maillard reaction). Starch hydrolyzes faster at acidic pH to promote the emission of carbonyl groups participating on to the Maillard reaction (BeMiller 1965). However, at lower pH (<pK_a of amino groups), less unprotonated amino acids are available which lead to a milder Maillard reaction (Martins et al. 2000). For example, the production of 5-hydroxymethylfurfural was gradually diminished as the pH increased from 4 to 10, since 1,2-enolization pathway was hindered by more protonated form of glutamic acid (Kavousi et al. 2015).

6.4.4 Frying Type

The conventional frying methods include stir frying, pan frying, and deep frying, which are distinguished by the amount of oil used. Although the volatiles produced in all the frying types resemble each other, the highest release of volatiles, especially the characteristic fried volatile compounds (e.g., hexanal and 2,4-decadienal), is found higher in the case of deep frying than in pan and stir frying (Peng et al. 2017). Deep-fried flavor, typically composed of a number of volatiles, is described as oily, grassy, fruity, buttery, nutty, burnt, and fishy odor (Nayak et al. 2016).

Today, consumers are more focused toward healthy diet and are seeking for food products with extraordinary quality in terms of nutrition, appearance, functionality, safety, and sensory properties. Vacuum frying attracts huge attentions owing to the use of lower temperatures (90°C–150°C) under pressure <6.65 kPa, least contact with oxygen, and lower oil absorption to develop fried products with less oil content and high retention of inherent colors and flavors (Moreira 2014). It is the only frying technique that can be applied on delicate food sources having high sugar content (e.g., fruit and vegetables). As established by some researchers, vacuum frying depicted superior protections to the atmospheric frying of canola oil (Aladedunye and Przybylski 2009) and high-oleic-acid sunflower oil (Nazarbakhsh et al. 2014) to retard oxidative deterioration, which was assessed by total polar compounds, fatty acids composition, anisidine value, and antioxidative activity. Chicken nuggets vacuum fried at 150°C for 6 minutes obtained better organoleptic characteristics (including color, oiliness, crispness, and overall acceptance) compared to the deep-fried ones. But the juiciness and the lightness (*L* value) in color were undesirable due to the loss of water content in the vacuum-fried chicken nuggets. In this study, little information was available on the flavor except oiliness, which was not significantly affected by the frying type. An opposite remark was delineated by Yang et al. (2012) on the sweet potato snacks fried using rice bran oil carried out by atmospheric frying (at 170°C for 2.5 minutes) and vacuum frying (at 90°C under 20 kPa). Vacuum-fried snacks had improved color and flavor acceptability, whereas the deep fat–fried snacks were crispier. Troncoso et al. (2009) observed that the sensory attributes, particularly flavor quality and physicochemical characters of deep-fried potato slices, were significantly better than vacuum frying. Therefore, the detailed measurement of volatile profile is still required for the feasibility analysis of vacuum frying in the food manufacturing.

Air frying is another novel alternative technique to deep-fat frying, which is developed to fulfill the requirements of consumers for the fried foods with reduced fat content and comparable sensory features as that of conventional frying. An emulsion of oil droplets in hot air is created when food products are fried in the air-frying chamber, and as a result, the product have a homogeneous contact with hot air and oil to be dehydrated and a crust layer structure of fried products forms (Andres et al. 2013). Air frying reduces fat content in the fried food up to 90% in comparison with deep-fried ones (Sansano et al. 2015). Yang et al. (2017) found that the variety and total amount of aroma-active compounds of air-fried pork loin were only 1/5 and 1/2 of these compounds as in the deep-fried one, in which 2-methyl-3-propylpyrazine, 2-pentylfuran, 6-methyl-2-ethylpyrazine, 2,5-dimethylpyrazine, 2,6-dimethylpyrazine, and trimethylpyrazine were not detected in the air-fried samples. This might be attributed to the lower rate of Maillard reaction and lipid oxidation as initial reducing sugar and frying oil contents were increased and decreased, respectively (Sansano et al. 2015). Caetano et al. (2018) presented similar findings on sweet potato chips fried in canola oil, in which the deep-fried chips showed much better sensory acceptability to induce greater purchase intent by consumers than the air-fried chips. Although air-fried products exhibit lower fat and moisture contents with superior texture properties (e.g., crispiness), weaker aroma features indeed hinder the extension of air frying in the fast food and food production industries. Hence, a better scientific understanding to enhance organoleptic sensation of air-fried foods is a matter of thought to provide sufficient evidences to substitute deep-fat frying.

Overall, compared to all other frying types, deep-fat frying is still a superior thermal treatment that provides a great quality of sensory characteristics (e.g., color, aroma, and texture), because of the usage of oils in large quantities, short cooking time, dehydrated environment, fast heat transfer, complex chemical reactions, and a very attractive deep-fried flavor (Ghidurus et al. 2010). Most operational recommendations on deep-fat frying are formerly concentrated on the reduction of off-flavors resulting from excessive lipid oxidation. Przybylski and Eskin (1988) advised that flushing nitrogen (15 minutes) or carbon dioxide (5 minutes) to frying oils prior to frying can significantly prevent oil oxidation to reduce rancid flavor. Stevenson et al. (1984) suggested that a replenishment of frying oil by 15%–25% capacity of the fryer should be done to limit the application of silicones (an antifoaming agent) and reduce off-flavors. However, practical measures and concerns on the modification of major chemical reactions (e.g., lipid oxidation and Maillard reaction) are rarely investigated to optimize deep-fried flavor.

6.5 Technologies for Flavor Detection and Characterization

Advanced technologies [e.g., gas chromatography-mass spectrometry (GC-MS)] of food production at a molecular level have been developed over the past decades. As a useful advanced technique in the field of instrumental analysis, chromatography has been considered a powerful tool to conduct qualitative and quantitative analyses of complex compounds such as lipids, carbohydrates, and proteins. The application of chromatography in lipid analysis has been extensively recommended to characterize the volatile flavor components and has continually improved according to the requirements of composition analysis (Bosque-Sendra et al. 2012). Due to the nature of the constituents of lipids, high-performance liquid chromatography (HPLC) is commonly used to directly measure TAGs (Lísa et al. 2009), and gas chromatography (GC) is applied to measure FAs after the derivatization of TAGs (Mitchell et al. 2009).

Furthermore, the combination of reliable sample pretreatment, the proper detector, and an appropriate data analysis method makes these techniques more effective and rapid in their measurement of the lipid composition. Among the above-mentioned products, some are expected to be generated and others are undesirable because of their adverse effects on the nutritional properties and quality of the frying oil and the final fried food. These products with potential side effect in frying oil and fried food have become the research topics due to the large consumption of fried food. Because of the complex features of these products and chromatographic techniques, chromatographic methods can be used to monitor the changes in the composition and evaluate the real-time quality of frying oil and fried foods during or at the end of a deep-fat frying process.

In view of the complex constituents of frying oil, different reliable sample pretreatments have been conducted. The complexity and homogeneity of the above-mentioned products formed during the frying process can be determined due to certain similarities among these products. First, the constituents of frying oil are usually divided into polar compounds and nonpolar compounds based on the polarity of the molecule. Oxidized decomposition compounds, hydrolysis products, oxidized TAG monomers, oxygenated TAG polymers, and oxidized sterols are the representative polar compounds. The nonpolar compounds include the unchanged TAGs, cyclic compounds, *trans* isomers, TAG polymers without oxygenated functional groups, and nitrogen-containing compounds such as heterocyclic amines. According to this classification, absorption chromatography is commonly utilized to initially separate the frying oil, and the separated portions are then analyzed by the appropriate chromatographic techniques. Second, based on the similarity of functional groups contained in the products, a chromatographic technique is specifically adopted to measure the targeted compounds after suitable sample pretreatments. The typical detection methods for each type of product are shown in Table 6.2.

Table 6.2 Commonly Used Detection Techniques and Corresponding Detection Conditions for Analysis of Reaction Products Formed during Food-Frying or Oil-Heating Processes

Main Products	*Samples*	*Detection Techniques*	*Chromatographic Parameters*	*Target Analytes*	*References*
Non-aldehyde volatile compounds	Soybean oil (SO), frying wheat dough (WD), and frying chicken breast meat	GC-MS	Helium as the carrier gas, and the flow rate of 1 mL/min, DB-5MS column (30 m×0.25 mm×0.25 µm). The temperature was isothermal for 2 minutes at 40°C, was raised to 160°C at a rate of 5°C/min, isothermal for 2 minutes at 160°C, was raised to 250°C at 10°C/min, and then maintained at 250°C for 10 minutes. The temperatures of the transfer line, the ion source, and the quadrupole mass filter were set at 250°C, 230°C, and 150°C, respectively. The mass spectrum was acquired over the m/z range 30–450 in the full-scan mode by electron ionization with an ionization energy of 70 eV	Alkanes, alkenes, alkynes, alcohols, ketones, nitrogen-containing volatiles (NCVs)	Zhang et al. (2018)

(*Continued*)

Table 6.2 (*Continued*) Commonly Used Detection Techniques and Corresponding Detection Conditions for Analysis of Reaction Products Formed during Food-Frying or Oil-Heating Processes

Main Products	*Samples*	*Detection Techniques*	*Chromatographic Parameters*	*Target Analytes*	*References*
Volatile flavor compounds	Corn products	Gas chromatography ion mobility spectrometry (GC-IMS)	Helium as the carrier gas, DB-5MS capillary column (60 mm×0.32 mm×1.0 μm), the flow rate of 0.8 mL/min. The injection temperature was 250°C. EI ionization was 70 eV, and the ion source temperature was 230°C. The GC temperature was programmed from 40°C (holding for 2 minutes) to 180°C at 5°C/min, then to 250°C at a rate of 10°C/min, and held at 250°C for 10 minutes. The full-scan mode between 35 and 450 m/z was conducted for qualitative analysis, and the SIM model was used for quantitative analysis	n-Hexanal, 1-octene-3-ol, decylaldehyde, and 2-pentylthiazol	Zhang et al. (2022)

(*Continued*)

Table 6.2 (*Continued*) Commonly Used Detection Techniques and Corresponding Detection Conditions for Analysis of Reaction Products Formed during Food-Frying or Oil-Heating Processes

Main Products	*Samples*	*Detection Techniques*	*Chromatographic Parameters*	*Target Analytes*	*References*
Aroma compounds	Raw shiitake mushrooms (*Lentinula edodes*)	High-resolution gas chromatography olfactometry	Fused-silica capillary columns: DB-FFAP and DB-5 (both 30 m×0.25 mm i.d.; 0.25 μm film thickness). The sample (1.0 μL) was injected cold on-column at 40°C using helium as a carrier gas at a flow rate of 1.2 mL/min. The oven temperature was programmed from 40°C, held for 2 minutes, increased with 6°C/min to 230°C (for the DB-FFAP column or to 240°C for the DB-5 column), and held for 5 minutes	3-Hydroxy-4,5-dimethylfuran-2(5H)-one, 1,2,4,5-tetrathiane, 4-hydroxy-2,5-dimethylfuran-3(2H)-one, phenylacetic acid, 3-(methylthio)-propanal, and trans-4,5-epoxy-(E)-2-decenal	Schmidberger and Schieberle (2020)

(*Continued*)

Table 6.2 (*Continued*) Commonly Used Detection Techniques and Corresponding Detection Conditions for Analysis of Reaction Products Formed during Food-Frying or Oil-Heating Processes

Main Products	*Samples*	*Detection Techniques*	*Chromatographic Parameters*	*Target Analytes*	*References*
Volatile compounds	Soybean oil	GC-MS/MS	HP-5MS silica capillary column (30 m×0.25 mm×0.25 μm), the flow rate of helium (99.999% purity) was 1.0 mL/min, the initial temperature of 45°C was held for 2 minutes, then raised to 180°C at ramp rate of 3°C/min, followed by increasing to 240°C at 10°C/min and maintained for 7 minutes. The mass spectrometer parameters were as follows: electron impact mode, 70 eV; ion source temperature, 230°C; and mass to charge ratio, 50–550 m/z	Aldehydes, ketones, alcohols, acids, hydrocarbons, and other volatiles (such as furanic compounds)	Liu et al. (2018)

(*Continued*)

Table 6.2 (*Continued*) Commonly Used Detection Techniques and Corresponding Detection Conditions for Analysis of Reaction Products Formed during Food-Frying or Oil-Heating Processes

Main Products	*Samples*	*Detection Techniques*	*Chromatographic Parameters*	*Target Analytes*	*References*
Volatile compounds	Fried shallot oil	GC-MS	A 7890B gas chromatograph with SPME and a 5977B mass selective detector (MSD), DB-WAX (30 m×0.25 mm×0.25 μm). The carrier gas was helium continuously injected at a flow rate of 1 mL/min. The column was maintained at an initial temperature of 60°C for 3 minutes, then the temperature programmed to 80°C at a rate of 5°C/min and held for 7 minutes, rising to 100°C at the rate of 3°C/min, held for 7 minutes, then increased 3°C/min to 150°C, held for 8 minutes, finally increased at 10°C/min to 230°C where it was maintained for 5 minutes. MS conditions were as follows: ion source temperature 230°C; EI ionization source; ionization energy 70 eV; transmission line temperature 280°C; Interface temperature 250°C; the mass scanning range was 30–450 m/z; and the solvent delay was 1 minute	Hexanal, (E)-2-heptenal, (E)-2-octenal, dipropyl disulfide, 2-ethyl-3,5-dimethylpyrazine, and 1-octen-3-ol	Tian et al. (2021)

(*Continued*)

Table 6.2 (*Continued*) Commonly Used Detection Techniques and Corresponding Detection Conditions for Analysis of Reaction Products Formed during Food-Frying or Oil-Heating Processes

Main Products	*Samples*	*Detection Techniques*	*Chromatographic Parameters*	*Target Analytes*	*References*
Polar compound (oxidized TAG-degraded compounds)	Olive oil and sunflower oil	GC-FID	HP Innowax fused-silica capillary column (30 m×0.25 mm×0.25μm), split ratio: 40:1; carrier gas: hydrogen at 1.1 mL/min; temperature program: 90°C for 2 minutes, increased to 240°C at 4°C/min and held for 20 minutes; injector temperature: 250°C; detector temperature: 250°C with hydrogen at 40 mL/min and air at 450 mL/min; auxiliary gas: nitrogen at 45 mL/min	Heptanoic, octanoic, 8-oxo-octanoic, 9-oxononanoic, octanedioic, and nonandioic acids	Velasco et al. (2005)
Nonpolar compounds	Sunflower oil and high-oleic acid sunflower oil fried with frozen pre-fried foods	GC-FID	HPLC: reversed-phase C18 column (250×7 mm; 0.25 μm); GC-FID: fused-silica capillary column (50 m×0.33 mm×0.25 μm); HPLC: a mobile phase consisted of acetone/acetonitrile (90:10, v/v) at 4 mL/min; GC-FID: split ratio: 50:1; carrier gas: helium; temperature program: from 60°C to 190°C at 20°C/min	Cyclic fatty acid monomers	Romero et al. (2006)

6.6 Odor Impacts of Fried Volatiles

Odor activity value (OAV, a ratio of volatile concentration in the food to its odorant threshold in a media) is universally applied for the judgment of odor sensitivity and the contribution of individual odorant to a given food aroma. It is expected that the volatile with higher OAV (at least >1) is a more important contributor to the corresponding flavor compared to the odorant threshold value. Buttery and Ling (1972) concluded that the most important odorant was methional (with the highest OAV) in the fried potato flavor, followed by phenylacetaldehyde, 2- and 3-methylbutanal, 3-ethyl-2,5-dimethyl-pyrazine, 2,4-decadienal, 1-penten-3-one, and hexanal (with the lowest OAV). Nevertheless, the absence of methional in the imitated flavor of French fries was not detected by sensory panels, due to the flavor-masking effect by other odorants (e.g., 2-ethyl-3,5-dimethylpyrazine, 3-ethyl-2,5-dimehylpyrazine, and 2-methylbutanal) (Wagner and Grosch 1998). Indeed, the relationship between flavor perception by consumers and volatile chemicals is complicated and also dependent on some minor important parameters (e.g., chemical stability, partition coefficient, volatility, and molecular weight) rather than threshold value and OAV (Jousse et al. 2002). For instance, molecular weights of volatiles are negatively correlated with the sense of odor, in which formaldehyde provides a strong sting and pungency odor, but aldehydes with >7 carbon atoms have lower odor intensities (Stier 2000). Therefore, it is essential to comprehensively consider various elements (e.g., OAV, threshold value, concentration, and molecular weight) to define the sensation efficiency of volatiles.

6.7 Comparison of Frying with Other Heating Treatments

Yang and coworkers (2017) estimated the effect of baking and deep frying on the formation of aroma compounds in pork loins and noticed that a number of significant pyrazines (e.g., 2,5-dimethylpyrazine, 2,6-dimethylpyrazine, 2-methyl-3-propylpyrazine) that expressed roasted and nutty odors were missing in the baked pork loin as compared with deep-fried pork loin. Shi et al. (2017) claimed that flavor profiles (e.g., roasted, beany, woody, and bitterness odors) of deep-fried peanuts were similar as roasted peanuts, but the roasted odor of fried peanuts lasted longer during storage by a great exposure of inner parenchyma cells during frying.

6.8 Potential Harmful Products in Fried Flavor

Butanal, pentanal, hexanal, heptane, pentanol, 2-hexenal, heptanal, 1-octen-5-ol, 2-pentylfuran, and 2-decenal provide off-odors in deep-fat frying (Prevot et al. 1988). Some of volatile compounds formed in deep-fat frying, such as 1,4-dioxane, benzene, toluene, and hexylbenzene, do not contribute to desirable flavor and are toxic compounds (Negroni et al. 2001). Many studies have announced that deep-fat frying favors the generation of undesirable toxicological relevant compounds [e.g., acrylamide, acrolein, glycidamide, a, b-unsaturated aldehydes, and heterocyclic aromatic amines (HAAs)] by violent chemical reactions at high temperatures to potentially threaten human health. Mutagenic and carcinogenic HAAs are most commonly studied potential toxic chemicals originating from thermal-processed, protein-rich products. They include thermic HAAs (resulting from creatine, free amino acids, and monosaccharides through the Maillard reaction at 150°C–250°C) and pyrolytic HAAs (possibly resulting from amino acids pyrolysis at temperature >250°C) (Jagerstad et al. 1998; Matsumoto et al. 1981). In the 25 identified HAAs, 2-amino-3,8-dimethylimidazo[4,5-f]quinoxaline [MeIQx, possible carcinogenic to human (Groups 2B)], 2-amino-1-methyl-6-phenylimidazo[4,5-b]pyridine (PhIP, Groups 2B), and 2-amino-3-methylimidazo[4,5-f]quinoline [IQ, probable carcinogenic to human (Group 2A)] are the most abundant HAAs exposed in the cooked foods (IARC 1993). A number of researches observed that the formation of HAAs drastically is promoted as the frying temperature and time increased (Shabbir et al. 2015). In addition, acrolein (Group 3: not classifiable to carcinogenic toxicants) and furan (Group 2B) generated during deep-fat frying via multiple pathways (e.g., carbohydrates and amino acids degradations, lipid oxidation, and Maillard reaction) also have potential adverse effects on human health (IARC 1993). Since the toxicity of aldehydes (best-known deep-fried volatiles) is highly dependent on their abilities to introduce inter- and/or intramolecular cross-linkages to modify proteins and nucleic acids and interfere cell metabolisms (Witz 1989), further establishments on regulations and classification of aldehydes will be fundamental to provide a clear guidance on deep-fat frying.

In current, majority of researchers intended to lower frying temperatures, applied pretreatments (e.g., microwaving and edible coating), and limited the involvement of oxygen (by vacuum fryers), in order to prevent the production of these food-borne toxicants (Hosseini et al. 2016; Thurer and Granvogl 2016; Yang et al. 2016; Zamora et al. 2016; Zhang et al. 2012), but the intensity of fried flavor drastically declined that decreased the sensory quality of deep-fried products. For example, the total content of HAAs was decreased by 70% in the air-fried pork loin, but its sensory satisfaction was greatly lowered in comparison with deep-fried one (Yang et al. 2016). Nevertheless, Haskaraca et al. (2014) did not find any detectable effect on the HAAs by adding green tea extract into the fried chicken meat, but lowering the number of frying cycles led to a reduction of HAA contents in the grass carp with a negative organoleptic response (Wang et al. 2015).

6.9 Conclusions

The hydrolysis, oxidation, and polymerization of oil are common chemical reactions in frying oil and produce volatile or non-volatile compounds accountable for fried flavor. Factors influencing the flavor of fried products mainly include frying temperature, time, pH, moisture, fatty acid composition of oil, amino acid, antioxidants, and frying type. High frying temperature, more number of frying, greater contents of free fatty acids, polyvalent metals, and unsaturated fatty acids of oil decrease the oxidative stability and flavor quality of oil. Food components have significant influences on the formation of volatiles during deep-fat frying. They affect the intensity and complexity of reactions to produce more volatiles under the conditions of high temperature and occurrence of oxygen. However, further hydroperoxidation, Maillard reaction, and thermal degradation could take place between food components and production intermediates to expedite the release of unstable volatiles. Regulating frying by-products and intermediate materials as well as balancing frying conditions to sustainably maintain ideal fried flavor along with simultaneously minimizing the production of toxicological substances is technically challenging. However, it is vitally essential to get substantially improved food quality and safety augmenting the consumer acceptance and economic benefits.

References

Adams, A., V. Kitryte, R. Venskutonis, and N. De Kimpe. 2011. Model studies on the pattern of volatiles generated in mixtures of amino acids, lipid-oxidation-derived aldehydes, and glucose. *Journal of Agricultural and Food Chemistry* 59 (4): 1449–56. doi: 10.1021/jf104091p.

Aladedunye, F. A., and R. Przybylski. 2009. Protecting oil during frying: A comparative study. *European Journal of Lipid Science and Technology* 111 (9): 893–901. doi: 10.1002/ejlt.200900020.

Ammawath, W., Y. B. C. Man, S. Yusof, and R. A. Rahman. 2001. Effect of variety and stage of fruit ripeness on the physicochemical and sensory characteristics of deep-fat-fried banana chips. *Journal of the Science of Food and Agriculture* 81 (12): 1166–71. doi: 10.1002/jsfa.922.

Andres, A., A. Arguelles, M. L. Castello, and A. Heredia. 2013. Mass transfer and volume changes in French fries during air frying. *Food and Bioprocess Technology* 6 (8): 1917–24. doi: 10.1007/s11947-012-0861-2.

Asaaf, S., Y. Hadar, and C. G. Dosoretz. 1997. 1-Octen-3-ol and 13-hydroperoxylinoleate are products of distinct pathways in the oxidative breakdown of linoleic acid by Pleurotus pulmonarius. *Enzyme and Microbial Technology* 21 (7): 484–90. doi: 10.1016/S0141-0229(97)00019-7.

Bansal, G., W. B. Zhou, P. J. Barlow, P. S. Joshi, H. L. Lo, and Y. K. Chung. 2010. Review of rapid tests available for measuring the quality changes in frying oils and comparison with standard methods. *Critical Reviews in Food Science and Nutrition* 50 (6): 503–14.

Bazina, N., and J. B. He. 2018. Analysis of fatty acid profiles of free fatty acids generated in deep-frying process. *Journal of Food Science and Technology* 55 (8): 3085–92. doi: 10.1007/s13197-018-3232-9.

BeMiller, J. N. 1965. Acid hydrolysis and other lytic reactions of starch. In *Starch: Chemistry and technology, volume I fundamental aspects*, ed. R. L. Whistler, and E. F. Paschall, 495–520. New York: Academic.

ben Hammouda, I., F. Freitas, S. Ammar, M. D. R. G. D. Silva, and M. Bouaziz. 2017. Comparison and characterization of volatile compounds as markers of oils stability during frying by HS-SPME-GC/MS and chemometric analysis. *Journal of Chromatography B* 1068: 322–34.

Bosque-Sendra, J. M., L. Cuadros-Rodríguez, C. Ruiz-Samblás, and A. P. de la Mata. 2012. Combining chromatography and chemometrics for the characterization and authentication of fats and oils from triacylglycerol compositional data—A review. *Analytica Chimica Acta* 724: 1–11.

Buttery, R. G., and L. C. Ling. 1972. Characterization of non-basic steam volatile components of potato chips. *Journal of Agricultural and Food Chemistry* 20 (3): 698–700. doi: 10.1021/jf60181a068

Buttery, R. G. 1989. Importance of lipid-derived volatiles to vegetable and fruit flavor. In *Flavor chemistry of lipid foods*, ed. D. B. Min, and T. H. Smouse, 156–65. Champaign, IL: American Oil Chemists Society.

Caetano, P. K., F. A. D. Mariano-Nasser, V. Z. de Mendonca, K. A. Furlaneto, E. R. Daiuto, and R. L. Vieites. 2018. Physicochemical and sensory characteristics of sweet potato chips undergoing different cooking methods. *Food Science and Technology* 38 (3): 434–40. doi: 10.1590/1678-457x.08217.

Cao, G., D. Ruan, Z. Chen, Y. Hong, and Z. Cai. 2017. Recent developments and applications of mass spectrometry for the quality and safety assessment of cooking oil. *Trends in Analytical Chemistry* 96: 201–11. doi: 10.1016/j.trac.2017.07.015.

Chang, C., G. Wu, H. Zhang, Q. Jin, and X. Wang. 2020. Deep-fried flavor: Characteristics, formation mechanisms, and influencing factors. *Critical Reviews in Food Science and Nutrition*, 60 (9): 1496–1514. doi: 10.1080/10408398.2019.1575792

Chang, S. S., R. J. Peterson, and C. T. Ho. 1978. Chemical-reactions involved in deep-fat frying of foods. *Journal of the American Oil Chemists' Society* 55 (10): 718–27. doi: 10.1007/BF02665369.

Choe, E., and D. B. Min. 2007. Chemistry of deep-fat frying oils. *Journal of Food Science* 72 (5): R77–86.

Chun, H. K., and C. T. Ho. 1997. Volatile nitrogen-containing compounds generated from maillard reactions under simulated deep-fat frying conditions. *Journal of Food Lipids* 4 (4): 239–44. doi: 10.1111/j.1745-4522.1997.tb00096.x.

Chung, T. Y., J. P. Eiserich, and T. Shibamoto. 1993. Volatile compounds identified in headspace samples of peanut oil heated under temperatures ranging from 50 to 200 °C. *Journal of Agricultural and Food Chemistry* 41 (9): 1467–70.

Chyau, C., and J. Mau. 2001. Effects of various oils on volatile compounds of deep-fried shallot flavouring. *Food Chemistry* 74 (1): 41–6. doi: 10.1016/S0308-8146(00)00336-8.

Dobarganes, C., G. Marquez-Ruiz, and J. Velasco. 2000. Interactions between fat and food during deep-frying. *European Journal of Lipid Science and Technology* 102 (8–9): 521–8. doi: 10.1002/1438-9312(200009)102:8/9.

Dominguez, R., M. Gomez, S. Fonseca, and J. M. Lorenzo. 2014. Effect of different cooking methods on lipid oxidation and formation of volatile compounds in foal meat. *Meat Science* 97 (2): 223–30. doi: 10.1016/j.meatsci.2014.01.023.

Dunkel, A., M. Steinhaus, M. Kotthoff, B. Nowak, D. Krautwurst, P. Schieberle, and T. Hofmann. 2014. Nature's chemical signatures in human olfaction: A foodborne perspective for future biotechnology. *Angewandte Chemie International Edition* 53 (28): 7124–43.

Frankel, E. N., K. Warner, and K. J. Moulton. 1985. Effects of hydrogenation and additives on cooking-oil performance of soybean oil. *Journal of the American Oil Chemists' Society* 62 (9): 1354–8. doi: 10.1007/BF02545957.

Ghidurus, M., M. Turtoi, G. Boskou, P. Niculita, and S. Stan. 2010. Nutritional and health aspects related to frying (I). *Romanian Biotechnological Letters* 15 (6): 5675–82.

Gibis, M. 2016. Heterocyclic aromatic amines in cooked meat products: Causes, formation, occurrence, and risk assessment. *Comprehensive Reviews in Food Science and Food Safety* 15 (2): 269–302. doi: 10.1111/1541-4337.12186.

Hammouda, I. b., F. Freitas, S. Ammar, M. D. R. Gomes Da Silva, and M. Bouaziz. 2017. Comparison and characterization of volatile compounds as markers of oils stability during frying by HS-SPME-GC-MS and chemometric analysis. *Journal of Chromatography B* 1068–9: 322–34.

Haskaraca, G., E. Demirok, N. Kolsarici, F. Oz, and N. Ozsarac. 2014. Effect of green tea extract and microwave pre-cooking on the formation of heterocyclic aromatic amines in fried chicken meat products. *Food Research International* 63: 373–81. doi: 10.1016/j.foodres.2014.04.001.

Ho, C., and Q. Chen. 1994. Lipids in food flavours – An overview. *ACS Symposium Series* 558: 2–14.

Hosseini, H., M. Ghorbani, N. Meshginfar, and A. S. Mahoonak. 2016. A review of frying: Procedure, fat, deterioration progress and health hazards. *Journal of the American Oil Chemists' Society* 93 (4): 445–66. doi: 10.1007/s11746-016-2791-z.

International Agency for Research on Cancer (IARC). 1993. IARC monographs on the evaluation of carcinogenic risks to humans. No. 56. Some naturally occurring substances: Food items and constituents. Heterocyclic aromatic amines and mycotoxins. Lyon: International Agency for Research on Cancer.

Jagerstad, M., K. Skog, P. Arvidsson, and A. Solyakov. 1998. Chemistry, formation and occurrence of genotoxic heterocyclic amines identified in model systems and cooked foods. *Zeitschrift fur €Lebensmittel-Untersuchung und -Forschung* 207 (6): 419–27. doi: 10.1007/s002170050355.

Jousse, E., T. Jongen, W. Agterof, S. Russell, and P. Braat. 2002. Simplified kinetic scheme of flavor formation by the maillard reaction. *Journal of Food Science* 67 (7): 2534–42.

Katsuta, I., M. Shimizu, T. Yamaguchi, and Y. Nakajima. 2008. Emission of volatile aldehydes from DAG-rich and TAG-rich oils with different degrees of unsaturation during deep-frying. *Journal of the American Oil Chemists' Society* 85 (6): 513–9. doi: 10.1007/s11746-008-1236-8.

Kavousi, P., H. Mirhosseini, H. Ghazali, and A. A. Ariffin. 2015. Formation and reduction of 5-hydroxymethylfurfural at frying temperature in model system as a function of amino acid and sugar composition. *Food Chemistry* 182: 164–70. doi: 10.1016/j.foodchem.2015.02.135.

Lísa, M., H. Velínská, and M. Holcapek. 2009. Regioisomeric characterization of triacylglycerols using silver-ion HPLC/MS and randomization synthesis of standards. *Analytical Chemistry* 81 (10): 3903–10.

Liu, Y., Y. Wang, P. Cao, and Y. Liu. 2018. Combination of gas chromatography-mass spectrometry and electron spin resonance spectroscopy for analysis of oxidative stability in soybean oil during deep-frying process. *Food Analytical Methods* 11 (5): 1485–1492.

Liu, H., and P. J. White. 1992. High-temperature stability of soybean oils with altered fatty acid composition. *Journal of the American Oil Chemists' Society* 69 (6): 533–7. doi: 10.1007/BF02636104.

Lu, F., G. K. Kuhnle, and Q. F. Cheng. 2018. The effect of common spices and meat type on the formation of heterocyclic amines and polycyclic aromatic hydrocarbons in deep-fried meatballs. *Food Control* 92: 399–411. doi: 10.1016/j.foodcont.2018.05.018.

Ma, R. X., T. Gao, L. Song, L. Zhang, Y. Jiang, J. L. Li, X. Zhang, F. Gao, and G. H. Zhou. 2016. Effects of oil-water mixed frying and pure-oil frying on the quality characteristics of soybean oil and chicken chop. *Food Science and Technology* 36 (2): 329–36. doi: 10.1590/1678-457X.0092.

Mandin, O., S. C. Duckham, and J. M. Ames. 1999. Volatile compounds from potato-like model systems. *Journal of Agricultural and Food Chemistry* 47 (6): 2355–9. doi: 10.1021/jf981277þ.

Martin, F. L., and J. M. Ames. 2001. Comparison of flavor compounds of potato chips fried in palmolein and silicone fluid. *Journal of the American Oil Chemists' Society* 78 (8): 863–6. doi: 10.1007/s11746-001-0356-2.

Martins, S. I. F. S., W. M. F. Jongen, and M. A. J. S. van Boekel. 2000. A review of maillard reaction in food and implications to kinetic modelling. *Trends in Food Science and Technology* 11 (9–10): 364–73. doi: 10.1016/S0924-2244(01)00022-X.

Matsumoto, T., D. Yoshida, and H. Tomita. 1981. Determination of mutagens, amino-a-carbolines in grilled foods and cigarette smoke condensate. *Cancer Letters* 12 (1–2): 105–10. doi: 10.1016/0304-3835(81)90045-8.

McClements, D. J., and E. A. Decker. 2007. Lipids. In *Food chemistry*, ed. S. Damodaran, K. L. Parkin, and O. R. Fennema, 155–216, 4th ed. Boca Raton: CRC Press.

Merkle, S., E. Giese, N. Dietz, K. Lösche, and J. Fritsche. 2016. Development of technofunctional-sensory characterization of virtually TFA free deep-frying fats for artisan bakery products. *European Journal of Lipid Science and Technology* 118 (12): 1827–38. doi: 10.1002/ejlt.201500527.

Min, D. B., and J. M. Boff. 2002. Lipid oxidation of edible oil. In *Food lipids*, ed. C. C. Akoh, and D. B. Min, 335–63, 2nd ed. New York: Marcel Dekker Inc.

Mitchell, T. W., H. Pham, M. C. Thomas, and S. J. Blanksby. 2009. Identification of double bond position in lipids: From GC to OzID. *Journal of Chromatography B* 877 (26): 2722–35.

Miyagi, A., and Y. Ogaki. 2014a. Chemical processes in peanut under thermal treatment. *Journal of Food Measurement and Characterization* 8 (4): 305–15. doi: 10.1007/s11694-014-9191-6.

Miyagi, A., and Y. Ogaki. 2014b. Sensory preferences among general Japanese consumers and physicochemical evaluation of deep-fried peanuts. *Journal of the Science of Food and Agriculture* 94 (10): 2030–9. doi: 10.1002/jsfa.6521

Molina-Garcia, L., C. S. P. Santos, S. C. Cunha, S. Casal, and J. O. Fernandes. 2017. Comparative fingerprint changes of toxic volatiles in low PUFA vegetable oils under deep-frying. *Journal of the American Oil Chemists' Society* 94 (2): 271–84.

Moreira, R. G. 2014. Vacuum frying versus conventional frying – An overview. *European Journal of Lipid Science and Technology* 116 (6): 723–34. doi: 10.1002/ejlt.201300272.

Mottram, D. S. 1994. Flavor compounds formed during the Maillard reaction. In *Thermally generated flavors*, ed. T. H. Parliament, M. J. Morello, and R. J. McGorrin, 104–26. Washington, DC: ACS Publications.

Nawar, W. W. 1998. Volatile components of the frying process. *Grasas Aceites* 49 (3–4): 271–4. doi: 10.3989/gya.1998.v49.i3-4.727.

Nayak, P. K., U. Dash, K. Rayaguru, and K. R. Krishnan. 2016. Physiochemical changes during repeated frying of cooked oil: A review. *Journal of Food Biochemistry* 40 (3): 371–90. doi: 10.1111/jfbc.12215.

Nazarbakhsh, V., H. Ezzatpanah, B. Tarzi, and M. Givianrad. 2014. Chemical changes of canola oil during frying under atmospheric condition and combination of nitrogen and carbon dioxide gases in the presence of air. *Journal of the American Oil Chemists' Society* 91 (11): 1903–9. doi: 10.1007/s11746-014-2539-6.

Negroni, M., A. D'Agostin, and A. Arnoldi. 2001. Effects of olive, canola, and sunflower oils on the formation of volatiles from the Maillard reaction of lysine with xylose and glucose. *Journal of Agricultural and Food Chemistry* 49 (1): 439–45.

Oluwaniyi, O. O., O. O. Dosumu, and G. V. Awolola. 2010. Effect of local processing methods (boiling, frying and roasting) on the amino acid composition of four marine fishes commonly consumed in Nigeria. *Food Chemistry* 123 (4): 1000–6. doi: 10.1016/j.foodchem.2010.05.051.

Peng, C., C. Lan, P. Lin, and Y. Kuo. 2017. Effects of cooking method, cooking oil, and food type on aldehyde emissions in cooking oil fumes. *Journal of Hazardous Materials* 324: 160–7. doi: 10.1016/j.jhazmat.2016.10.045.

Perkins, E. G. 2007. Volatile odor and flavor components formed in deep frying. In *Deep frying: Chemistry, nutrition, and practical applications*, ed. M. D. Erickson, 51–6, 2nd ed. Urbana: AOCS Press.

Petersen, K. D., G. Jahreis, M. Busch-Stockfisch, and J. Fritsche. 2013. Chemical and sensory assessment of deep-frying oil alternatives for the processing of French fries. *European Journal of Lipid Science and Technology* 115 (8): 935–45. doi: 10.1002/ejlt.201200375

Pokorny, J. 1980. Effect of browning reactions on the formation of flavour substances. *Die Nahrung* 24 (2): 115–27.

Pokorny, J. 1981. Browning from lipid-protein interaction. *Progress in Food & Nutrition Science* 5 (1–6): 421–8.

Pokorny, J. 1989. Flavor chemistry of deep fat frying in oil. In *Flavor chemistry of lipid foods*, ed. D. B. Min, and T. H. Smouse, 113–5. Champaign, IL: AOCS Press.

Pokorny, J., and A. Kolakowska. 2011. Lipid-protein and lipid-saccharide interactions. In *Chemical, biological, and functional aspects of food lipids*, ed. Z. E. Sikorski, and A. Kolakowska, 277–90, 2nd ed. Boca Raton, FL: CRC Press.

Prevot, A., J. L. Perrin, G. Laclaverie, P. Auge, & J. L. Coustille. 1990. A new variety of low-linolenic rapeseed oil; characteristics and room-odor tests. *Journal of the American Oil Chemists' Society* 67 (3): 161–164. doi: 10.1007/BF02539617

Przybylski, R., and N. A. M. Eskin. 1988. A comparative study on the effectiveness of nitrogen or carbon dioxide flushing in preventing oxidation during the heating of oil. *Journal of the American Oil Chemists' Society* 65 (4): 629–33. doi: 10.1007/BF02540692.

Ramirez, M. R., M. Estevez, D. Morcuende, and R. Cava. 2004. Effect of the type of frying culinary fat on volatile compounds isolated in fried pork loin chops by using SPME-GC-MS. *Journal of Agricultural and Food Chemistry* 52 (25): 7637–43. doi: 10.1021/jf049207s.

Romero, A., S. Bastida, and F. J. Sánchez-Muniz. 2006. Cyclic fatty acid monomer formation in domestic frying of frozen foods in sunflower oil and high oleic acid sunflower oil without oil replenishment. *Food and Chemical Toxicology* 44 (10): 1674–81.

Saguy, I. S., and D. Dana. 2003. Integrated approach to deep fat frying: Engineering, nutrition, health and consumer aspects. *Journal of Food Engineering* 56 (2–3): 143–52.

Sansano, M., M. Juan-Borras, I. Escriche, A. Andres, and A. Heredia. 2015. Effect of pretreatments and air-frying, a novel technology, on acrylamide generation in fried potatoes. *Journal of Food Science* 80 (5): T1120–8. doi: 10.1111/1750-3841.12843.

Schieberle, P. and W. Grosch. 1991. Potent odorants of the wheat bread crumb. *Zeitschrift für Lebensmittel Untersuchung und Forschung* 192 (2): 130–135.

Schmidberger, P. C., and P. Schieberle. 2020. Changes in the key aroma compounds of raw shiitake mushrooms (Lentinula edodes) induced by pan-frying as well as by rehydration of dry mushrooms. *Journal of Agricultural and Food Chemistry* 68 (15): 4493–506.

Shabbir, M. A., A. Raza, F. M. Anjum, M. R. Khan, and H. A. R. Suleria. 2015. Effect of thermal treatment on meat proteins with special reference to heterocyclic aromatic amines (HAAs). *Critical Reviews in Food Science and Nutrition* 55 (1): 82–93.

Shi, X. L., J. P. Davis, Z. T. Xia, K. P. Sandeep, T. H. Sanders, and L. O. Dean. 2017. Characterization of peanuts after dry roasting, oil roasting, and blister frying. *Lebensmittel-Wissenschaft & Technologie* 75: 520–8. doi: 10.1016/j.lwt.2016.09.030.

Stevenson, S. G., M. Vaisey-Genser, and N. A. M. Eskin. 1984. Quality control in the use of deep frying oils. *Journal of the American Oil Chemists' Society* 61 (6): 1102–8. doi: 10.1007/BF02636232.

Stier, R. F. 2000. Chemistry of frying and optimization of deep-fat fried food flavour – An introductory review. *European Journal of Lipid Science and Technology* 102 (8–9): 507–14. doi: 10.1002/1438-9312(200009)102:8/9<507::AID-EJLT507>3.3.CO;2-M.

Thurer, A., and M. Granvogl. 2016. Generation of desired aroma-active as well as undesired toxicologically relevant compounds during deep-frying of potatoes with different edible vegetable fats and oils. *Journal of Agricultural and Food Chemistry* 64 (47): 9107–15. doi: 10.1021/acs.jafc.6b04749.

Tian, P., P. Zhan, H. Tian, P. Wang, C. Lu, Y. Zhao, ... & Y. Zhang. 2021. Analysis of volatile compound changes in fried shallot (Allium cepa L. var. aggregatum) oil at different frying temperatures by GC–MS, OAV, and multivariate analysis. *Food Chemistry* 345: 128748.

Troncoso, E., F. Pedreschi, and R. N. Zuniga. 2009. Comparative study of physical and sensory properties of pre-treated potato slices during vacuum and atmospheric frying. *LWT - Food Science and Technology* 42 (1): 187–95. doi: 10.1016/j.lwt.2008.05.013.

Uriarte, P. S., and M. D. Guillen. 2010. Formation of toxic alkylbenzenes in edible oils submitted to frying temperature influence of oil composition in main components and heating time. *Food Research International* 43 (8): 2161–70. doi: 10.1016/j.foodres.2010.07.022.

Van Ba, H., I. Hwang, D. Jeong, and A. Touseef. 2012. Principle of meat aroma flavors and future prospect. In *Latest research into quality control*, ed. I. Akyar, 145–76. Rijeka: IntechOpen.

van der Klis, F., M. H. van den Hoorn, R. Blaauw, J. van Haveren, and D. S. van Es. 2011. Oxidative decarboxylation of unsaturated fatty acids. *European Journal of Lipid Science and Technology* 113 (5): 562–71.

Velasco, J., S. Marmesat, O. Berdeaux, G. Márquez-Ruiz, and C. Dobarganes. 2005. Quantitation of short-chain glycerol-bound compounds in thermoxidized and used frying oils. A monitoring study during thermoxidation of olive and sunflower oils. *Journal of Agricultural and Food Chemistry* 53 (10): 4006–11.

Vichi, S., L. Pizzale, L. S. Conte, S. Buxaderas, and E. LopezTammames. 2003. Solid-phase microextraction in the analysis of virgin olive oil volatile fraction: Modifications induced by oxidation and suitable markers of oxidative status. *Journal of Agricultural and Food Chemistry* 51 (22): 6564–71. doi: 10.1021/jf030268k.

Wagner, R. K., and W. Grosch. 1998. Key odorants of French fries. *Journal of the American Oil Chemists' Society* 75 (10): 1385–92. doi: 10.1007/s11746-998-0187-4.

Wang, Y., T. Hui, Y. W. Zhang, B. Liu, F. L. Wang, J. K. Li, B. W. Cui, X. Y. Guo, and Z. Q. Peng. 2015. Effects of frying conditions on the formation of heterocyclic amines and trans fatty acids in grass carp (Ctenopharyngodon idellus). *Food Chemistry* 167: 251–7.

Warner, K., P. Orr, and M. Glynn. 1997. Effect of fatty acid composition of oils on flavor and stability of fried foods. *Journal of the American Oil Chemists' Society* 74 (4): 347–56.

Warner, K., P. Orr, L. Parrott, and M. Glynn. 1994. Effects of frying oil composition on potato chip stability. *Journal of the American Oil Chemists' Society* 71 (10): 1117–21. doi: 10.1007/BF02675905.

Whitfield, F. B., and D. S. Mottram. 1992. Volatiles from interactions of maillard reactions and lipids. *Critical Reviews in Food Science and Nutrition* 31 (1–2): 1–58.

Witz, G. 1989. Biological interactions of a, b-unsaturated aldehydes. *Free Radical Biology and Medicine* 7 (3): 333–49. doi: 10.1016/0891-5849(89)90137-8.

Yang, J. H., H. Y. Park, Y. S. Kim, I. W. Choi, S. S. Kim, and H. D. Choi. 2012. Quality characteristics of vacuum-fried snacks prepared from various sweet potato cultivars. *Food Science and Biotechnology* 21 (2): 525–30. doi: 10.1007/s10068-012-0067-4.

Yang, Y., I. Achaerandio, and M. Pujola. 2016. Influence of the frying process and potato cultivar on acrylamide formation in French fries. *Food Control* 62: 216–23. doi: 10.1016/j.foodcont.2015.10.028.

Yang, Z. M., R. Lu, H. L. Song, Y. Zhang, J. N. Tang, and N. Zhou. 2017. Effect of different cooking methods on the formation of aroma components and heterocyclic amines in pork loin. *Journal of Food Processing and Preservation* 41 (3): 1–8. doi: 10.1111/jfpp.12981.

Yaylayan, V. A., and A. Huyghues-Despointes. 1996. Retro-aldol and redox reactions of a Madori compounds: Mechanistic studies with various labeled D-[13C] glucose. *Journal of Agricultural and Food Chemistry* 44 (3): 672–81. doi: 10.1021/jf9502921.

Yin, H. Y., L. B. Xu, and N. A. Porter. 2011. Free radical lipid peroxidation: Mechanisms and analysis. *Chemical Reviews* 111 (10): 5944–72.

Zamora, R., and F. J. Hidalgo. 2005. Coordinate contribution of lipid oxidation and Maillard reaction to the nonenzymatic food browning. *Critical Reviews in Food Science and Nutrition* 45 (1): 49–59.

Zamora, R., I. Aguilar, M. Granvogl, and F. J. Hidalgo. 2016. Toxicologically relevant aldehydes produced during the frying process are trapped by food phenolics. *Journal of Agricultural and Food Chemistry* 64 (27): 5583–9. doi: 10.1021/acs.jafc.6b02165

Zhang, Q., A. S. M. Saleh, J. Chen, and Q. Shen. 2012. Chemical alterations taken place during deep-fat frying based on certain reaction products: A review. *Chemistry and Physics of Lipids* 165 (6): 662–81. doi: 10.1016/j.chemphyslip.2012.07.002

Zhang, Q., W. Qin, D. Lin, Q. Shen, and A. S. M. Saleh. 2015. The changes in the volatile aldehydes formed during the deep-fat frying process. *Journal of Food Science and Technology* 52 (12), 7683–96.

Zhang, Q., C. Wan, C. Wang, H. Chen, Y. Liu, S. Li, D. Lin, D. Wu, and W. Qin. 2018. Evaluation of the non-aldehyde volatile compounds formed during deep-fat frying process. *Food Chemistry* 243: 151–61. doi: 10.1016/j.foodchem.2017.09.121.

Zhang, N., B. Sun, X. Mao, H. Chen, and Y. Zhang. 2019. Flavor formation in frying process of green onion (Allium fistulosum L.) deep-fried oil. *Food Research International* 121: 296–306.

Zhang, K., L. Gao, C. Zhang, T. Feng, and H. Zhuang. 2022. Analysis of volatile flavor compounds of corn under different treatments by GC-MS and GC-IMS. *Frontiers in Chemistry* 10: 1–11.

Zhang, Y. G., and C. T. Ho. 1989. Volatile compounds formed from thermal interaction of 2,4-decadienal with cysteine and glutathione. *Journal of Agricultural and Food Chemistry* 37 (4): 1016–20.

Chapter 7

Influence of Frying on Food Macromolecule Constituents

Jessica Pandohee
Telethon Kids Institute

Johra Khan
College of Applied Medical Sciences, Majmaah University

Aaliya Ali, Parneet Kaur, and Saurabh Kulshreshtha
Shoolini University of Biotechnology and Management Sciences

Mubasshira Shaikh and Priya Sundarrajan
St. Xavier's College (Autonomous)

Ritee Basu, Sukanya Dasgupta, and Ayesha Noor
Vellore Institute of Technology (VIT)

DOI: 10.1201/9781003329244-7

7.1 Introduction

Food processing includes heating, frying, and cooling under different moisture conditions which may cause structural changes in food at granular and molecular levels. Different processing conditions have different effects on structure and accessibility of starch which may have impact on digestion (Ziaiifar et al. 2008). One of the oldest ways of preparing food is frying. Its appeal arises from the convenience and quickness, as well as sensory qualities such as a distinct taste and flavor (Dobarganes, Márquez-Ruiz, and Velasco 2000). It is a low-cost, high-speed technique of continuous mass and heat transfer that alters sensory as well as nutritional properties as a consequence of intricate dynamics between oil and food (Bordin et al. 2013a). The method is defined as the interaction between oil and food at elevated temperatures that also cooks as well as dehydrates the food, and alters both its physical and chemical properties, including the denaturation of protein and the production of color and flavor through the Maillard reaction (Debnath et al. 2012).

Frying is considered one of the oldest and most popular methods of food processing due to its ease and speed, as well as the unique flavor and taste it adds to food. To understand the connection between good health and diet, it is necessary to appreciate the effect of processing on food components and relationships between structures, behavior, and their interactions in the gastrointestinal tract. Frying works as a preservative as it causes thermal destruction of microorganisms and enzymes, and reduction of water content on the food surface. The properties of food nutrients change depending upon the type of oil, volume to surface ratio of oil, temperature of oil, heating processes, length of frying food material in oil, and the materials used for frying processes. Long exposure of food to frying and oil at high temperatures can produce highly toxic and oxidized products harmful to human health (Bordin et al. 2013). This chapter is intended to elucidate the findings of research on the impact of frying on macromolecules.

7.2 Carbohydrates and Indigestible Polysaccharides

Dietary carbohydrates are the biggest source of energy in adult diet. A study by Subar et al. (1998) reported carbohydrates as the biggest contributor (more than 60%) to the energy of adult diet in the US. Carbohydrates found in plants are classified into two broad classes based on their impact on human health and their structure. The first class is stored carbohydrates such as starch, sugar, and oligosaccharides, and the second class includes cell wall polysaccharides derived mainly from plants. Polysaccharides are complex carbohydrates such as starches and dietary fibers (Rudrapal et al. 2022). Dietary fibers are indigestible polysaccharides in plant cell wall such as cellulose, hemicellulose, pectin, mucilage, gums from plants, and algal polysaccharides (Khan et al. 2021).

Among the plant-based carbohydrates, starch is the major form of carbohydrate and a significant source of calories for human and animals. Biophysical properties of starch have a great impact on food properties and texture (Han and BeMiller 2007). The digestion of starch takes place in small intestine, whereas the resistant starch and polysaccharides remain undigested. The latter are fermented in colon by microbiota resulting in production of short chain fatty acids. Plant starch is a polymer of D-glucose, made of amylose and amylopectin. Amylopectin is comprised of large chain glucose ranging from 10,000 to 100,000 branch point at α-1,6 linkages on each 20–25 glucose units (Albuquerque et al. 2012). The α-1,6 linkages are about 5% of the total glycosidic bonds present in amylopectin. Starch granules of various structures and compositions are found in both forms of native starch in different proportions. The size of starch granules affects the surface to volume ratio causing increase in surface area availability for enzyme hydrolysis. Frying transfers heat and mass to food that changes sensory and nutritional characters of food due to complex interaction between food carbohydrates and oil. In frying, oil acts as the heat-transferring medium (Debnath et al. 2012).

7.2.1 Changes in Starch and Indigestible Polysaccharides after Frying

The starch present in plant products (e.g., amylose and amylopectin) dissolves in water. When heated in oil during frying, starch gelatinization occurs. Gelatinization of starch takes place after the denaturation of carbohydrates, protein, lipids, and water at very high temperatures (Albuquerque et al. 2012). For instance, potato contains a large amount of carbohydrate, and when potatoes are fried, starch granules are gelatinized. The firm structure of potato carbohydrates is lost after 1–2 minutes of frying, and on further frying, a crispy crust is formed which contains least water content due to which the gelatinization is not very strong and starch granules retain their crystalline structure. Some studies found that frozen food such as French fries had no change on starch structure in comparison to fresh French fries (Tynek et al. 2001). They also reported that frying increases the percentage of resistant starch significantly and partially help in the formation of amylose–lipid complex and increase the fiber content of food. Upon frying, polysaccharides form a solid film on the surface of the food and this process prevents fat migration into food and loss of water (Morales, Rios, and Aparicio 1997).

During deep frying, many chemical reactions occur. Some of them are hydrolysis, polymerization, oxidation, and isomerization resulting in production of free fatty acids, alcohol, lactone, aldehyde, ketone, monoglyceride and diglyceride, trans isomers, epoxy compounds, dimmers, monomer, and oligomer. The unsaturation of fatty acids affects the oxidative stability of oil or fat during frying. Minor macromolecules such as free fatty acids, monoacylglycerols, diacylglycerols, and metals negatively affect frying stability of oil. A study by Blumenthal (1991) reported that reused oil develops better characteristics in fried food products with carbohydrates due to the polar compounds that infuse to food surface. The reused oil increases flavor and increases the contact between oil and water on the surface of food. It also helps in catalyzing heat transfer; however, to maintain the quality of product, the oil should be replaced from time to time (Aladedunye and Przybylski 2013).

The chemical structure of starch (carbohydrates) changes dramatically as a result of heat transfer from hot oil during frying. The heat breaks the glycosidic bonds in polysaccharide chain resulting in degradation of starch. Gokoglu et al. (2004) showed no significant relation between microstructure of starch (crystals or granules) and the source of thermal transfer. The pure starch degradation results into dehydrated oligomers of glucose as found in toasted bread. The Maillard reaction occurs during frying, and when frying bread, sugar and oligomers are produced as a result of the Maillard reaction. A study by McDougall et al. (1996) reported physical and chemical changes in natural starch products. Frying causes degradation and solubilization of cell wall polymer leading to cell breaking or separation resulting in increase in accessibility and food surface area. Frying causes depolymerization of pectin and causes the degradation of sensitive natural starch products at a temperature around 50°C (Mishra, Hardacre, and Monro 2012). The pH of the mixture also has an impact on frying. Frying above pH 7 causes demethylation, depolymerization, and chemical b elimination, while at pH 4.5, frying causes acid hydrolysis. Frying also changes conformational properties of pectin; similarly, β-glucan on heating or frying at high temperature causes depolymerization and interrupts in their beneficial effects on human health. Depolymerization of pectin in vegetables and β-glucan in bread after frying is similar to depolymerization by enzymes and difficult to differentiate between the two (Miao et al. 2014).

7.2.2 Nutritional Changes

The nutrients present in food comprise lipids, proteins, vitamins, and minerals all get affected by the process of frying. Although it is difficult to estimate the nutritional effects, it is well known that the type of oil and the technique of frying affect the nutrition of food. The main problem in the consumption of fried food is increase in fat content of food, which also increases the energy intake due to absorption and retention of oil. Fried food contributes more than 75% of calories in diet. In frying, the reuse of oil or fat also changes the nutritional value of the food and its digestion. Many hazardous compounds, which have been identified as potential carcinogens including carbonyl compounds and few aldehydes, produced from linoleic acid such as 4-hydroxy-2-transnonenal, have been shown to be cytotoxic. The process of frying results in the production of trans fatty acids (non-conjugated carbon–carbon double bond compound) which has been reported to cause cardiovascular diseases. During frying, the formation of trans fatty acids depends on three important factors including method of measurement of trans fatty acids, frying materials, and frying conditions.

Protein is another important nutrient that is also affected during frying. Some studies reported increase in the concentration of protein content after frying as in sardines. The increase may be due to formation of new products that is similar in structure to protein, which influences the determination of protein in fried products. Some studies also found that frying of protein causes degradation of amino acids causing reduction in protein concentration. The lowest amount of protein content was found while frying in palm oil. Similarly in fish frying, using palm oil reported decrease in lysine concentration by 17%, and in fish fried after 24 hours of treatment, the concentration reduces up to 25%. Other than lysine no change was found in histidine, valine, leucine, threonine, methionine, phenylalanine, arginine, isoleucine, serine, glutamic acid, aspartic acid, alanine, cysteine, proline, and tyrosine. Steiner-Asiedu et al. (1991) on the contrary found no difference between fresh and fried fish amino acid contents.

The mineral content of fried food was found to have no effect on frying. Some studies even reported that frying helps in preservation of minerals, especially when it is done at very high temperature ranging from 160°C to 190°C. A study by Ozeren and Ersoy (2009) reported no decrease in concentration of minerals (Zn, Na, K, Ca, and Fe) after frying, whereas in the case of trout fillets, all the minerals were found to increase significantly while frying.

Vitamins are also essential requirements though required in very small amounts. All vitamins are thermo-sensitive, and their oxidation depends on the temperature of frying process. The vitamins of group B (B6, thiamine, riboflavin, and niacin) as well as vitamin C are most frequently affected by frying. A similar study by Manorama and Rukmini (1991) on loss of carotene on frying using palm oil reported that β-carotene is the most lost type of carotene, whereas the other carotene types even after loss from food are found in oil. Due to oxidation in frying, retinol, tocopherols, and carotenoids are destroyed due to which flavor and color of oil changes. The oxidation of tocopherols in frying helps prevent antioxidant effects of tocopherols. Jimenez-Monreal et al. (2009) studied the effect of home cooking (frying) on antioxidants of vegetables and reported a loss of up to 50% of antioxidants in garlic, 30%–40% in Swiss chard, 5% in cauliflower, and 30% in pepper on the basis of ABTS radial scavenging capacity.

7.3 Lipids

Deep-fat frying of food is an important preparation process that has grown in popularity recently. The deep-frying process, however, exposes the cooking fat to elements that significantly change the way that it is structurally composed. At high temperatures, say 180°C, several complex reactions take place resulting in the formation of products like triacylglycerol polymers, triacylglycerol dimers, and oxidized triacylglycerols. Hydrolytic alteration is brought on by food-derived moisture, oxidation is brought on by ambient oxygen contacting the oil, and hydrolytic alteration is brought on by the temperature (180°C) at which the processes take place (Arroyo et al. 1996).

There is no precise definition of lipids. The Association of Official Analytical Chemists (AOAC) methods or other trustworthy and acceptable practices have traditionally been utilized by the U.S. Food and Drug Administration (FDA) to extract components having lipid properties from food. To assess caloric fatty acids, the FDA altered its definition from one based on solubility to "total lipid fatty acids expressed as triglycerides." The caloric values of fatty acids are influenced by their solubility and size. This is significant for items that benefit from it, such as Benefat/Salatrim, and thus these by-products would be scrutinized individually. A special methodology is required to quantify the caloric fatty acids in food items that include sucrose polyesters. Foods that contains vinegar (4.5% acetic acid) create a challenge due to the lack of detailed methodologies to remove polar fatty acids or decrease acetic acid content. Thus these will be categorized as having 4.5% fat (Akoh and Min 2002).

7.3.1 Nutritional Changes

7.3.1.1 Maillard Reactions and Caramelization-Induced Changes of Color, Flavors, and Taste

The caramelization, Maillard reaction, and evaporation of surface water produced changes in the food's surface and gave fried food its characteristic texture. Maillard is the name of the primary process that transforms sugars during frying. It includes carbonyl groups or other aldehydes, sugar ketones, free amino groups of amino acids, peptides, and proteins. At frying temperatures, a number of intermediate intermediates, such as Amadori products or pre-melanoidins, quickly polymerize to create dark-colored molecules. Some hazardous chemicals, such as acrylamide, were generated during the Maillard reaction. According to a recent study, the amount of acrylamide in potato chips increased as treatment duration and temperature increased. Water activity has little effect on this molecule; in fact, research has shown that acrylamide production increases as water activity declines. In addition to Maillard and caramelization, frying oil can also engage in non-enzymatic browning through the interaction of lipid oxidation products with amines, amino acids, and proteins (Bordin et al. 2013).

7.3.1.2 Changes to the Lipid Content of Foods

The increasing temperature and increased oxygen content may speed up the oxidation of some of the lipids present in fried foods. Due to the many factors that might affect nutrition, such as the type of oil, history of thermal treatment, and amount consumed, nutrition impacts can be difficult to determine. The selection of frying oil must take into account both its nutritional benefits and its technological characteristics because of the incorporation of oil. Due to oil absorption and retention, the fat content of food rises, indicating an increase in caloric intake. When the procedure is carried out utilizing recycled oils or fats, the digestibility of the fat is also altered. Trans-fatty acids can also be produced during the frying process, depending on three primary variables: the frying environment, the frying medium, and the trans-fatty acids testing techniques (Bordin et al. 2013). By calculating the relative proportions of unsaturated, polyunsaturated, and saturated fatty acids, the effect of the fatty acid composition on lipid oxidation is assessed (Crosa et al. 2014).

7.3.1.3 Changes in Protein Content

While the quantity of protein in the diet is represented by its nutritional value, the quality of the protein taken and utilized by the organism is reflected by its functional content. Heat treatment can alter the protein content of food by reducing the amount of protein and destroying some amino acids. As we all know, dehydration occurs during the frying process, and therefore the concentration effect causes the protein content to increase. There is no consensus in the literature about the loss of essential amino acids. However, as lysine is the first amino acid implicated in the Maillard reaction, it is anticipated that it will be lost throughout the frying process (Bordin et al. 2013). Fried meals usually contain proteins, but in varying amounts. Proteins act as emulsifying and film-forming agents, offer structure, aid in flavor development, participate in browning events, and improve water-holding capacity (Soriguer et al. 2003).

7.3.1.4 Content of Minerals

Mineral content does not appear to have significantly decreased. Studies have shown that frying, particularly at high temperatures and for a brief period of time, effectively preserves minerals. According to one study, the amount of minerals such as Na, Ca, Mg, K, P, Fe, and Zn in rainbow trout fillets increased dramatically when they were being fried. With the exception of K and Mg in mussels and shrimp, respectively, the frying procedure greatly increases the concentration of macro- and microelements as well as heavy metals (Na, Mg, Ca, Cu, Fe, Zn, Pb, Cd, and Hg) (Bordin et al. 2013).

7.3.1.5 Vitamin Content Variations and Antioxidant Deterioration

Vitamin molecules are temperature-sensitive. Their oxidation is influenced by the food's interior temperature and the frying procedure.

Typically, vitamin C has the highest thermo-sensitivity. Thiamine, riboflavin, niacin, and B6 are the B-group vitamins most frequently impacted by the procedure. One of the researchers found that frying fruits in soybean oil caused a minor loss of vitamin C. High heat, enzymatic oxidation during preparation, and prolonged frying are the causes of vitamin loss. Total beta-carotene was destroyed during the initial frying process, according to Manorama and Rukmini's investigation, but a mixture of other carotenes, despite minor losses, was still present in the oil. Beta-carotene is lost during frying, perhaps in part as a result of the compound's oxidation, as evidenced by the discovery of some carotene oxidation products in heated palm oil. According to Kourimska and Gordon, canola oil's tocopherol concentration degraded more quickly than that of other oils. The effect of frying using several oil samples (canola, vegetable, and soya oils) on the vitamins, minerals, and protein contents of *Colocasia esculenta* was studied. The findings of this study demonstrated that almost all vitamins and nutrients were lost during the frying of cocoyam. For instance, vitamin A was completely lost in canola oil and reduced in other oils, while vitamin D was raised in soy oil and vegetable oil (Omotosho et al. 2015).

7.3.1.6 Alterations in Starch and Indigestible Polysaccharides

In food, starch is the most prevalent polysaccharide. Starch gelatinization results from the dissolution of amylose and amylopectin in water with heat, generating a polymer network. Carbohydrates, proteins, lipids, and water are all involved in the gelatinization process, which follows the denaturation of globular proteins at high temperatures. Changes in starch are crucial when frying raw potato goods because heated oil quickly gelatinizes starch granules, causing the hard structure of raw potatoes to disappear in 1–2 minutes and the fried chips to soften. The consumer much enjoys the firm, crispy crust that forms on the surface of the fried particles after additional heating. The gelatinization is not powerful on the surface of chips, where the water content is significantly lower than in deeper layers, allowing the starch granules to partially retain their crystalline structure. According to studies on dietary fiber content, frying significantly increased the percentage of resistant starch, which is partly attributed to the formation of amylose–lipid complex, which increases fiber content. However, cooking frozen French fries had no effect on the starch composition when compared to fresh samples. When making French fries, high-molecular-weight non-starch polysaccharides (dietary fiber) break down, especially when using low-specific gravity tubers. However, in certain instances, dietary fiber increased while the food was being fried, maybe as a result of the formation of melanoidins or other indigestible chemicals (Bordin et al. 2013).

7.3.1.7 Migration of Minor Compounds

Due to oil performance and quality, minor food chemicals that leached into the frying oil were easily changed. The following groups stand out among the primary substances influencing the physical and chemical characteristics of used frying fats and oils:

1. Emulsifiers and phospholipids are examples of amphiphilic substances that can cause early foaming.
2. Depending on whether they have an antioxidant or prooxidant impact, lipid-soluble vitamins, and trace metals have the ability to either slow down or speed up oil oxidation when they are released from food during frying.
3. Vegetable frying oils include cholesterol from fatty animal foods and can be absorbed into non-fatty foods during later frying processes.
4. Pigments and Maillard browning substances alter the frying oil's sensitivity to oxidation and help it darken.
5. Phenolic chemicals in the dish or additional spices make frying oil more stable.
6. Particular odd flavors may come from volatile molecules found in foods with strong flavors, such as fish or onions (Márquez-Ruiz and Velasco 2000).

Breaded or battered foods may contribute surface coating particles to the fat, which could cause burning, in addition to the migration of lipids from the food into the frying oil (Márquez-Ruiz and Velasco 2000).

7.3.2 Production of Toxic Compounds or Degradation of Lipids

There are numerous oxidation products produced as a result of the reaction between oxygen and unsaturated lipids, known as lipid peroxidation. Lipid hydroperoxides are lipid peroxidation's main by-products (LOOH). Malondialdehyde, propanal, hexanal, and 4-hydroxynonenal are only a few of the many aldehydes that can result from lipid peroxidation. The most deadly consequence of lipid peroxidation is 4-hydroxynonenal, while malondialdehyde seems to be the most mutagenic (Ayala, Muñoz, and Argüelles 2014). Unsaturated fatty acid–rich culinary oils are frequently found to contain lipid oxidation products, such as cytotoxic and genotoxic aldehydes, as well as their lipid hydroperoxide precursors, epoxy-fatty acids, and many other secondary or even tertiary lipid oxidation products, as a result of the peroxidative degradation of unsaturated fatty acid, particularly polyunsaturated fatty acids (Grootveld et al. 2020). It has been consistently discovered that the toxicity of aldehydes in a homologous series increases with increasing chain length, suggesting that the aldehyde's length may be significant. It implies a connection between toxicity and lipophilicity (Esterbauer, Muskiet, and Horrobin 1993).

Malondialdehyde and 4-hydroxynonenal: Due to their relatively high production levels and high reactivity as the "second messengers of free radicals," 4-hydroxyalkenals are the most important products. One of the most hazardous by-products of lipid peroxides is 4-hydroxynonenal, which has received extensive scientific attention since the 1990s, in particular. The fast interactions of 4-hydroxynonenal with thiols and amino groups can account for its high toxicity. Reactive aldehydes, particularly 4-hydroxynonenal, are harmful consequences of lipid peroxidation with long-lasting biological impacts, particularly through covalently changing macromolecules (Ayala, Muñoz, and Argüelles 2014).

While being cooked, oils break down into volatile and non-volatile decomposition products (Figure 7.1). Foods fried in deteriorated oils could contain a lot of breakdown by-products, which could have a negative impact on the food's flavor, texture, color, and safety. Thermal polymers do not directly alter flavor, even though volatile chemicals

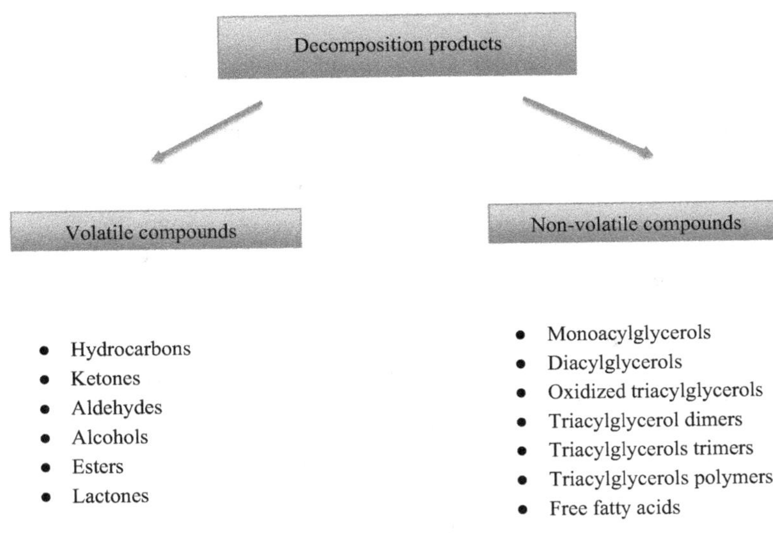

Figure 7.1 Decomposition products formed during frying.

are predominantly responsible for flavor—both positive and negative (Akoh and Min 2002).

The primary reaction by-products during lipid oxidation are hydroperoxides. Despite having no taste or smell, they do respond to a number of secondary degradation products from diverse chemical families. Particularly, volatile organic compounds, which include aldehydes, alkanes, alcohols, esters, and epoxides, are produced when hydroperoxides are broken down and are typically thought of as the secondary degradation products of lipid oxidation. In light of this, tertiary volatile organic compounds such as the methyl ketones of 2,4-alkadienals can also be produced via lipid oxidation. The majority of these volatile organic compounds are aroma-active and can alter the sense of smell significantly. Some volatile organic compounds not only have unpleasant sensory characteristics, but they can also be harmful to human health. This holds true particularly for the unsaturated 2,4-alkadienals and 2-alkenals.

Used frying oil contains polymeric triacylglycerols, oxidized triacylglycerol derivatives, cyclic compounds, and breakdown products as non-volatile degradation by-products. To form polar and nonpolar high-molecular-weight compounds, two or more triacylglycerol molecules must condense. The majority of the triacylglycerols in the oil that hasn't been polymerized are present, along with some of their oxidized derivatives. Mono- and diacylglycerols, partial glycerols, chain-scission products, triacylglycerols with cyclic and/or dimeric fatty acids, and partial glycerols without fatty acids are only a few examples of additional non-volatile products that are included in this. The additional classification of degradation products includes monomeric, unmodified, modified, oxidized, cyclized, isomerized, and fragmented fatty acid esters, as well as polar and nonpolar polymeric fatty acid methyl esters and polar and nonpolar polymeric fatty acid methyl esters (Akoh and Min 2002).

Triacylglycerol Toxicology: Hydroperoxides, epoxides, and hydroperoxides are the first triacylglycerol products. Following ingestion, oxidized triacylglycerol transits to the colon and interconverts to oxidized phospholipids in chylomicrons. LDL absorbs the oxidized phospholipids and oxidized triacylglycerol, converting them into oxidized LDL. Due to the rise in reactive oxygen species when the oxidized LDL reaches the liver, it is not or is the only slowly digested reactive oxygen species. As a result, both atherogenesis and the formation of fatty liver are influenced by oxidized triacylglycerol (Zeb 2019).

Junk food contains harmful chemicals such as trans-fats: Trans fats are vegetable oils that have undergone partial hydrogenation; they extend food's shelf life and give it a crisper texture. These trans fats are the worst ingredients we eat through meals (Bhaskar 2012).

7.4 Proteins

7.4.1 Chemical Modification of Proteins Due to Frying

Maillard (non-enzymatic browning) is regarded as the major significant process in food browning. This is the primary process that affects the amino acids, free amino acid group, proteins, and peptides during frying. This process results in changes to the proteins' composition (Bordin et al. 2013). Zhou et al. (2019) reported the influence of the frying procedure on the characteristics of gluten in order to establish a theoretical foundation for making nutritious wheat-based fried dishes (Zhou et al. 2019). The primary constituents of wheat, like dietary fiber, protein, and starch, interact with one another and impact product quality throughout the frying process. Wheat protein is an essential component in wheat, accounting for around 10%–15% of the overall weight. Unextracted protein, gliadin, albumin, glutenin, and globulin are the five components of wheat protein. Gluten protein is mostly made up of glutenin and gliadin, and contributes to approximately 80%–85% of overall wheat protein. The extensibility and elasticity of the gluten protein are the distinguishing characteristics which set it apart from the other cereal proteins, and also serve as the foundation for the production of a wide variety of foods that are based on wheat. The formation of the gluten network structure is facilitated by SH/SS exchange reactions that take place between glutenin and gliadin as the temperature rises. Furthermore, it is also inferred that post frying, the gluten was degraded and unfolded, devising the structure easily available for aggregation. The gluten network appeared to be incompact post frying, as per secondary and tertiary structural studies. The predominant disulfide bonds and hydrophobic interactions make up the gluten network. Frying distorted the many noncovalent bonds associated with gluten formation. Therefore, the flexibility of the peptide bond was decreased by the elevated frying temperature, resulting in a weaker gluten network (Zhou et al. 2019).

7.4.2 Nutritional Changes

The nutritional value of dietary protein is determined by a concoction of quantity and quality of food. A protein's functional content ingested and utilized by an individual is regarded as quality, and the protein content of the food is represented by quantity (Bordin et al. 2013). The nutrition studies show that browning reactions and the severe loss of nutritive benefits are caused by any or all of the following: (i) a reduction in the accessibility of certain amino acids (AA) once subjected to heat being maintained in an environment that contains reducing sugars, (ii) reduction in the ability of proteins to be digested when heated in the presence of carbohydrates, and (iii) the potential generation of toxic compounds or metabolic repressors (Gray 1981). Heat treatment can significantly decrease protein content and degrade certain AA, as well as alter the quality of protein composition in food (Bordin et al. 2013). But a study carried out by Zhang et al. (2013) reported that the significant percentage of protein increased in the grass carp fillet after being fried. In another research, the fried sardine's protein levels were greater than the uncooked ones. This might be as a result of the frying processes producing new compounds that resemble protein. The palm oil–fried samples decreased the AA content (Zhang et al. 2013). Frying process reduces the bioavailability of lysine in fish fillets by around 17% and by 25% when the fish oil is used for prolonged frying for 48 hours, and it may be due to interactions between both the carbonyl molecule and amino group of lysine. Regardless of this, there was no discernible impact of the frying practices on the amino acid content including methionine, proline, serine, histidine, tyrosine, lysine, cysteine, valine, aspartic acid, glycine, leucine, valine, isoleucine, threonine, phenylalanine, arginine, glutamic acid, and alanine. An analysis of the amino acid content of raw, cooked, and fried fish revealed that there was no variation in the composition of specific amino acid due to frying (Bordin et al. 2013). Since frying is a form of dehydration, the concentration of protein content rises throughout the frying process. Lysine, nevertheless, is the first amino acid to take part in the Maillard reaction, and it is hypothesized that it will be lost during frying. Labile nutrients' alterations can result from frying. Therefore, it is crucial to recognize, assess, and understand how frying affects nutrition and in return affects human health (Bordin et al. 2013).

7.4.3 Production of Toxic Compounds or Degradation of Proteins

The primary goal of frying is to create high-quality meals that are well-liked by consumers. However, not all palatable meals are safe. In animal models, fat deposits and changes in serum biochemistry were frequently observed post consumption of high amount of fried food. Therefore, the novel hazardous toxic molecule formed during frying piqued the interest of researchers (Zeb 2019). Several classes of compounds are produced during the frying process (Figure 7.2). Several fried foods contain high concentrations of the potential carcinogen acrylamide, which forms during the heating process of some foods in low humidity conditions and at temperatures greater than 120°C, such as roasting, cooking, or frying. The primary mechanism for acrylamide synthesis in foods is associated with the Maillard reaction,

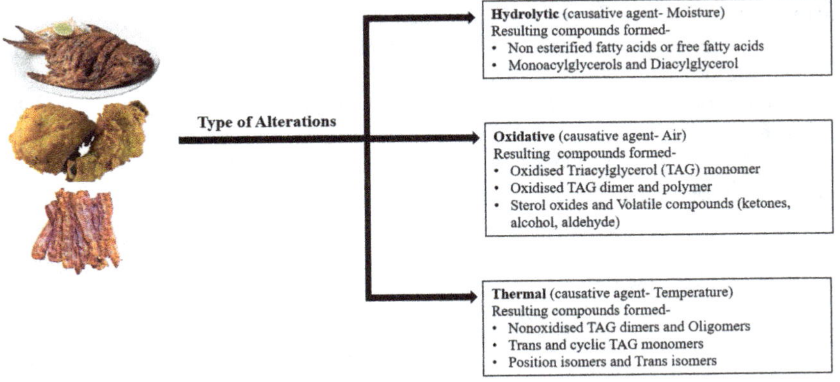

Figure 7.2 Predominant classes of compounds that are produced during the frying process.

specifically the amino acid asparagine. The level of acrylamide elevates as the duration of frying increases (Hosseini et al. 2016). Additionally, studies found that frying at a temperature of 175°C reduced the production of acrylamide (Hosseini et al. 2016). The epoxidation of acrylamide to glycidamide through Cytochrome P450, a hemoprotein with enzyme activity, is crucial for the genotoxic activity of acrylamide. In addition to the reacting with enzymes and hemoglobin, glycidamide can also interact with DNA, leading to cancer and point mutations (Zeb 2019). Aldehyde acrolein is a carcinogenic substance. It has the ability to cross-link proteins in foods through formation of adducts with lysine, cysteine residues, and histidine. The production of acrolein adducts in soy protein results in the decreased solubility as well as surface hydrophobicity, which in turn leads to increased protein aggregation (Zeb 2019). Fried foods and the oils used to fry them are two of the most prominent sources of acrolein. Nevertheless, it was found that the amount of acrolein that is present in frying oil and in fried foods is not the same (Ou et al. 2020). The mutagenic potential of heterocyclic aromatic amines, which are produced during the cooking of meat and fish, is about 100 times greater than that of aflatoxin and 2,000 times greater than that of benzo α-pyrene (Zaghi et al. 2019). Animal studies conducted by Haskaraca et al. (2014) revealed that the animals contain increased levels of heterocyclic aromatic amines and the unregulated frying conditions of fast foods may be responsible for this predicament (Haskaraca et al. 2014). These compounds are produced from free AA and creatine that are usually found in meals of animal origin, which are high in protein. According to research, some oils might also promote the production of heterocyclic aromatic amines (Zaghi et al. 2019). Omojola et al. (2015) investigated the formation of heterocyclic aromatic amines in drake breasts while they were being prepared using a variety of cooking methods, such as panfrying (180°C without fat or oil, for 10 minutes), deep-frying in soy oil (180°C, for 10 minutes), grilling with liquefied gas (200°C, for 20 minutes), and roasting in an oven (200°C, for 20 minutes), all while managing the meat core temperature of 80°C (Omojola et al. 2015). Therefore, to avert toxicity and ecological degradation, frying oils should be discarded appropriately.

7.5 Nucleic Acids

7.5.1 Chemical Modification of Nucleic Acids Due to Frying

Frying requires heating up edible oil or fat and simply utilizing the hot oil to cook food during deep frying. It is one of the quickest, oldest, and simplest ways of food preparation. When oil is heated to high temperatures in the presence of moisture and air, it behaves like a heat-transferring substance. Frying process is one of the competent cooking methods due to its high temperature and fast heat transfer (Sanibal n.d.). The method preserves the food by thermally destroying microbes and enzymes, and decreasing water activity on the food's surface (Fellows 2009). Several chemical processes take place in this environment, such as the oxidation, hydrolysis, cyclization, and polymerization of unsaturated fatty acids. It is well known that frying lowers the protein quality and causes some amino acids to be destroyed. Heat treatment can alter the quality of the protein content in food by reducing the amount of protein and destroying particular amino acids (Henry 1998).

Protein quality can suffer from excessive heat in the absence of water; tryptophan, arginine, methionine, and lysine, for instance, may be harmed in fish proteins (Henry 1998). Heat treatment can cause a variety of chemical processes, such as breakdown, threonine and serine dehydration, sulfur loss from cysteine, cysteine and methionine oxidation, and the cyclization of glutamic and aspartic acids and threonine (Mauron 1972).

Because of interactions between the amino group of lysine and carbonyl compounds, deep-fat frying has been shown to reduce the amount of lysine that is readily available in fish fillets by around 17%, and by 25% when the fish oil has been used for continuous frying for 48 hours (Oluwaniyi, Dosumu, and Awolola 2010). In general, the concentration impact of frying, which is also a process of dryness, causes the content of protein to increase. There is a disagreement in the literature on essential amino acid loss in relation to it. Lysine is the first amino acid implicated in the Maillard reaction; hence, it should be lost throughout the frying process (Bordin et al. 2013).

7.5.2 Nutritional Changes

One of the oldest methods for processing of food is frying. It is also considered as an efficient cooking method as it preserves food by the thermal destruction of microorganisms, inactivation of enzymes, and reduction of water activity in food (Fellows 2009). Frying process causes various physical, chemical changes which impacts the composition of the food, flavor, texture and color development (Bordin et al. 2013). There are various undesirable changes that occur due to frying such as loss of nutritive contents of vitamins and minerals. A large amount of oil is absorbed by food during frying which leads to 30%–50% of fat in the final product (Kosseva and Kennedy 2003). The nutritional effects are not easily estimated due to the variety of factors influencing it. The influencing factors include type of oil used, thermal treatment given, and the portion of the food which is retained (Singh, Gamlath, and Wakeling 2007). The lipid content which is present in the food gets oxidized, and the consumption of fried food leads to the increase in the intake of the calories. Trans-fatty acids are also produced by frying which are unsaturated fatty acids which increase the risk of cardiovascular diseases. The degrees of trans-fatty acids formation depend on frying material, conditions of frying, and methods for the measurement of trans-fatty acids (Tsuzuki, Matsuoka, and Ushida 2010). In a study, it was found that in grass carp fillet, the protein content increased after frying (Zhang et al. 2013). In a recent study, it was also found that the protein content of fried sardine was higher than that of unfried sardine, which may be due to formation of new products similar to proteins during frying process. Although the process of frying reduced the content of the amino acids, no significant effect was found on the reduction of amino acids. The fats and fried oils that are absorbed by food become the part of our diet. High temperature and oxidation lead to changes in vitamins of fried foods (Oke et al. 2018). During heating, the oxidation of unsaturated fatty acids leads to loss of vitamin E. The vegetable frying oils are a source of vitamin E. In a recent study, it was found that there was decrease in vitamin E content from 4.6 to 4.9 mg/100 g in chicken nuggets after frying (Simonne and Eitenmiller 1998). It has also been observed that deep frying of vegetables leads to decrease in beta-carotene twice as high as in shallow frying (Ghidurus et al. 2010). Stir frying leads to the better retention of vitamin C, and niacin content increases during frying (Ren et al. 2020). Minerals also play an essential role in the nutritional quality of food. The minerals are preserved in the food by frying at high temperature. It has been found that mineral losses in potatoes and beef are 1% and 26%, respectively, when deep fried. In a recent study, it was found that there is increase in the macro- and microelements and heavy metals due to frying in mussels and shrimps (Anna and Kamila 2013). The quality of the oil used for frying did not induce any significant change in the bioavailability of calcium (Pérez-Granados, Vaquero, and Navarro 2000). There is no effect of frying on the digestibility of proteins. There is a change in the structure of nutrients such as proteins, vitamins, and antioxidants due to the process of frying. Hence, it is important to estimate the effect of frying on the nutritional quality of the food.

7.6 Conclusion

The process of frying has provided several advantages in the food industry. The extermination of unwanted microbes and extension of shelf-life are examples of benefits of frying that have tremendously helped with distribution of food. Moreover, this chapter has discussed the effects of frying on the nutrients of food. Despite the enhancement in taste, texture, and flavor of fried food, several reports of the over-consumption of fried food have emphasized their negative impacts on health. Of note, the production of dangerous chemicals has been emphasized in most studies.

References

Akoh, Casimir C., and David B. Min. 2002. *Food Lipids: Chemistry, Nutrition, and Biotechnology*. Marcel Dekker, Inc.

Aladedunye, Felix A., and Roman Przybylski. 2013. "Minor Components in Oils and Their Effects on Frying Performance." *Lipid Technology* 25 (4): 87–90.

Albuquerque, Tânia Gonçalves, Ana Sanches-Silva, Lèlita Santos, and Helena S. Costa. 2012. "An Update on Potato Crisps Contents of Moisture, Fat, Salt and Fatty Acids (Including Trans-Fatty Acids) with Special Emphasis on New Oils/Fats Used for Frying." *International Journal of Food Sciences and Nutrition* 63 (6): 713–717.

Anna, Czech, and Stachyra Kamila. 2013. "Effect of Processing Treatments (Frozen, Frying) on Contents of Minerals in Tissues of 'Frutti Di Mare.'" *International Journal of Food Science & Technology* 48 (2): 238–45.

Arroyo, R., F. J. Sánchez-Muniz, C. Cuesta, F. J. Burguillo, and J. M. Sánchez-Montero. 1996. "Hydrolysis of Used Frying Palm Olein and Sunflower Oil Catalyzed by Porcine Pancreatic Lipase." *Lipids* 31 (11): 1133–1139. https://doi.org/10.1007/BF02524287.

Ayala, Antonio, Mario F. Muñoz, and Sandro Argüelles. 2014. "Lipid Peroxidation: Production, Metabolism, and Signaling Mechanisms of Malondialdehyde and 4-Hydroxy-2-Nonenal." *Oxidative Medicine and Cellular Longevity*. https://doi.org/10.1155/2014/360438.

Bhaskar, Rajveer. 2012. "Junk Food: Impact on Health." *Journal of Drug Delivery and Therapeutics* 2 (3). https://doi.org/10.22270/JDDT.V2I3.132.

Blumenthal, Micheel M. 1991. "A New Look at the Chemistry and Physics of Deep-Fat Frying." *Food Technology (Chicago)* 45 (2): 68–71.

Bordin, Keliani, Mariana Tomihe Kunitake, Keila Kazue Aracava, and Carmen Silvia Favaro Trindade. 2013. "Changes in Food Caused by Deep Fat Frying - A Review." *Archivos Latinoamericanos de Nutricion* 63 (1): 5–13.

Crosa, Maria Jose, Veronica Skerl, Mónica Cadenazzi, Laura Olazábal, Roberto Silva, Gabriela Suburú, and Marina Torres. 2014. "Changes Produced in Oils during Vacuum and Traditional Frying of Potato Chips." *Food Chemistry* 146 (March): 603–607. https://doi.org/10.1016/J.FOODCHEM.2013.08.132.

Debnath, Sukumar, Navin K. Rastogi, A. G. Gopala Krishna, and B. R. Lokesh. 2012. "Effect of Frying Cycles on Physical, Chemical and Heat Transfer Quality of Rice Bran Oil during Deep-Fat Frying of Poori: An Indian Traditional Fried Food." *Food and Bioproducts Processing* 90 (2): 249–56. https://doi.org/10.1016/j.fbp.2011.05.001.

Dobarganes, Carmen, Gloria Márquez-Ruiz, and Joaquín Velasco. 2000. "Interactions between Fat and Food during Deep-Frying." *European Journal of Lipid Science and Technology* 102 (8–9): 521–28. https://doi.org/10.1002/1438-9312(200009)102:8/9<521::aid-ejlt521>3.0.co;2–a.

Ersoy, Beyza, and Akif Özeren. 2009. "The effect of cooking methods on mineral and vitamin contents of African catfish." Food chemistry 115 (2): 419–422.

Esterbauer, H., F. Muskiet, and D. F. Horrobin. 1993. "Cytotoxicity and Genotoxicity of Lipid-Oxidation Products." *The American Journal of Clinical Nutrition* 57 (5): 779S-786S. https://doi.org/10.1093/AJCN/57.5.779S.

Fellows, Peter J. 2009. *Food Processing Technology: Principles and Practice.* Elsevier.

Ghidurus, Mihaela, Mira Turtoi, George Boskou, Petru Niculita, and Visilica Stan. 2010. "Nutritional and Health Aspects Related to Frying (I)." *Romanian Biotechnological Letters* 15(6): 5675–82.

Gokoglu, Nalan, Pinar Yerlikaya, and Emel Cengiz. 2004. "Effects of Cooking Methods on the Proximate Composition and Mineral Contents of Rainbow Trout (Oncorhynchus mykiss)." *Food Chemistry* 84 (1): 19–22.

Gray, J. I., and I. D. Morton. 1981. "Some Toxic Compounds Produced in Food by Cooking and Processing." *Journal of Human Nutrition* 35 (1981): 5–23.

Grootveld, Martin, Benita C. Percival, Justine Leenders, and Philippe B. Wilson. 2020. "Potential Adverse Public Health Effects Afforded by the Ingestion of Dietary Lipid Oxidation Product Toxins: Significance of Fried Food Sources." *Nutrients* 12 (4): 974. https://doi.org/10.3390/NU12040974.

Han, Jung-Ah, and James N BeMiller. 2007. "Preparation and Physical Characteristics of Slowly Digesting Modified Food Starches." *Carbohydrate Polymers* 67 (3): 366–374.

Haskaraca, G., E. Demirok, N. Kolsarici, F. Öz, and N. Özsaraç. 2014. "Effect of Green Tea Extract and Microwave Pre-Cooking on the Formation of Heterocyclic Aromatic Amines in Fried Chicken Meat Products." *Food Research International* 63: 373–381. https://doi.org/10.1016/j.foodres.2014.04.001.

Henry, C. J. K. 1998. "Impact of Fried Foods on Macronutrient Intake, with Special Reference to Fat and Protein." *Grasas y Aceites* 49 (3–4): 336–339.

Hosseini, Hamed, Mohammad Ghorbani, Nasim Meshginfar, and Alireza Sadeghi Mahoonak. 2016. "A Review on Frying: Procedure, Fat, Deterioration Progress and Health Hazards." *JAOCS, Journal of the American Oil Chemists' Society* 93 (4): 445–466. https://doi.org/10.1007/s11746-016-2791-z.

Jiménez-Monreal, A. M., L. García-Diz, M. Martínez-Tomé, M. M. M. A. Mariscal, and M. A. Murcia. 2009. "Influence of Cooking Methods on Antioxidant Activity of Vegetables." *Journal of Food Science* 74 (3): H97–H103.

Khan, Johra, Prashanta Kumar Deb, Somi Priya, Karla Damián Medina, Rajlakshmi Devi, Sanjay G Walode, and Mithun Rudrapal. 2021. "Dietary Flavonoids: Cardioprotective Potential with Antioxidant Effects and Their Pharmacokinetic, Toxicological and Therapeutic Concerns." *Molecules* 26 (13): 4021.

Kosseva, Maria R., and John F. Kennedy. 2003. "Chemical and Functional Properties of Food Proteins-ZE Sikorski (Ed.); Technomic Publishing Co., Inc., Basel, 2001, X+ 490 Pp, ISBN 1 56676 960 4." *Carbohydrate Polymers* 4 (54): 531.

Manorama, R., and C. Rukmini. 1991. "Nutritional Evaluation of Crude Palm Oil in Rats." *The American Journal of Clinical Nutrition* 53 (4): 1031S–1033S.

Márquez-Ruiz, Gloria, and Joaquín Velasco. 2000. "Interaction between Fat and Food During Deep-Frying." *Article in European Journal of Lipid Science and Technology* 102. https://doi.org/10.1002/1438-9312(200009)102:8/93.0.CO;2-A.

Mauron, J. 1972. "Influence of Industrial and Household Handling on Food Protein Quality." *Protein and Amino Acid Functions* 11: 417.

McDougall, Gordon J., Ian M. Morrison, Derek Stewart, and John R. Hillman. 1996. "Plant Cell Walls as Dietary Fibre: Range, Structure, Processing and Function." *Journal of the Science of Food and Agriculture* 70 (2): 133–150.

Miao, YuTian, HuanJie Zhang, LuLu Zhang, SiJia Wu, YiJia Sun, Yu Shan, and Yuan Yuan. 2014. "Acrylamide and 5-Hydroxymethylfurfural Formation in Reconstituted Potato Chips during Frying." *Journal of Food Science and Technology* 51 (12): 4005–4011.

Mishra, Suman, Allan Hardacre, and John Monro. 2012. "Food Structure and Carbohydrate Digestibility." In *Carbohydrates-Comprehensive Studies on Glycobiology and Glycotechnology*, Chuan-Fa Chang (Ed.), 289–316. InTech. doi:10.5772/2702.

Morales, M. T., J. J. Rios, and R. Aparicio. 1997. "Changes in the Volatile Composition of Virgin Olive Oil During Oxidation: Flavors and Off-Flavors." *Journal of Agricultural and Food Chemistry* 45 (7): 2666–2673.

Oke, E. K., M. A. Idowu, O. P. Sobukola, S. A. O. Adeyeye, and A. O. Akinsola. 2018. "Frying of Food: A Critical Review." *Journal of Culinary Science & Technology* 16 (2): 107–27.

Oluwaniyi, O. O., O. O. Dosumu, and G. V. Awolola. 2010. "Effect of Local Processing Methods (Boiling, Frying and Roasting) on the Amino Acid Composition of Four Marine Fishes Commonly Consumed in Nigeria." *Food Chemistry* 123 (4): 1000–1006.

Omojola, A. Babatunde, Saheed A. Ahmed, Victoria Attoh-Kotoku, and Gs Ikani Wogar. 2015. "Effect of Cooking Methods on Cholesterol, Mineral Composition and Formation of Total Heterocyclic Aromatic Amines in Muscovy Drake Meat." *Journal of the Science of Food and Agriculture* 95 (1): 98–102. https://doi.org/10.1002/jsfa.6687.

Omotosho, O. E., O. C. Laditan, O. E. Adedipe, and J. A.O. Olugbuyiro. 2015. "Effect of Deep-Fat Frying on the Vitamins, Proximate and Mineral Contents of Colocasia Esculenta Using Various Oils." *Pakistan Journal of Biological Sciences* 18 (6): 295–299. https://doi.org/10.3923/PJBS.2015.295.299.

Ou, Juanying, Jie Zheng, Junqing Huang, Chi Tang Ho, and Shiyi Ou. 2020. "Interaction of Acrylamide, Acrolein, and 5-Hydroxymethylfurfural with Amino Acids and DNA." *Journal of Agricultural and Food Chemistry* 68 (18): 5039–5048. https://doi.org/10.1021/acs.jafc.0c01345.

Pérez-Granados, Ana M., M. Pilar Vaquero, and M. Pilar Navarro. 2000. "Intake of Unused Palm Olein and Palm Olein Used in Frying Does Not Affect Calcium and Phosphorus Bioavailability in Rats." *Journal of the Science of Food and Agriculture* 80 (9): 1379–1385.

Ren, Xue, Tiantian Tang, Xinfang Xie, Wenjun Wang, Xuanming Tang, Charles S. Brennan, Jie Zhang, and Zhidong Wang. 2020. "The Effects of Preparation and Cooking Processes on Vitamins and Antioxidant Capacity of Sour and Spicy Potato Silk." *International Journal of Food Science & Technology* 55 (11): 3475–83.

Rudrapal, Mithun, Shubham J. Khairnar, Johra Khan, Abdulaziz Bin Dukhyil, Mohammad Azam Ansari, Mohammad N. Alomary, Fahad M. Alshabrmi, Santwana Palai, Prashanta Kumar Deb, and Rajlakshmi Devi. 2022. "Dietary Polyphenols and Their Role in Oxidative Stress-Induced Human Diseases: Insights into Protective Effects, Antioxidant Potentials and Mechanism (s) of Action." *Frontiers in Pharmacology* 13. DOI:10.3389/fphar.2022.806470.

Sanibal, E. A. A. n.d. "Filho, JM (2002). Physical, Chemical and Nutritional Oils Subjected to the Frying Process." *Food Ingredient South America* 18 (3): 64–71.

Simonne, A. H., and R. R. Eitenmiller. 1998. "Retention of Vitamin E and Added Retinyl Palmitate in Selected Vegetable Oils during Deep-Fat Frying and in Fried Breaded Products." *Journal of Agricultural and Food Chemistry* 46 (12): 5273–77.

Singh, Shivendra, Shirani Gamlath, and Lara Wakeling. 2007. "Nutritional Aspects of Food Extrusion: A Review." *International Journal of Food Science & Technology* 42 (8): 916–929.

Soriguer, Federico, Gemma Rojo-Martínez, M. Carmen Dobarganes, José M. García Almeida, Isabel Esteva, Manuela Beltrán, M. Soledad Ruiz De Adana, et al. 2003. "Hypertension Is Related to the Degradation of Dietary Frying Oils." *The American Journal of Clinical Nutrition* 78 (6): 1092–1097. https://doi.org/10.1093/AJCN/78.6.1092.

Steiner-Asiedu, Matilda, Kåre Julshamn, and Øyvind Lie. 1991. "Effect of Local Processing Methods (Cooking, Frying and Smoking) on Three Fish Species from Ghana: Part I. Proximate Composition, Fatty Acids, Minerals, Trace Elements and Vitamins." *Food Chemistry* 40 (3): 309–321.

Subar, Amy F., Susan M. Krebs-Smith, Annetta Cook, and Lisa L. Kahle. 1998. "Dietary Sources of Nutrients among US Children, 1989–1991." *Pediatrics* 102 (4): 913–923.

Tsuzuki, Wakako, Akiko Matsuoka, and Kaori Ushida. 2010. "Formation of Trans Fatty Acids in Edible Oils during the Frying and Heating Process." *Food Chemistry* 123 (4): 976–982.

Tynek, Maria, Zdzisllawa Hazuka, Roman Pawlowicz, and Marta Dudek. 2001. "Changes in the Frying Medium During Deep-Frying of Food Rich in Proteins and Carbohydrates." *Journal of Food Lipids* 8 (4): 251–261.

Zaghi, Aline Nalon, Sandra Maria Barbalho, Elen Landgraf Guiguer, and Alda Maria Otoboni. 2019. "Frying Process: From Conventional to Air Frying Technology." *Food Reviews International* 35 (8): 763–777. https://doi.org/10.1080/87559129.2019.1600541.

Zeb, Alam. 2019. "Toxicity of Food Frying." *Food Frying*, no. Ldl: 365–406. https://doi.org/10.1002/9781119468417.ch12.

Zhang, Jinjie, Dan Wu, Donghong Liu, Zhongxiang Fang, Jianchu Chen, Yaqin Hu, and Xingqian Ye. 2013. "Effect of Cooking Styles on the Lipid Oxidation and Fatty Acid Composition of Grass Carp (Ctenopharynyodon Idellus) Fillet." *Journal of Food Biochemistry* 37 (2): 212–219. https://doi.org/10.1111/j.1745-4514.2011.00626.x.

Zhou, Ruoxin, Juan Sun, Haifeng Qian, Yan Li, Hui Zhang, Xiguang Qi, and Li Wang. 2019. "Effect of the Frying Process on the Properties of Gluten Protein of You-Tiao." *Food Chemistry* 310: 125973. https://doi.org/10.1016/j.foodchem.2019.125973.

Ziaiifar, Aman Mohammad, Nawel Achir, Francis Courtois, Isabelle Trezzani, and Gilles Trystram. 2008. "Review of Mechanisms, Conditions, and Factors Involved in the Oil Uptake Phenomenon During the Deep-Fat Frying Process." *International Journal of Food Science & Technology* 43 (8): 1410–1423.

Chapter 8

Physical and Chemical Parameters as Qualitative Indicators of Used Frying Oils

Siddharth Vishwakarma and Shubham Mandliya
Indian Institute of Technology Kharagpur

Chandrakant Genu Dalbhagat
National Institute of Technology Rourkela

Chirasmita Panigrahi
Indian Institute of Technology Madras

DOI: 10.1201/9781003329244-8

8.1 Introduction

Frying is one of the oldest and most popular cooking methods because it produces pleasing tastes and appearance in a product. It is as easy and quick way of cooking because the operation involves a few minutes of submerging a food item in hot oil. For frying, a variety of fatty materials in solid, semi-solid, or liquid form at room temperature are utilized. However, liquid vegetable oil is widely used for cooking due to its proven nutritional benefits. Palm olein, maize oil, sunflower oil, soybean oil, safflower oil, cottonseed oil, peanut oil, rapeseed oil, and canola oil are some of the refined, bleached, and deodorized oils which are extensively used for cooking (Bansal et al. 2010).

Although frying is a basic method of cooking, it includes the most complicated reactions in relation with food science. It involves continuous heating of oil at high temperature (160°C–180°C) in contact with air, metal container, and moisture, which sometimes resulted in both thermal and oxidative breakdown of the oil (Esfarjani et al. 2019). Furthermore, chemical events such as hydrolysis, cyclization, and polymerization arise during the deep-frying process. These chemical interactions produce various volatile and non-volatile compounds which induce undesirable changes in the oil during repeated frying operation. The volatile breakdown products change and influence the flavor of the fried dish. Non-volatile products remain in the oil and are absorbed in meals, with an effect on the physical properties of the oil used (Gertz 2000). As a result, it is necessary to analyze the oil quality by developing some chemical and physical indicators that can help in determining at which point the reused oil needs to be discarded. These indicators are used to create methods for determining the quality of fried oil, such as chemical, physical, and analytical procedures. However, understanding the physical and chemical features associated with the quality factor of fried oil is important.

The purpose of this chapter is to cover the various physical and chemical quality indicators of used frying oil as well as its method of determination. Some standard and analytical methods are also described related to these parameters.

8.2 Major Components Deciding Quality of Used Fried Oils

During deep frying of food, various physical and chemical changes in the oil takes place. However, these changes are mostly due to three major chemical reactions, viz. hydrolytic reaction, oxidative alterations, and thermal reactions occurred during deep frying. The transfer of moisture from foods into the oil causes hydrolytic reactions. The oil is always in contact with air, and movement of oil during frying causes oxidative changes which resulted in production of decomposition products like free radicals, ketones, acids, hydrocarbons, etc. High-temperature thermal changes result in the creation of cyclic monomers, dimers, and polymers (Paul, Mittal, and Chinnan 1997).

Total polar materials (TPM) and polymers were formed as a result of the aforementioned chemical processes. These two factors are the most essential in determining whether or not the oil should be discarded (Zainal and Isengard 2010). The TPM are the oil's combined degradation products, and the degraded components other than the nonpolar portion are referred to as the polar fraction. This portion is divided into two groups: polymers and decomposition products. The term "polymers" refers to all degradation products with a molecular weight greater than the triglycerides in the original oil (fresh oil is almost entirely composed of triglycerides), whereas "decomposition products" refer to degradation products with a molecular weight less than the triglycerides (Bansal et al. 2010). The TPM in frying oil and fried foods may be hazardous to human health. As a result, several nations have imposed a TPM limit in frying oil ranging from 24% to 30%. TPM levels in frying oil must be less than 24% in Germany; otherwise, the oil must be discarded (Zainal and Isengard 2010).

So, the TPM can be used for determining the quality of oil and should be measured before re-using the fried oil. However, it is often analyzed using silica gel column chromatography and gravimetric analysis. This analysis is time consuming and needs large amounts of solvents (Cascant, Garrigues, and de la Guardia 2017). But, various studies showed the correlation of TPM with physical and chemical parameters of fried oil (Gertz 2000; Bansal et al. 2010). Some of these parameters are easily determined, and they are also employed in the development of analytical procedures. Thus, studying these characteristics is important, and it is covered in detail in the next sections of this chapter.

8.3 Types of Quality Indicator of Used Fried Oils

The quality of fried oils can be found out using subjective and quantitative approach. Sometimes in practice, the subjective way of judging oil quality is done by assessing the color, odor, excessive foaming, smoking, and tasting the fried product (Gertz 2000). However, each of these parameters is influenced by a variety of other factors in addition to the quality of the frying oil. Thus, quantitative assessment of characteristics linked to oil quality is a more reliable approach. The quantitative measurement is primarily determined by the physical and chemical methods listed in Figure 8.1.

While the chemical indications described in Figure 8.1 can be a more accurate technique to detect frying oil degradation. The volatile chemicals created by chemical reactions during the frying process help to increase the peroxide value (PV) of the oil. As the oil is heated, the PV rises over time. Furthermore, iodine value (IV) is utilized to determine the appropriateness of the oils. In addition, high-IV oil is bad for lipid oxidation and formation of hydroperoxides between unsaturated fatty acids and oxygen. In addition, free fatty acids (FFA), polymeric triglycerides, anisidine value (AnV), and polymerized and oxidized material (POM) are often used as indicators of frying oil quality, but are not definitive. At the moment, measuring TPC is considered the most widely used approach for evaluating oil quality since it determines the overall chemical deterioration occurring in the oil (Khaled, Aziz, and Rokhani 2015).

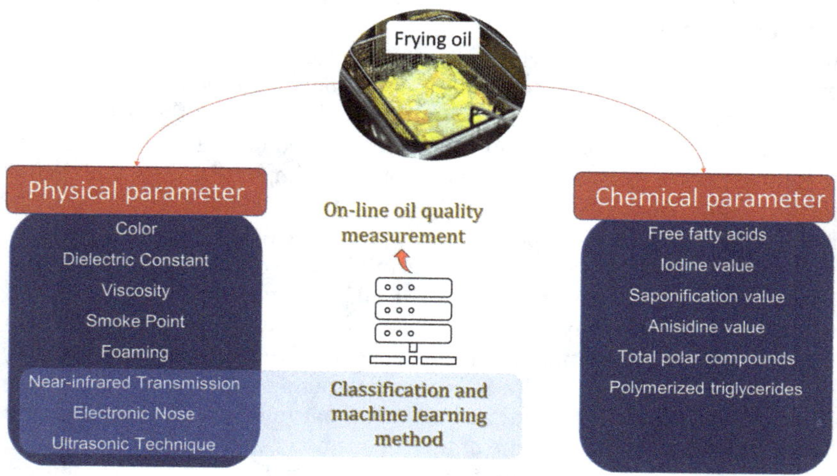

Figure 8.1 Physical and chemical parameters related to frying oil quality.

8.4 Physical Quantitative Quality Indicator of Used Fried Oils

As the frying process continues, the physical properties of the oils get changed with time. These variations are darkening, thickening, and foaming which act as quality indicators. Smoke point, dielectric constant, viscosity, and other physical measures also determine the oil quality which are described in brief in Table 8.1. Moreover, changes in the physical characteristics of frying oils are used as an indication of oil quality in restaurants and food services. For instance, it is good to remove frying oil if it gets black, emits excessive smoke, and has strong odor and a greasy consistency, or if a durable foam layer of the significant thickness is seen (Esfarjani et al. 2019)

8.4.1 Color

The subjective assessment of oil color is the most straightforward technique to evaluate oil quality. The color of an oil becomes objectionable much sooner than the flavor or odor of the oil. Color qualities are connected to other oil quality criteria to some extent, as seen by their association with several oil quality metrics. Based on the significant link between the color index and the TPC, a chromametric approach for the quick assessment of oils was developed by Xu (2003). The polymers, TPC, and FFAs have an inverse effect on the color of frying oil. Absorption spectroscopy in the ultraviolet and visible range of 200–700 nm is another method for detecting color changes in oils (Bansal et al. 2010). However, the acceptability of oil based on color is still in doubt since the color of the deteriorated oil is heavily dependent on various factors such as the original color of the oil, the type of food being fried, the type of oil, and contact time with light and heat. For example, oil darkening is caused by the caramelization of food particles and the release of oil-soluble compounds (Bansal et al. 2010; Esfarjani et al. 2019).

8.4.2 Dielectric Constant

Dielectric constant (DC) was previously considered the most suitable quality measure in industrial frying (Smith et al. 1986). As the oil degrades, the amount of polar molecules rises, directly increasing the DC. Furthermore, the polymer content of deteriorated oil causes a significant shift in its relative permittivity and, eventually, the DC of the oil, making it a useful oil indicator (Rubalya Valantina et al. 2017). Furthermore, the DC is linked with heating time of oil. So, early in the 20th century, several instruments were

Table 8.1 Physical Parameters of Used Fried Oil with Its Measuring Instruments and Advantages

Type of Parameter	*Correlated Oil Components*	*Instrument Used*	*Instrument Manufacturer*	*Advantages of Being Measured*	*References*
Color	Free fatty acids, polar compounds	Chromameter	Minolta CR300, Minolta CameraCo Ltd, Osaka, Japan	Rapid, convenient and reliable	Xu (2003)
Dielectric constant	Heating time, density, polar compounds, saturated fatty acids	Capacitive sensor	KOKUYO-KDK Electric Corporation Limited	Non-destructive method	Rubalya Valantina et al. (2017)
Viscosity	Saturation of oil	Viscometer	SV-10 Vibro Viscometer, A&D Company Limited, Japan	Simple and inexpensive way	Khaled, Aziz, and Rokhani (2015)
Near-infrared transmission spectroscopy	Polar components	FTNIR spectrometer	Bruker Optik, GmbH, Ettlingen, Germany	Simple, no complex pre-treatment, rapid and environment friendly	Chen et al. (2015)
Near-infrared transmission spectroscopy	Carbonyl value	MPA-TM FTNIR system	Bruker Optics, Germany	Simple operation, rapid determination, and no hazardous chemicals required	Wang et al. (2014)

(Continued)

Table 8.1 (*Continued*) Physical Parameters of Used Fried Oil with Its Measuring Instruments and Advantages

Type of Parameter	*Correlated Oil Components*	*Instrument Used*	*Instrument Manufacturer*	*Advantages of Being Measured*	*References*
E-nose	Peroxide value	Electronic nose	SPEC Sensors, Newark, CA, USA	Simple handling, real-time measurement, green analytical methods	Wasik, Majchrzak, and Wojnowski (2021)
E-nose	Peroxide value, acid value, oxidation in oil	PEN3 portable E-nose	Win Muster Airsense Analytics, Inc., Schwerin, Germany	Time-saving and non-destructive	Xu et al. (2016)
Ultrasonic technique	Free fatty acid level and total polar compounds	Ultrasonic transducer	A309S-SU Model, Panametrics, Olympus	For real-time quality control	Izbaim et al. (2010)

developed to detect DC, such as the Northern Instruments Corporation's Food Oil Sensor (FOS) model NI-20 and the Center for Chemical Information Technology's "CapSens 5000." Through its sensitive bridge circuit, the FOS model can detect minute dielectric changes in oils (Bansal et al. 2010). Studies have shown a relationship between polar components measured by column chromatography and FOS readings. In this study, oil quality testing techniques using FOS were found to be the most practical (Smith et al. 1986).

Aside from the oil quality parameter, the DC is impacted by a variety of external parameters such as water, salt, and minerals. These components have an effect on the polarity of a sample and can provide inaccurate information about the oil's quality. However, using filtered test samples before measurement can help alleviate some of these issues. There are still some problems with the DC meter, such as the long warm-up time, the need for daily calibration, and a substantial effect from the kind of oil (Bansal et al. 2010). Unheated coconut oil samples, for example, had a FOS value of 5.6 higher than 4.0 FOS units, equivalent to 27% TPC, which is the recommended standard for rejecting frying oils (Gertz 2000).

8.4.3 Viscosity

The extent of deterioration of oil can be estimated through viscosity measurement but not as conclusive remark (Khaled, Aziz, and Rokhani 2015). Oil viscosity generally increases with frying time due to oil polymerization. Also, oils mostly behave as Newtonian fluids, meaning that their viscosity is independent of shear rate (Bansal et al. 2010). Many devices have been developed to measure the viscosity of frying oil, which has a direct or indirect relationship with oil quality. Previously, Kress-Rogers, Gillatt, and Rossell (1990) created an instrument for oil-fired boilers that included a probe produced by GEC Marconi. The instrument consists of two vibrating steel tubes powered by an internally mounted piezoelectric crystal and an internal thermocouple. This instrument measurement revealed a strong relationship between viscosity and oxidized and polymerized decomposition products.

The frying time and frequency have a considerable impact on the viscosity of the oils, which is heavily impacted by frying temperature instead of frying medium. Increased frying time produces a rise in the viscosity of vegetable oils, which may be attributed oil polymerization (Bansal et al. 2010). Microfluidic sensors have recently gained popularity due to its use in various food safety tests, such as pesticide residues, food additives, heavy metals, and formaldehyde concentrations. These sensors are tiny, efficient, rapid, portable, and cost-effective, need few samples, and are simple to integrate. They may be used to check the frying oil quality. The microfluidic devices produce W/O droplets, and due to degradation of oil and its adulteration, the droplet size varies with changes in oil viscosity and surface tension. Deng et al. (2018) employed the device in olive oil quality testing, and by comparing droplet size and adulteration rate/TPM, the device and method proved to be effective in evaluating frying oils, especially olive oil.

8.4.4 Near-Infrared Transmission Spectroscopy

Near-infrared (NIR) transmission spectroscopy may be applied for oil quality evaluation subjected to frying. In comparison to traditional analytical procedures, NIR transmission spectroscopy is a quick, non-destructive, cost-effective, and environmentally friendly technology. The NIR does not directly test oil quality, but rather analyses characteristics such as PV, iodine value, total polar chemicals, polymerized triacylglycerols, and acid value. Several authors discovered a strong association between NIRS analytical parameters and several traditional analytical parameters showed the benefits of using this method for detecting oil deterioration (Gertz and Behmer 2014). The NIR spectroscopy was effectively used for measuring the TPM and FFA in soybean frying oil.

The calibration models need to be developed through regression in each NIR spectroscopy. Further, these models are validated and then used for predicting the required parameter value. However, the development of calibration models is mostly done for a definite wavelength range in NIR spectroscopy (Wang et al. 2014).

8.4.5 Electronic Nose

The Electronic Nose (E-nose), also known as the Artificial Olfactive System (AOS), is a fast analytical system used to evaluate oil deterioration during thermoxidation (Upadhyay, Sehwag, and Mishra 2017). This device is made up of arrays of various chemical sensors that detect volatile chemicals and odors in the headspace above the oil sample. Pattern recognition algorithms related to statistical approaches and artificial neutral networks are effective tools for evaluating sensor array responses. Conducting polymers, quartz microbalance sensors, metal oxide semiconductors, and bulk acoustic waves are the most often used commercial sensors (Innawong, Mallikarjunan, and Marcy 2004).

Several research have already been conducted on E-nose for oil quality determination. For measuring the quality of virgin olive oil like rancidity, Aparicio et al. (2000) developed an E-nose system which is related to conducting polymers. When compared to the findings of a trained panel test, they discovered that the E-nose was very close to sensory results. Metal oxide-based sensors were utilized in the systems of Yang, Han, and Noh (2000) and Muhla et al. (2000). To detect rancidity in soybean oil, a portable device which is commercially available was employed by Yang, Han, and Noh (2000) containing six metal oxide sensors. Muhla et al. (2000) investigated the degradation of frying fats with their system containing three gas sensors.

The E-nose has various advantages, including the fact that it does not require sample pre-treatment and provides immediate results. The study of Savarese et al. (2007) revealed the capacity of E-nose to discern among oils with varying levels of deterioration caused by longer frying operation. They discovered E-nose to be quite efficient and has the potential to be used as an alternative to conventional analytical approaches for oil quality monitoring during deep frying.

The E-nose was recently used to do a non-invasive online evaluation of the quality of frying oils during deep frying of products. The researchers discovered that analyzing frying fumes is a legitimate method for determining the oil deterioration levels where E-noses work effectively. It was feasible for checking the oil quality online using this technology. Further, using the same device, the linear regression model developed from support vector machine was used to accurately predict the PV of the oil. The evaluation of the parameters was done within 3 minutes which is near to real-time evaluation of oil quality. This made this technology useful to be applied on deep frying in the food industry (Wasik, Majchrzak, and Wojnowski 2021).

8.4.6 Ultrasonic Technique

Ultrasonic technology is increasingly used in the food industry. This method is quick, non-invasive, and inexpensive. Diagnostic ultrasonography and power ultrasound are the two broad types of ultrasound. The oil quality is usually measured by diagnostic ultrasound by analyzing the changes in the attenuation coefficient, signal velocity, and other factors that vary depending on the oil quality. There are two types of diagnostic ultrasounds, viz. pulse-echo and pulser-receiver, used for testing the oil quality as shown in Figure 8.2. The working of these instruments includes the transmission of signal from the oil and detection of change in transmitter signal attributes which is related to the oil quality. The velocity of an ultrasonic signal is highly sensitive to molecular structure and intermolecular interactions, making it appropriate for identifying the composition, structure, and physical condition of oil, as well as detecting a foreign substance (Zarezadeh, Aboonajmi, and Ghasemi Varnamkhasti 2021).

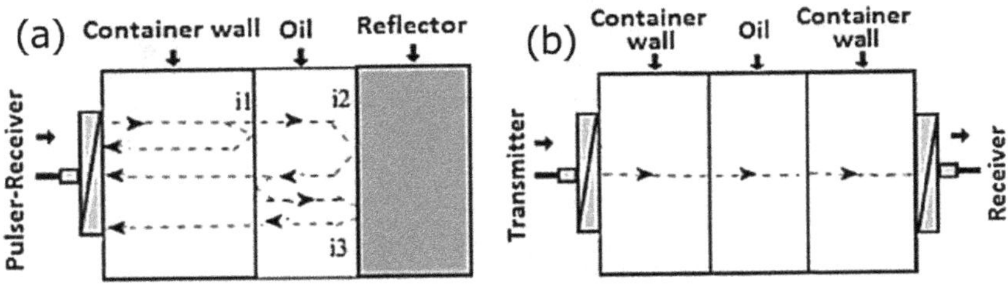

Figure 8.2 Diagnostic ultrasound system in oil quality testing (a) pulser-receiver and (b) pulse-echo. Adapted from Zarezadeh, Aboonajmi, and Ghasemi-Varnamkhasti 2022.

So, oils can be evaluated using ultrasonic velocity. This metric is strongly related to chemical changes such as FFA levels and TPC. So, this parameter can be effectively used for analyzing the oil deterioration levels and for monitoring changes in oil during deep frying. Furthermore, the real-time evaluation of oil during frying can be designed using ultrasonic methods in the food industries (Izbaim et al. 2010).

8.4.7 Smoke Point

During frying operation, the fat is converted into glycerol and fatty acids with a series of breakdown processes. The smoke is generated due to acrolein which is a breakdown product of glycerol. The increase in the amount of smoke is proportional to the temperature and low-molecular-weight decomposed product content in the oil. The smoke can also be formed at lower temperatures due to accumulation of breakdown products during repetitive use of oils. A smoke point of 170°C has been established as the legal limit for oil disposal in countries such as Austria and Germany (Bansal et al. 2010).

8.4.8 Foaming

Foaming is another physical feature that may be used to assess oil quality after repeated application. Compounds having a transitional hydrophilic–lipophilic stability are recognized to be accountable for oil foaming which improves the flexibility of oil/air contact. Foaming may be used to assess oil degradation such as thermal denaturation of fatty acids, polar chemicals, and oxidative polymerization. However, deterioration of oil due to prolonged frying affects the strength of foaming. It was established that the more degraded the oil, the greater the foam height (Hosseini et al. 2016; Li et al. 2020). Several studies have examined the effect of triglyceride content on foam formation and have developed techniques to evaluate foam under actual frying conditions (Bansal et al. 2010)

8.5 Chemical Methods of Measuring Quality of Used Fried Oils

The chemical methods are a more reliable way of measuring the oil quality. Free fatty acids, iodine value, saponification value, anisidine value, total polar compounds, and polymerized triglycerides are the chemical parameters informed about frying oil quality (Table 8.2).

8.5.1 Free Fatty Acids

The FFA content is regarded as an important component in determining the quality and cost of frying oils. At prolonged and high-temperature frying operations, FFA are produced as a hydrolysis product due to oxidation of oil and fat. Initially, standard procedures for determining FFA were proposed by the Association of Official Analytical Chemists (AOAC), American Oil Chemists' Society (AOCS), and the European Commission Regulation (EC). These procedures are similar in that they are based on oil titration. Here, the 50–100 mL of hot-neutralized 95% ethanol is taken to dissolve the oil or fat sample and later titrate against a strong base with addition of a small of amount of phenolphthalein indicator. Later, various procedures (Figure 8.3) are also introduced by many researchers for saving time and chemicals, and increasing accuracy and environment friendliness (Mahesar et al. 2014).

Vicentini-Polette et al. (2021) updated AOAC technique and utilized it analytically to determine solely the FFA in crude oils (Figure 8.4).

8.5.2 Iodine Value

The unsaturation in the oil or fats can be calculated using the IV value. The breakdown of double bonds by oxidation, scission, and polymerization is responsible for the decrease in IV. The heat treatment during frying generates oxidative rancidity in the oil, which can be measured using the IV. The breaking of double bonds causes drop in the IV value which is a sign of oxidation of the oil (Alireza et al. 2010). Only oils that do not include conjugated double bonds have an IV (Paul, Mittal, and Chinnan 1997).

Table 8.2 Chemical Parameters of Used Fried Oil with Its Method of Determination and Relation with Quality of Oil

Type of Parameter	*Relation with Oil Quality*	*Method Name*	*Major Chemical or Reagent Used*	*Deterioration Status of Oil*	*References*
Free fatty acids	Rancidity	AOCS, AOAC, EC	Ethanol, phenolphthalein	Increasing free fatty acid	Mahesar et al. (2014)
Iodine value	Unsaturation level	Hübl method, Wijs/Hanuš method, Kaufmann method, and Rosenmund–Kuhnhenn method	Carbon tetrachloride, potassium iodide, sodium thiosulfate solution	Decreasing iodine value	Chebet, Kinyanjui, and Cheplogoi (2016)
Saponification value	Ester linkages, average molecular weight of all fatty acids present	ISO method, AOCS	Potassium oxide, sulfuric or hydrochloric acid	Increasing saponification value	Wypych (2017)
Anisidine value	Carbonyl content	AOCS	Isooctane, glacial acetic acid	Increasing anisidine value	Moigradean, Poiana, and Gogoasa (2012)
Total polar compounds	Polar material	AOCS	-	Increasing total polar compounds value	Cascant, Garrigues, and de la Guardia (2017)
Polymerized triglycerides	Polymeric compounds	High-performance size-exclusion chromatography	-	Increasing polymerized triglycerides value	Khor et al. (2019)

AOCS, American Oil Chemists' Society; AOAC, Association of Official Analytical Chemists; EC, European Commission Regulations; ISO, International Organization for Standardization.

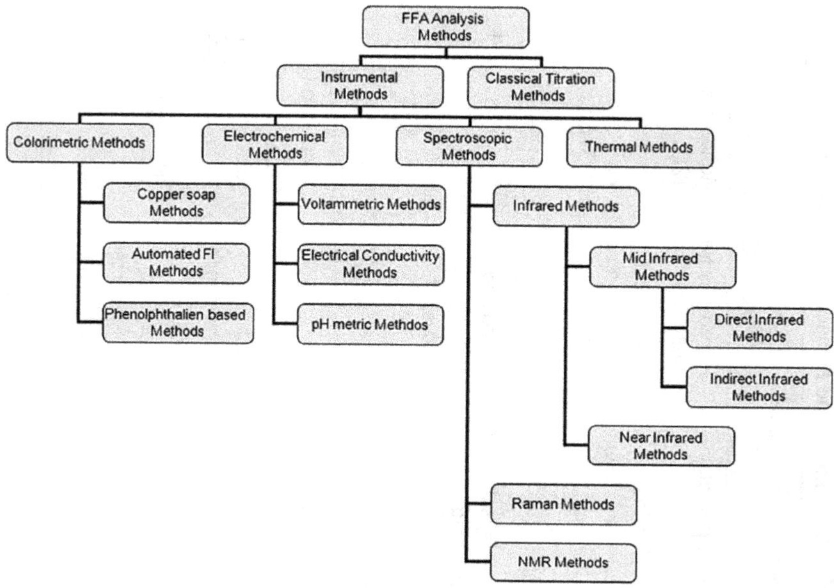

Figure 8.3 Various methods developed for the determination of FFA content. Adapted from Mahesar et al. 2014.

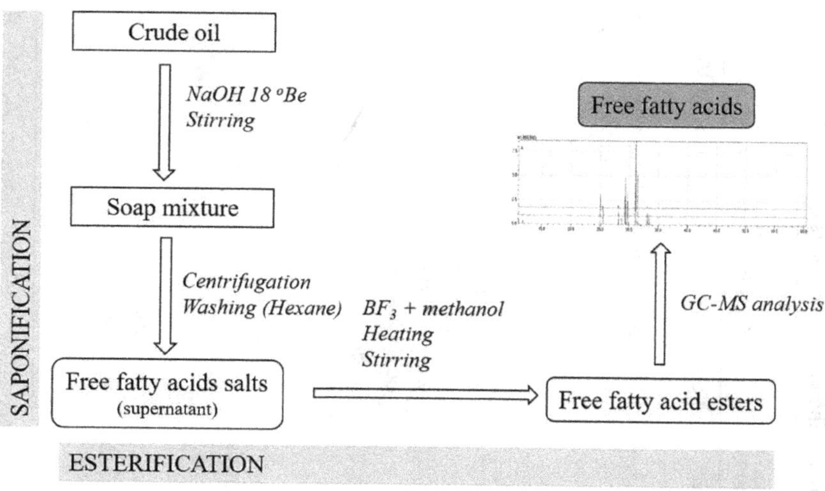

Figure 8.4 Schematic flowchart depicting the method for determining the free fatty acid. Adapted from Vicentini-Polette et al. 2021.

The IV determination techniques include the Hübl method, the Wijs/Hanu method, the Kaufmann method, and the Rosenmund–Kuhnhenn method. Chebet, Kinyanjui, and Cheplogoi (2016) described the Wijs technique, which is frequently used for determining IV. Approximately 5 g of oils in flask are dissolved in 20 mL of carbon tetrachloride, then adding approximately 25 mL of the Wijs reagent and heating in a fume chamber. The flask's contents are vigorously stirred before being placed in the dark for 1 hour. At the completion of this time, 20 mL of 10% aqueous potassium iodide is added in the mixture along with 100 mL of water. Excess iodine is titrated using a solution of 0.1 M sodium thiosulfate. Similar steps can be followed for blank test, and 1% starch is utilized as indicator. The IV is expressed as follows:

$$\text{Iodine Value}\left(\text{IV}\right)=\frac{12.69C\left(V_1-V_2\right)}{M} \tag{8.1}$$

where C is the concentration of sodium thiosulfate, V_1 is the volume of sodium thiosulfate used for blank, V_2 is the volume of sodium thiosulfate used for determination, and M is the mass of sample.

8.5.3 Saponification Value

Saponification value is a measure of the ester linkage content. The back titration technique using potassium oxide and 0.5 N sulfuric or hydrochloric with phenolphthalein indicator is used for the determination of saponification value. In this method, the sample is dissolved in 25 mL of alcoholic KOH solution and put in the steam bath for 1 hour. The amount of the sample must be decided so that almost half of the KOH is consumed in the process. The ISO standard defines the quantity of phthalate sample required for analysis in relation to its molecular weight. Later, the sample is again treated with 1 N alcoholic KOH (50 mL) and water (5 mL) before being refluxed for 1 hour. The sample is then titrated with 1 N HCl after the addition of phenolphthalein solution (Wypych 2017).

The saponification value, which indicates the presence of surfactants in the oil, rises as the oil deteriorates. Surfactants, on the other hand, make up a minor part of the overall decomposition products. As a result, the saponification value is slightly insufficient to describe the overall changes occurring in the degraded oil (Paul, Mittal, and Chinnan 1997).

8.5.4 Anisidine Index

The thermo-oxidative changes in the oil can be detected through the calculation of anisidine value (AnV) (Gertz and Behmer 2014). It can be determined using spectrophotometer through the standard AOCS method

(AOCS 1997). In this method, very little quantity of oil sample is diluted with isooctane and then used for absorption at 350nm as a blank using an ultraviolet/visible spectrophotometer. Later, a small amount of this solution is kept in a test tube, and the same amount of isooctane is taken in another test tube. A very small amount of p-anisidine solution is added into both tubes followed by measuring the absorbance of the sample test tube at 350nm after 10 minutes, and the second test tube was treated as reference. At last, the p-AnV is calculated through the standard formula (AOCS 1997; Esfarjani et al. 2019).

The AnV test quantifies secondary oxidation. The p-AnV detects high-molecular-weight saturated and unsaturated carbonyl compounds in triacylglycerols. Oils with a p-AnV less than 10 were deemed high grade; however another study defined high-quality oils as having an AnV less than 2 (Moigradean, Poiana, and Gogoasa 2012).

8.5.5 Total Polar Compounds

Oil degrades throughout the frying operation due to hydrolytic, oxidative, and thermal processes. Nonpolar triacylglycerols are converted into polar molecules. These compounds are known as TPM, whereas the nonpolar portion is made up of the remaining triacylglycerols and other low-polarity molecules. The ratio and amount of these two fractions are determined by the kind of frying oil used, storage conditions, and processing factors such as frying temperature, frying duration, and the item being fried. The TPM in frying oil and fried foods may be hazardous to human health. As a result, the oil degradation is effectively determined by measuring TPM content in oil. The repeated use of oil can generated polar oxidized, mono- di-, and polymerized degradation products, whose quantity is generally in relation with the TPM content. As a result, several nations have imposed a TPM limit in frying oil ranging from 24% to 30% as already discussed (Zainal and Isengard 2010; Gertz and Behmer 2014; Cascant, Garrigues, and de la Guardia 2017; Lioumbas, Zamanis, and Karapantsios 2013). TPM is typically determined using silica gel column chromatography and gravimetric analysis, in accordance with IUPAC- and AOAC-approved standard methods (AOCS 1997).

8.5.6 Polymerized Triglycerides

The heat treatment of oil resulted in various chemical reactions like oxidation, hydrolysis, and polymerization which leads to the formation of TPC with a higher molecular weight and higher polarity than regular untreated triacylglycerols. These polar parts can be separated from the oil, and the polymeric triglyceride (PTG) distribution is regarded as effective reliable indicator for measuring oil degradation level. It should be noted that oil samples with similar TPC concentrations can have different PTG distributions. Maximum amounts of 24% TPM and 12% PTG are recommended for oils (Petersen et al. 2013).

8.6 Application of Machine Learning in the Frying Oil Quality Determination

Multivariate data analysis and pattern recognition algorithms are effective operations that help for quick determination of oil quality. The primary goal of pattern recognition methods is to use the most appropriate data treatment to describe and characterize a group of objects or samples that display a specific characteristic or behavior (Jiménez-Carvelo et al. 2019).

Analytical techniques such as olfaction machines, colorimetry, spectroscopy, ultrasonic technology, viscosity measurement, and DC measurement can be used to assess oil quality. In general, oil quality results from these physical methodologies must be in accordance with results of more reliable chemical parameters like TPC, p-AnV, IV, and so on. As a result, after selecting the most significantly influencing factors, the development of a multivariable regression model with high prediction accuracy becomes essential. Many researchers performed this type of work effectively using classification and machine learning-based regression models (Zarezadeh, Aboonajmi, and Ghasemi Varnamkhasti 2021; Chen et al. 2015; Cascant, Garrigues, and de la Guardia 2017; Upadhyay, Sehwag, and Mishra 2017).

Artificial neural network, k-nearest neighbors (kNN), support vector machine (SVM), gradient boosting classifier, blockchain, etc. are the different tools that can be used for analyzing and classifying the data (Zarezadeh, Aboonajmi, and Ghasemi Varnamkhasti 2021). However, the principal component analysis (PCA) is the most basic step before proceeding to regression and classification models. Here, the data are converted from a high-dimensional space to a low-dimensional space. Because analysis in reduced dimension space is easier and quick, PCA is a kind of feature selection approach utilized in dimension reduction applications.

An online evaluation of oil quality during deep frying through an E-nose was developed using hierarchical cluster analysis (HCA) for grouping the variables and SVM algorithm for developing regression models. Wasik, Majchrzak, and Wojnowski (2021) found the possibility of predicting the PV of oil using SVM-based linear regression model with high degree of accuracy using the output data of E-nose. Zarezadeh, Aboonajmi, and Ghasemi Varnamkhasti (2021) converted four features into three features with negligible change in classification accuracy using PCA. The most accurate classification algorithms were the Naive Bayes, SVM approaches, and gradient boosting classifier. They showed that by extracting useful ultrasound features and training the machine with this data, the minute deterioration in quality of extra virgin olive oil can be identified with great accuracy.

The PCA is used to calibrate spectral intensity data at many frequencies to a reasonably small number of intensities for determining polar components in frying oils using Fourier-transform near-infrared (FTNIR) spectroscopy. As a result, it has produced optimal principal components while filtering out noise. The partial least square calibration is used for making a correlation between reference value and the FTNIR-predicted value of polar compounds content with the least error (Chen et al. 2015). Furthermore, many commonly used pattern recognition tools, such as soft independent modeling by class analogy, partial least squares-discriminant analysis, kNN, multivariate curve resolution-alternating least squares, and parallel factor analysis, can be used to classify the analytically measured oil quality data. Furthermore, utilizing SVM, classification and regression tree, and random forest, this data may be modeled with the reference data (TPC, AnV, FFA, etc.) (Jiménez-Carvelo et al. 2019).

The advance classification algorithms and machine learning (ML) tools have now decreased the burden of finding the most affecting measured data (analytically) and modeling it for predicting oil quality. These approaches have resulted in the development of online oil quality assessment instruments for the rapid identification of any harmful degradation of oil (Wasik, Majchrzak, and Wojnowski 2021).

8.7 Future Scope

The measurement of TPM among several chemical indicators is regarded as the most widely used approach for evaluating oil quality since it determines the total chemical deterioration occurring in the oil (Khaled, Aziz, and Rokhani 2015). However, it is measured using column chromatography, which requires skilled operators, numerous chemicals, and are time consuming, and costly (Lioumbas, Zamanis, and Karapantsios 2013). As a result, methodologies for measuring other parameter but related to TPM are in high demand, with a focus on rapid and low-cost testing of oil quality. The analytical technique can play a key role satisfying all these requirements (Deng et al. 2018). Many researchers have developed NIRS, E-nose, pore-based wicking sensor, dielectric constant sensors, ultrasonic based technique, etc. for analysing the oil quality (Upadhyay, Sehwag, and Mishra 2017; Chen et al. 2015; Kalogianni and Karastogiannidou 2015)

However, the majority of analytically developed methods are laboratory-based and must be translated for real-time oil quality measurement during frying. The simultaneous development of a framework of devices, sensors, classification, and regression algorithms can aid in the design of an online oil quality testing system (Wasik, Majchrzak, and Wojnowski 2021). Furthermore, in response to customer demand, a home-scale oil quality measurement system based on physical factors such as viscosity and color should be produced and linked to smartphones via the internet of things. These gadgets can assist in deciding when to discard frying oil and save money.

8.8 Conclusion

The measurement of oil quality during repeated usage of oil during frying is important for the safe and healthy consumption of fried meals. Many chemical parameters, such as FFA, IV, SV, AnV, TPM, and PTG, are utilized to determine the present oil quality. TPM was recognized as a standard parameter, with a value of less than 24% required for the oil to be accepted for further use. However, these procedures require a long time, effort, a lot of chemicals and solvents, and are often expensive. As a result, other physical methods such as color, dielectric constant, viscosity, smoke point, NIR, E-nose and ultrasonic technology can be utilized to determine oil quality quickly. The first four characteristics cannot be properly utilized to determine oil quality since they are impacted by various external circumstances such as the starting color of the oil, the type of food cooked, the oil variety, and the duration of exposure to light and heat. As a result, NIR, E-nose-based, and ultrasonic techniques have found use in measuring oil quality. These approaches required first classification of data in order to identify significantly influencing parameters, followed by model creation utilizing regression methods. Classification methods such as PCA and HCA, and regression methods such as SVM, PLS, ANN, and kNN are now employed in the development of analytical tools. These developments may pave the way for the implementation of more ML-based methodologies with the objective of designing online quality evaluation of oils during frying.

References

Alireza, S., C. P. Tan, M. Hamed, and Y. B. Che Man. 2010. “Effect of Frying Process on Fatty Acid Composition and Iodine Value of Selected Vegetable Oils and Their Blends.” *International Food Research Journal* 17 (2): 295–302.

American Oil Chemists' Society. *Official methods and recommended practices of the AOCS*. American Oil Chemists' Society, 1997.

Aparicio, Ramon, Silvia M. Rocha, Ivonne Delgadillo, and Maria Teresa Morales. 2000. "Detection of Rancid Defect in Virgin Olive Oil by the Electronic." *Journal of Agricultural and Food Chemistry* 48 (3): 853–60. https://doi.org/10.1021/jf9814087.

Bansal, Geeta, Weibiao Zhou, Philip J. Barlow, Pranav S. Joshi, Hui Ling Lo, and Youne Kow Chung. 2010. "Review of Rapid Tests Available for Measuring the Quality Changes in Frying Oils and Comparison with Standard Methods." *Critical Reviews in Food Science and Nutrition* 50 (6): 503–14. https://doi.org/10.1080/10408390802544611.

Cascant, Mari Merce, Salvador Garrigues, and Miguel de la Guardia. 2017. "Comparison of Near and Mid Infrared Spectroscopy as Green Analytical Tools for the Determination of Total Polar Materials in Fried Oils." *Microchemical Journal* 135 (November): 55–59. https://doi.org/10.1016/j.microc.2017.07.012.

Chebet, Josphine, Thomas Kinyanjui, and Peter K. Cheplogoi. 2016. "Impact of Frying on Iodine Value of Vegetable Oils before and after Deep Frying in Different Types of Food in Kenya." *Journal of Scientific and Innovative Research* 5 (5): 193–96. https://doi.org/10.31254/jsir.2016.5508.

Chen, Xiumei, Xiuzhu Yu, Yage Wang, Yandie Yang, and Jingya Zhang. 2015. "Determination of Polar Components in Frying Oils by Fourier-Transform Near-Infrared Spectroscopy." *Journal of Oleo Science* 64 (3): 255–61. https://doi.org/10.5650/jos.ess14227.

Deng, Ning, Ning Cao, Piaopiao Li, Yu Peng, Xiaomao Li, Liang Liu, Huayan Pu, et al. 2018. "Microfluidic Evaluation of Some Edible Oil Quality Based on Viscosity and Interfacial Tensions." *International Journal of Food Science & Technology* 53 (4): 946–53. https://doi.org/10.1111/ijfs.13667.

Esfarjani, Fatemeh, Khadijeh Khoshtinat, Aziz Zargaraan, Fatemeh Mohammadi-Nasrabadi, Yeganeh Salmani, Zahra Saghafi, Hedayat Hosseini, and Manochehr Bahmaei. 2019. "Evaluating the Rancidity and Quality of Discarded Oils in Fast Food Restaurants." *Food Science & Nutrition* 7 (7): 2302–11. https://doi.org/10.1002/fsn3.1072.

Gertz, Christian. 2000. "Chemical and Physical Parameters as Quality Indicators of Used Frying Fats." *European Journal of Lipid Science and Technology* 102 (8–9): 566–72. https://doi.org/10.1002/1438–9312(200009)102:8/9<566::aid-ejlt566>3.3.co;2-2.

Gertz, Christian, and Dagmar Behmer. 2014. "Application of FT-NIR Spectroscopy in Assessment of Used Frying Fats and Oils*." *European Journal of Lipid Science and Technology* 116 (6): 756–62. https://doi.org/10.1002/ejlt.201300270.

Hosseini, Hamed, Mohammad Ghorbani, Nasim Meshginfar, and Alireza Sadeghi Mahoonak. 2016. "A Review on Frying: Procedure, Fat, Deterioration Progress and Health Hazards." *Journal of the American*

Oil Chemists' Society 93 (4): 445–66. https://doi.org/10.1007/s11746-016-2791-z.

Innawong, Bhundit, Parameswarakumar Mallikarjunan, and Joseph E. Marcy. 2004. "The Determination of Frying Oil Quality Using a Chemosensory System." *Lebensmittel-Wissenschaft & Technologie* 37 (1): 35–41. https://doi.org/10.1016/S0023-6438(03)00122-1.

Izbaim, D., B. Faiz, A. Moudden, N. Taifi, and I. Aboudaoud. 2010. "Evaluation of the Performance of Frying Oils Using an Ultrasonic Technique." *Grasas y Aceites* 61 (2): 151–56. https://doi.org/10.3989/gya.087709.

Jiménez-Carvelo, Ana M., Antonio González-Casado, M. Gracia Bagur-González, and Luis Cuadros-Rodríguez. 2019. "Alternative Data Mining/Machine Learning Methods for the Analytical Evaluation of Food Quality and Authenticity – A Review." *Food Research International* 122 (August): 25–39. https://doi.org/10.1016/j.foodres.2019.03.063.

Kalogianni, Eleni P., and Calliope Karastogiannidou. 2015. "Development of a Rapid Method for the Determination of Frying Oil Quality Based on Capillary Penetration." *International Journal of Food Science and Technology* 50 (5): 1215–23. https://doi.org/10.1111/ijfs.12752.

Khaled, Alfadhl Yahya, Samsuzana Abd Aziz, and Fakhrul Zaman Rokhani. 2015. "Capacitive Sensor Probe to Assess Frying Oil Degradation." *Information Processing in Agriculture* 2 (2): 142–48. https://doi.org/10.1016/j.inpa.2015.07.002.

Khor, Yih Phing, Khai Shin Hew, Faridah Abas, Oi Ming Lai, Ling Zhi Cheong, Imededdine Arbi Nehdi, Hassen Mohamed Sbihi, Mohamed Mossad Gewik, and Chin Ping Tan. 2019. "Oxidation and Polymerization of Triacylglycerols: In-Depth Investigations towards the Impact of Heating Profiles." *Foods* 8 (10): 475. https://doi.org/10.3390/foods8100475.

Kress-Rogers, Erika, P. N. Gillatt, and J. B. Rossell. 1990. "Development and Evaluation of a Novel Sensor for the In Situ Assessment of Frying Oil Quality." *Food Control* 1 (3): 163–78. https://doi.org/10.1016/0956–7135(90)90008–Z.

Li, Xu, Xinyi Cheng, Gangcheng Wu, Jianhua Huang, Hui Zhang, Qingzhe Jin, and Xingguo Wang. 2020. "Individual and Combined Effects of Frying Load and Deteriorated Polar Compounds on the Foaming of Edible Oil." *Food Research International* 134 (August): 109206. https://doi.org/10.1016/j.foodres.2020.109206.

Lioumbas, John S., Angelos Zamanis, and Thodoris D. Karapantsios. 2013. "Towards a Wicking Rapid Test for Rejection Assessment of Reused Fried Oils: Results and Analysis for Extra Virgin Olive Oil." *Journal of Food Engineering* 119 (2): 260–70. https://doi.org/10.1016/j.jfoodeng.2013.05.037.

Mahesar, S. A., S. T. H. Sherazi, Abdul Rauf Khaskheli, Aftab A. Kandhro, and Siraj Uddin. 2014. "Analytical Approaches for the Assessment of Free Fatty Acids in Oils and Fats." *Anal. Methods* 6 (14): 4956–63. https://doi.org/10.1039/C4AY00344F.

Moigradean, Dian, Mariana Atena Poiana, and Ioan Gogoasa. 2012. "Quality Characteristics and Oxidative Stability of Coconut Oil during Storage." *Journal of Agroalimentary Processes and Technologies* 18 (4): 272–76.

Muhla, Mike, Hans Ullrich Demischa, Frank Beckerb, and Claus Dieter Koh. 2000. "Electronic Nose for Detecting the Deterioration of Frying Fat – Comparative Studies for a New Quick Test" 102 (8–9): 581–85. https://doi.org/10.1002/1438-9312(200009)102:8/9<581::AID-EJLT581>3.0.CO;2-N.

Paul, S., G. S. Mittal, and M. S. Chinnan. 1997. "Regulating the Use of Degraded Oil/Fat in Deep-fat/Oil Food Frying." *Critical Reviews in Food Science and Nutrition* 37 (7): 635–62. https://doi.org/10.1080/10408399709527793.

Petersen, Katharina Domitila, Gerhard Jahreis, Mechthild Busch-Stockfisch, and Jan Fritsche. 2013. "Chemical and Sensory Assessment of Deep-Frying Oil Alternatives for the Processing of French Fries." European *Journal of Lipid Science and Technology* 115 (8): 935–45. https://doi.org/10.1002/ejlt.201200375.

Rubalya Valantina, S., D. R. Phebee Angeline, S. Uma, and B. G. Jeya Prakash. 2017. "Estimation of Dielectric Constant of Oil Solution in the Quality Analysis of Heated Vegetable Oil." *Journal of Molecular Liquids* 238 (July): 136–44. https://doi.org/10.1016/j.molliq.2017.04.107.

Savarese, Maria, C. Parisini, E. De Marco, I. Battimo, S. Falco, and R. Sacchi. 2007. "Application of Electronic Nose to Monitor the Frying Process. A Preliminary Study." *Rivista Italiana Delle Sostanze Grasse* 84 (1): 33–39.

Smith, L. M., A. J. Clifford, C. L. Hamblin, and R. K. Creveling. 1986. "Changes in Physical and Chemical Properties of Shortenings Used for Commercial Deep-Fat Frying." *Journal of the American Oil Chemists' Society* 63 (8): 1017–23. https://doi.org/10.1007/BF02673790.

Upadhyay, Rohit, Sneha Sehwag, and Hari Niwas Mishra. 2017. "Electronic Nose Guided Determination of Frying Disposal Time of Sunflower Oil Using Fuzzy Logic Analysis." *Food Chemistry* 221 (April): 379–85. https://doi.org/10.1016/j.foodchem.2016.10.089.

Vicentini-Polette, Carolina Medeiros, Paulo Rodolfo Ramos, Cintia Bernardo Gonçalves, and Alessandra Lopes De Oliveira. 2021. "Determination of Free Fatty Acids in Crude Vegetable Oil Samples Obtained by High-Pressure Processes." *Food Chemistry: X* 12: 100166. https://doi.org/10.1016/j.fochx.2021.100166.

Wang, Yage, Xiuzhu Yu, Xiumei Chen, Yandie Yang, and Jingya Zhang. 2014. "Application of Fourier Transform Near-Infrared Spectroscopy to the Quantification and Monitoring of Carbonyl Value in Frying Oils." *Anal. Methods* 6 (19): 7628–33. https://doi.org/10.1039/C4AY00703D.

Wasik, Andrzej, Tomasz Majchrzak, and Wojciech Wojnowski. 2021. "On-Line Assessment of Oil Quality during Deep Frying Using an Electronic Nose and Proton Transfer Reaction Mass Spectrometry." *Food Control* 121: 107659. https://doi.org/10.1016/j.foodcont.2020.107659.

Wypych, George. 2017. "Typical Methods of Quality Control of Plasticizers." In *Handbook of Plasticizers*, edited by George Wypych, Third Edition, 85–109. ChemTec Publishing. https://doi.org/10.1016/B978-1-895198-97-3.50005-6.

Xu, Lirong, Xiuzhu Yu, Lei Liu, and Rui Zhang. 2016. "A Novel Method for Qualitative Analysis of Edible Oil Oxidation Using an Electronic Nose." *Food Chemistry* 202: 229–35. https://doi.org/10.1016/j.foodchem.2016.01.144.

Xu, Xin Qing. 2003. "A Chromametric Method for the Rapid Assessment of Deep Frying Oil Quality." *Journal of the Science of Food and Agriculture* 83 (13): 1293–96. https://doi.org/10.1002/jsfa.1535.

Yang, Y. M., K. Y. Han, and B. S. Noh. 2000. "Analysis of Lipid Oxidation of Soybean Oil Using the Portable Electronic Nose." *Food Science and Biotechnology* 9 (3): 146–50.

Zainal, and Heinz Dieter Isengard. 2010. "Determination of Total Polar Material in Frying Oil Using Accelerated Solvent Extraction." *Lipid Technology* 22 (6): 134–36. https://doi.org/10.1002/lite.201000019.

Zarezadeh, Mohammad Reza, Mohammad Aboonajmi, and Mahdi Ghasemi Varnamkhasti. 2021. "Fraud Detection and Quality Assessment of Olive Oil Using Ultrasound." *Food Science & Nutrition* 9 (1): 180–89. https://doi.org/10.1002/fsn3.1980.

Zarezadeh, Mohammad Reza, Mohammad Aboonajmi, and Mahdi Ghasemi-Varnamkhasti. 2022. "Applications of Ultrasound Techniques in Tandem with Non-Destructive Approaches for the Quality Evaluation of Edible Oils." *Journal of Food Science and Technology* 59 (8): 2940–50. https://doi.org/10.1007/s13197-022-05351-1.

Chapter 9

Promising Approaches to Reduce Oil Uptake of Deep Fat-Fried Foods

Chander Mohan and Subhadip Manik
ICAR-National Dairy Research Institute

Sheetal Thakur
Chandigarh University

Gurpreet Kaur
Mata Gujri College

Surbhi
Punjabi University

DOI: 10.1201/9781003329244-9

9.1 Introduction

All age groups of customers consume a significant amount of fried food products due to their distinct flavor profiles, palatability, affordability, accessibility, and availability. The process of frying, in which heat and mass are transferred in a lipid medium, is thought to be the simplest and oldest way to prepare food. During frying operation, there are numerous chemical and physical changes occur, including lipid oxidation, protein denaturation, moisture evaporation, starch gelatinization, and textural modification. However, one major problem with fried food is their high oil content. High intake of fried food is therefore concerning because of its link to high rates of health problems like obesity, high cholesterol, and high blood pressure. Therefore, one of the key drivers of the most recent research trends in this field, which encourages studies of ways to reduce the oil content of fried foods, seems to be the development of healthier fried products that contain less oil and absorb less oil during frying operations.

According to the available literature, three categories may be drawn from the different treatments used to lower the amount of oil that deep-fried foods absorb:

1. Product surface modification: use of edible coatings
2. Modifying the frying procedure: pre-frying, air frying, vacuum frying, and de-oiling after frying
3. Altering the frying medium

Because frying is mostly a surface phenomenon, altering the product's surface has been found to be the most effective method of reducing oil uptake.

9.2 Mechanism of Oil Absorption during Deep Fat Frying

Deep fat frying process is a complex process in which foods are cooked by immersing them in edible oil at a temperature above the boiling point of water. The process of deep fat frying involves heat and mass transfer, structural changes, and temperature gradient between the crust and core of the product. This phenomenon causes some quality changes such as oil absorption in fried foods. Higher frying temperature and time have been associated with lower oil absorption. On the contrary, recently there are some studies that found increased oil absorption as the frying temperature and time increase (Zhang et al., 2018). The type of pretreatment and structural variations in the samples are potential causes of the discrepancy.

The oil uptake phase is thought to be a complicated process involving a number of interconnected mechanisms that can be explained by the heat and mass transfer phenomenon, as shown in Figure 9.1. Heat and mass transfer happens simultaneously during the frying process, which causes the vaporization front to move and separate the humid core from the dehydrated crust (Farkas et al., 1996). Farkas et al. (1996) divided the frying process into four temporal stages based on the rate of heat and mass transfer: initial heating (surface reaches the boiling point of water), surface boiling (water vapor bubbles appear), falling rate (crust forms, and bubbling reduces), and bubble endpoint (bubbling is no longer observed). At the beginning of the frying process, heat is carried from the oil to the exterior of the fried food via convection, and subsequently heat is transferred by conduction from the exterior to the interior of the product. As a result, the moisture spreads from the center to the exterior and evaporates, allowing the oil to enter inside the food (Yagua & Moreira, 2011)

Moreira et al. (1997) demonstrated a physical connection between oil uptake and capillary pores, and described the oil uptake mechanism via capillary forces. At a temperature over the boiling point of water, dehydration starts to occur. The moisture vapor causes a selective weakening of cell adhesion, which helps generate capillary channels, sometimes referred to as capillary holes. The size and tortuosity of the fried food's pores have an impact on the oil absorption process, and the pores are mostly governed

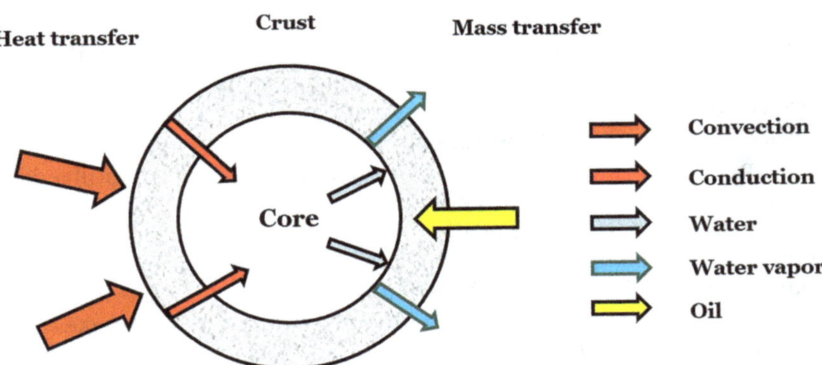

Figure 9.1 Diagrammatic representation of the simultaneous heat and mass transfers that occur during deep fat frying.

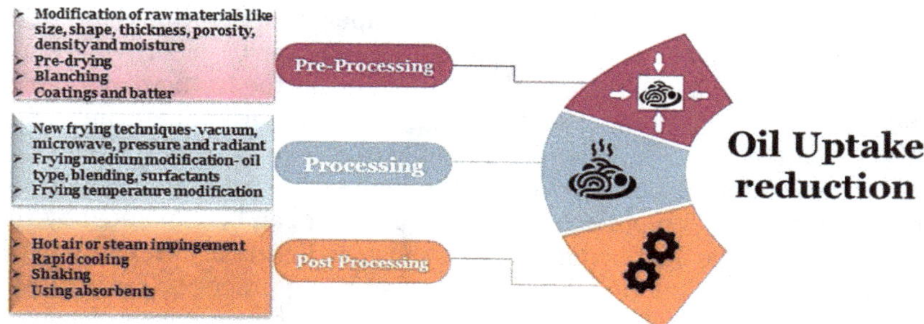

Figure 9.2 Strategies for oil uptake reduction in fried food products.

by surface-level breaks or collapses (Ziaiifar et al., 2008). Due to increased internal vapor pressure, there is a tendency of surface rupture as the product comes into contact with heated oil. When the pressure drops as a result of vapor escape, the oil enters the void field. Alternately, once water evaporation is complete, collapse could happen while the product is cooling (Ziaiifar et al., 2010). Figure 9.2 illustrates various strategies for oil uptake reduction. These are frying parameters, including pretreatment, frying procedures, post-frying techniques, frying product composition, and oil quality, that affect oil uptake during the frying process.

9.3 Modification of Food Surface

As mentioned earlier, the surface phenomena govern both moisture loss and oil uptake; therefore, altering the surface would substantially aid in optimizing the frying operations by reducing the oil uptake (Huse et al., 1998). Intake of oil during deep fat frying of coated products is significantly influenced by the microstructure and surface irregularities of batter coatings (Dueik et al., 2012). There is a strong correlation between the food surface roughness and oil absorption.

In the previous decade, edible coatings have been utilized to minimize oil uptake in deep fat-fried food products with effectiveness. Identifying the perfect formulation design, film-forming polymer characteristics, and component combinations to enhance adhesion and active qualities required more extensive studies. Due to its good thermogelling properties, hydrocolloid coatings can decrease excessive oil uptake while remaining invisible and having no negative effects on the sensory qualities of fried foods. Additionally, coating fried foods can enhance their nutritional qualities, increase crispiness, and improve palatability (Zeng et al., 2016).

9.3.1 Edible Films and Coatings

In the late 1960s, it was suggested to use edible films and coatings, initially to extend the shelf life of meat and later to improve the quality of fresh and frozen foods. A thin layer with a film thickness typically not exceeding 0.3 mm is referred to as an edible film. Three macromolecules, namely, polysaccharides, proteins, and lipids, as well as combinations of these, are used to produce edible coatings. It is applied to the surface of food as a substitute for natural protective coverings (Huber & Embuscado, 2009). The best indication of its safety is that it is edible. One aspect that must be taken into account is that it must be acceptable to customers while keeping the original organoleptic characteristics and not be detectable to the tongue of the consumer. Edible coatings build barriers to prevent the transport of solutes, oxygen, and moisture (Guilbert et al., 1996). Edible coatings are used in fried foods primarily for two reasons: (i) to create a thin-film surface coating that could be helpful in reducing oil uptake during and after frying operation, and (ii) the second reason is related to its technological functions, which include improving quality attributes like adherence, viscosity, and crispiness (Ajo, 2017; Norizzah et al., 2016). The edible coatings and films can prevent water in the food from evaporating during the first frying period and the food medium from absorbing oil. The most essential feature of an edible coating is that it can obstruct oil absorption and the evaporation of water in fried foods. The ability of an edible coating to prevent fried food from absorbing oil and from having water evaporate off of it is its most crucial quality. The mechanism of oil absorption with edible coatings is connected to the movement of oil in the voids and heat transfer. Thermogelling, crosslinking, and altered hydrophobicity are potential explanations for the phenomenon. In the first stage of the frying process, water escapes from the crust, and in the second stage, water from the product's core migrates to the crust. As a result, a weak crust forms with unfilled pores that are available to be filled with migratory fat from the frying medium. This, unfortunately, offers flavors and fatty nutrients. The amount of oils that are absorbed into food when it is being fried depends on the original moisture level of the food's surface, as well as its size and form. It is widely accepted that both water condensation and evaporation can absorb oils (Mellema, 2003). Accordingly, pre-drying food is a frequent method for lowering the intake of fat (Krokida et al., 2001). It is possible to generate a stronger surface with fewer voids during the frying process, which keeps water and steam from escaping the porous surface. As a result, the sample retains an acceptable amount of moisture, and the surface permeability is decreased, preventing oil from replacing the water (Funami et al., 1999). Contrarily, in the case of non-coated products, water in the crust evaporates through the pores during the frying process and is replaced by the oil, resulting in an oil-concentrated crust. According to Daraei et al. (2014), deep frying does not completely absorb the oil from the surface of the fried product. When samples are taken out of the frying medium, the residual oil may get into the product. Due to bigger magnitudes of positive gauge pressure that are created inside the food for a longer period of time in uncoated food, during the first 60 seconds of frying, quick and intense evaporation of water from the product occurs (Lalam, Sandhu, Takhar, Thompson, & Alvarado, 2013). Additionally, coatings can shield food surfaces from oil absorption by altering the hydrophobicity of the surface (Annapure, Singhal, & Kulkarni, 1999). Punctures may form on a coated surface if there is insufficient binding between the polymer and the food surface. Due to strong capillary forces, this may result in a greater uptake of oil (Moreira, Sun, & Chen, 1997). Less water will evaporate and less oil will be absorbed if the edible coating is compact. The fact that hydrocolloid is in the solid state at or below the critical gelling temperature is the primary property of thermal gelation. This indicates that the migration of oil into and water out of the product is prevented when the item is coated and submerged in hot frying oil. This is a result of the development of a reversible thermal gel that traps water.

9.3.1.1 Application Methods

There are some technical difficulties when coating food before frying. In the initial stage of preparation, film-making elements (polymer, plasticizers, emulsifications, and functional compounds) are dissolved at all times. The film-forming solution can then be heated or not, crosslinked, or in the case of proteins, and the pH can be adjusted depending on the primary polymer properties. Plasticizers are typically applied to the polymer after it has fully dissolved. By lowering the number of active centers available for stiff polymer–polymer interactions or the polymer's macromolecular mobility, these low-molecular-weight components enhance the free volume between the polymer chains. If the plasticizer and polymer are sufficiently compatible in terms of polarity, hydrogen bonding, dielectric constant, and solubility properties, successful plasticization will result (Vieira et al., 2011). Most commonly used plasticizers for edible coating are glycol, sorbitol, and glycerol. Each plasticizer has the potential to alter the final performance of the final coating; thus, it is important to understand its mechanism of action. According to Tavera and coworkers (2012), sorbitol and methylcellulose work better together to reduce fat in potato chips than glycerol. The two main methods for coating fried foods are dipping and spraying. The choice of the method is influenced by thickness, texture, smoothness, and applicability of the coating. Dipping involves immersing samples for a specific period of time in a coating solution. Food samples are fried after the extra liquid has been drained off of them. In the spraying method, there are multiple factors that have impacts on coating such as apparent viscosity of the film-forming solutions, the consistency coefficient, and flow behavior index. The spraying method involves the following steps: first, food samples are placed on rotating platform, the spraying is done using a commercial gun, and then it is dried and fried. Hydrophilic materials are preferred over hydrophobic materials because the later have more chances to get rupture during frying due to evaporation of water. Coatings that exhibit high viscosity at low concentration are usually carbohydrate based and are much easily applicable than those with low viscosities, where high concentration is required such as proteins. In the case of protein-based coatings, thickeners such as gums can be added to increase the viscosity. There are a range of hydrocolloids that have been found to be effective in reducing oil uptake. Most commonly used hydrocolloids are discussed in the following section.

9.3.1.1.1 Methylcellulose Methylcellulose (MC) exhibits some desirable properties like thermogelling and thickening characteristics that have been relevant in reducing oil uptake in fried foods (Dueik et al., 2014). MC helps in retention of moisture in the food during frying. One of the reasons for the performance of MC is attributed to thermogelation phenomenon which takes place at a temperature more than 60°C (Figure 9.3).

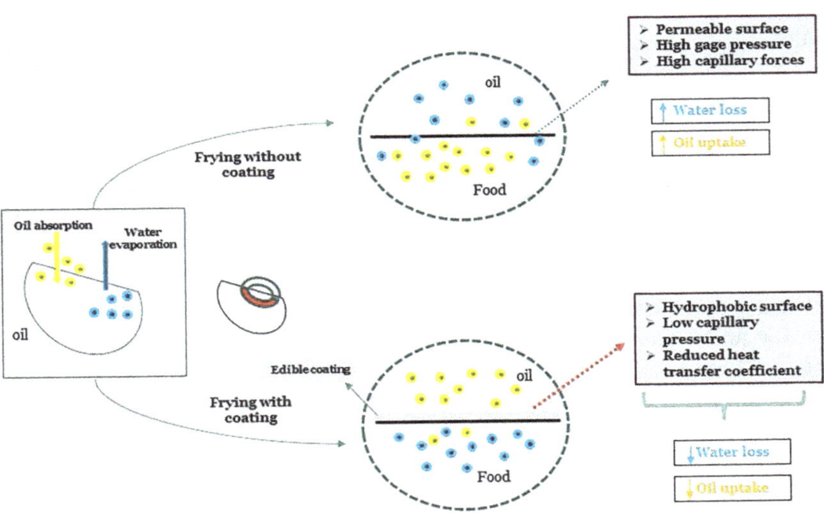

Figure 9.3 Mechanism of oil reduction in deep fat-fried foods with coatings (Kurek & Scetar, 2017).

According to Galus and Kadzińska (2015), the induced coating prevents moisture and oil from migrating between the core, crust, and frying oil. Similar to methylcellulose, hydroxypropyl methylcellulose is a cellulose ether that has the distinct capabilities of forming a structural gel when exposed to high temperatures and having its gelation properties be reversible when cooled (Suárez et al., 2008). When used as part of a batter system in a deep fat frying process, its gel formed during the first step of heating can act as a barrier against moisture loss during the final phase of frying when coated food items are exposed to high temperature (Aukkanit, Naowakul, & Jamsang, 2019). During the cooling process, this helps lower oil absorption into pores (Albert et al., 2009; Amboon, Tulyathan, & Tattiyakul, 2012; Klangmuang & Sothornvit, 2018). Because of this distinctive quality, HPMC is efficient at reducing the amount of oil absorbed during frying (Chen, Kuo, Peng, & Chen, 2009).

9.3.1.1.2 Carboxymethyl Cellulose Carboxymethyl cellulose (CMC) is one of the effective hydrocolloids reported to reduce oil uptake by deep fat-fried foods (Priya et al., 1996). However, the type of food and the nature of constituent formulation influence the efficiency of CMC in reducing oil uptake. CMC was found to be more effective than other hydrocolloids such as xanthan and guar gum in reducing oil uptake during the frying of potato chips. It was also found that efficiency of CMC enhanced when used synergistically with pectin. Due to its capacity to interact with calcium in the product and its ability to build thick, resistant structures, pectin was likely successful in the formulation that inhibits the absorption of oil.

9.3.1.1.3 Xanthan Gum The first natural biopolymer to be produced at an industrial scale was xanthan gum, a microbial polysaccharide that is manufactured in industries from carbon sources by fermentation by the gram-negative bacterium Xanthomonas campestris (Kim, Min, Lee, & Lee, 2012; Habibi & Khosravi-Darani, 2017). It is one of the most widely utilized hydrocolloids in both commercial and residential settings for batter and breading coatings to reduce the amount of oil absorbed by deep-fried foods (Alimi et al., 2014).

Recently, it was shown that combining xanthan gum with carrageenan reduced the quantity of oil absorbed in fried potato chip-based pellets (Ajo, 2017). Banana fritters, chicken nuggets, and other foods that are fried in deep fat have shown to absorb less oil when xanthan gum is added (Norizzah et al., 2016).

Table 9.1 Use of Various Hydrocolloids in Reducing Oil Uptake

Food Products	*Coatings*	*Concentration (%)*	*Plasticizer*	*Percentage Reduction in Oil Uptake*	*References*
Pastry mix	Hydroxypropyl cellulose (HPC)	1.5;2	No	51	Williams and Mittal (1999)
Cereal products	Microcrystalline cellulose (MCC)	0.5–5	Glycerol, PEG	14	Albert and Mittal (2002)
Potato balls	Hydroxypropyl methyl cellulose (HPMC)	2	Polyethylene glycol	61.4	Mallikarjunan et al. (1997)
Potato	Carboxymethyl cellulose	1	No	21.2	Garakhany et al. (2008)
Potato pallet chips	Corn starch	2	Glycerol	44.3	Angor et al. (2013)
Cereal products	Kappa-Carrageenana-konjac blend	2	Glycerin, PEG	54	Albert and Mittal (2002)
Potato chips	Okra	1	Glycerol	45	Archana et al. (2016)
Potato chips	Okra+Carrageenan	1	Glycerol	45	Archana et al. (2016)
Shrimps	Salep	1.5	No	29	Karimi and Kenari (2016)

9.4 Novel Frying Technologies

9.4.1 Vacuum Frying

Modern food-processing industries are always searching for novel technologies to manufacture healthy meals that are safe to eat. Furthermore, effectiveness in manufacturing is increasing and energy consumption is at a minimal range (Tao & Sun, 2015). Vacuum frying (VF) is considered as one of the patronizing frying techniques because of its ability to replace conventional frying methods, producing fried products of appropriate quality and adjusting to current health trends. The term "vacuum" refers to conditions where the pressure is less than atmospheric pressure. In this method, items are deep-fried in a confined environment at reduced air pressures—typically under 50 Torr or 6.65 kPa. Thus, vacuum condition lowers the boiling point of water and makes it possible for frying the items at lower temperature as compared with traditional frying (Su et al., 2016). Five unique steps make up the VF process: depressurization, frying, de-oiling, cooling, and pressure. Samples are inserted into the frying chamber's headspace during the depressurization procedure, and the pressure around them drops to a predetermined level. Different temperatures are used for the frying process at different periods of time. The items are fried in this process through the transfer of heat and mass. Without disturbing the vacuum, the de-oiling step is performed; once it is finished, a pressurization step takes place, returning the system to atmospheric air pressure. Then, the fried foodstuffs are removed from the frying chamber and cooled to normal temperature when the unit returns to air pressure (Yagua & Moreira, 2011).

The key characteristics of VF are minimal temperature range (90°C–120°C) and very less oxygen exposure, which decrease oil uptake and avoid the production of hazardous chemicals (Maity et al., 2014). Additionally, VF process with low frying pressure accelerates transport of air into porous structure and prevents the product from absorbing oil and thus results in less fat absorption than deep fat frying process (Sharanabasava, 2018). De-oiling is employed before pressurization to lessen surface oil before product is removed from the frying chamber, which is another unique advantage of VF. According to Yagua and Moreira (2011), use of de-oiling system before pressurization reduced surface oil amount on fried food by 87% and also produced higher-quality fried meals than deep fat-fried foodstuffs. In order to identify the precise advantages of vacuum technology, Dueik and coworkers (2010) compared vacuum- and atmospheric-fried treated carrot slices. This study showed that less than 50.5% of oil was absorbed on basis of dry matter by carrot slices during VF technique as compared with the atmospheric one and also there were a retention of around 90% of trans-α-carotene and 80% of trans-β-carotene. Furthermore, the study showed retention of raw carrot color. Perhaps, a lot of research has been conducted to make a detailed comparison between vacuum and atmospheric frying on food products such as apple (Mariscal & Bouchon, 2008), potato (Dueik & Bouchon, 2011), carrot (Dueik et al., 2010), and sweet potato (Da Silva & Moreira, 2008). They all claimed that compared to traditional frying methods, the VF process yields fried foodstuffs with a lower oil content. Additionally, VF method assists to lower development of acrylamide in foodstuffs that were fried in reduced temperature and vacuum environment (García-Segovia et al., 2016). In comparison to deep fat frying, which produced 1,000 μg/kg of acrylamide, VF produced very less (250 μg/kg) of it (Belkova et al., 2018). Similar to this, Granda et al. (2004) examined the amounts of acrylamide in VF and deep fat frying. According to the analysis, lowering temperature in VF in the range of 140°C–125°C decreased acrylamide generation by 63%, while 51% was reduced in deep fat frying when the temperature was lowered from 180°C to 165°C.

9.4.2 Microwave Frying

Food-processing industries use microwave (MW) heating for a variety of purposes such as pasteurization, blanching, gelatinization, sterilization, roasting, tempering, dehydration, and baking (Neetoo, 2014). MW refers to electromagnetic waves generated with the frequency range of 300 MHz to 300 GHz (Singh & Heldman, 2001). The International Telecommunications Union has approved a particular range of electromagnetic spectrum for use in households and industry. Industrial MW is often from 915 MHz or 2.45 GHz, while domestic MW is typically in the frequency of 2.45 GHz (Guo et al., 2017).

Many items can be heated rapidly and effectively with MW energy; extensive research has been done on the use of MW for frying. Dipolar rotation and ion conduction are two different sources of energy that are used in microwave frying of food (Ramaswamy & Tang, 2008). Polar molecules rotate in an electromagnetic field as part of the dipolar rotation mechanism, and direction of their rotation is dependent on their frequency. Polar molecules continually rotate, which causes them to clash and generate heat. Positive ions migrate in the direction of electric field, while negative ions flow in the opposite direction into the ionic conduction mechanism. By changing the direction of field, the ions can be diverted. As a result of friction caused by this motion—heated by the food—they collide with the other molecules. Thermal energy is transported from inside to outside of meal during microwave-frying process, causing water to boil inside food, increasing difference in vapor pressure between center and outside, and eliminating moisture from the food. Sansano et al. (2018) examined performance measures of deep fat frying, microwave frying, and traditional frying of potatoes. Investigators came to the conclusion that microwave cooking accelerated dryness, and cutting time is required for deep fat frying to retain the same moisture content considerably. According to this study, microwave frying might be an alternative to deep fat frying of fried foodstuffs due to its lower acrylamide and oil content. While utilizing varied MW energies, Oztop et al. (2007) examined the impact of MW frying process on the quality of fried potatoes and came to the conclusion that the moisture content was greatly reduced. By contrast, as the friction duration and microwave power increase, the color, oil content, and hardness of potatoes increased. The oil quality used in conventional frying and the microwave was compared by Aydinkaptan et al. (2017). They noted that when compared to traditional frying, microwave frying showed low oil degradation and thus reduced artifact creation.

9.4.3 Microwave Vacuum Frying (MVF)

Common issues with the microwave-frying process include uneven heating, loss of color and flavor, wet surface, significant loss of moisture, and firm texture (Sumnu & Sahin, 2005). Inventors found a different heating source in conjunction with a MW heating system, such as suction as well as ultrasonic power, to get by the problems with microwave frying. MVF has been identified as a potential method for enhancing the frying process and increasing the quality of fried items in comparison to MW frying and VF (Su et al., 2017). MVF has grown in popularity recently because of its adequate heating rate, uniformity in heating, ease of operation, and low cost. This technique combines MW energy with VF process to provide water molecules more strength to diffuse externally and create paths for the moisture diffusion (Sham et al., 2001). During comparative studies with MW frying and VF, MVF has been seen as a more promising technique to improve frying process and raise the quality of fried items (Su et al., 2017). Compared to VF, this produces less porous regions thus reducing the contact between the dry and oil parts during the pressurizing and cooling phases (Su et al., 2016). In the MVF method, a small quantity of porous regions limits oil absorption in the final fried food products, and thus it minimizes the use of oil and improves the quality of fried potato chips. Su et al. (2016) conducted a comparison of MVF and VF technologies. Their results revealed that compared to VF samples, MVF significantly reduced the oil intake of potato chips (ranging from 39.14 to 29.35 g oil per 100 g in dry matter basis), improved crunchiness, and maintained a more natural color.

Quan et al. (2014) reported that reduced MW power increases crispiness and color in potato chips, while reducing the rate of oil absorption, as well as the frying time. Jumras et al. (2020) investigated the effect of MV and conventional frying (VF) conditions on banana chips and found that the MVF condition used less energy and reduced the frying time by 33%. Subsequent research has demonstrated that the low temperature and short processing time of MVF could reduce degradation and absorption of content (Al Faruq et al., 2022). Chu and Hsu (2001) conducted an assessment of the effect of heat treatment on the quality of fried shallots and their frying oils and found that quick evaporation was beneficial. According to Gharachorloo et al. (2010), frying oil had a higher induction duration and less polar compounds, and a more number of saturated fatty acids are produced during the MVF process than during deep frying. The scientists also noted that the shorter frying duration and lower heat treatment used in the MVF method resulted in the production of peroxides that were not broken down by oxidation chain reactions.

9.4.4 Ultrasonic Microwave Vacuum Frying (UMVF)

The quality, speed, and safety of food items may all be improved with the use of ultrasound – a cutting-edge technology (Mason et al., 2011). Ultrasound has been used in the food-processing companies for fermentation, crystallization, extraction, dehydration, filtration, and microbial and enzymes inactivation (Gallo et al., 2018). Additionally, using ultrasonic as a pretreatment procedure before drying or frying has demonstrated an impressive influence on moisture reduction and color retention (Dehghannya, Naghavi, & Ghanbarzadeh, 2016). Recently, scientists have experimented with the use of ultrasound in conjunction with another heating source such that ultrasound might serve as heat source in a chamber for the purpose of frying or drying. Investigations combining ultrasound with other thermal sources to dry or fry plant-based materials are limited. Ultrasonic energy with a frequency from 20 to 100kHz propagates through a medium, resulting in a series of compressions and rarefactions that generate intense shear forces which may lead to medium disintegration (Park et al., 2014). The pressing and releasing of a sponge is analogous to the generation of ultrasonic waves which, when they impinge on a material's surface, can create energy through the alternating processes of compression (positive pressure) and rarefaction (negative pressure). This 'sponge effect' of ultrasound makes it possible to produce small channels in porous materials and remove moisture (Yao, 2016). Ultrasound can also lead to cavitation, which is the accumulation of liquid during rarefaction, and then its collapse during compression, resulting in micromechanical shocks which have the potential to cause cellular lysis (Neetoo, 2014). The bubbles created by this process can also increase the speed of microjet formation and mass transfer (Baeghbali et al., 2019). According to Su et al. (2018), the utilization of ultrasound at low frequencies has caused a change in physicochemical qualities of materials, including disruption of cellular structure, which has resulted in improved mass transfer, decreased production time, and energy savings. In a number of cases, it has also led to a better quality of the finished product. With its ability to operate at low temperatures and consuming less amount of time, UMVF is considered a hybrid technology that can enhance the quality of fried foodstuffs. Relevant features of UMVF include enhanced natural color preservation, energy savings, and improving evaporation rate (Carrillo-López & colleagues, 2017). Devi et al. (2018) assessed surface topography of fried mushroom chips through diverse frying techniques. Comparing VF and MVF, they observed that UMVF-fried chips had minor damage to the cell wall with a firm and uniform surface. This lesser damage and stiffness in the surface resulted in higher hardness, which decreased the oil absorption by 16%–20%. Similarly, Islam et al. (2019) discovered that the usage of ultrasound reduced oil uptake and water activity of fried edamame in comparison to the MVF and VF approaches. This experiment also revealed that the color, crispiness, chlorophyll content, and vitamin C levels of fried foods were all increased by the UMVF treatment. Su et al. (2018) analyzed the efficacy of the UMVF and MVF processes when conducted at low temperatures. The inclusion of ultrasound reduced the processing time by 36.4%–54.5%, and used 34.9%–48.3% less energy than in the MVF process. In addition to frying meals with greater quality, the combined impacts of MW and ultrasonic also contribute to the improved physicochemical and qualitative features of frying oil when compared to conventional frying techniques. Sun et al. (2019) conducted a study to determine if the use of ultra-high-molecular-weight fraction (UMVF) and low-field nuclear magnetic resonance (LF-NMR) improve the thermal stability of frying oil when compared to molecular weight fraction (MVF). The findings suggested that the usage of the two aforementioned techniques can prevent the increase in polar compounds, leading to a higher thermal stability.

9.4.5 Air Frying

Air frying is another alternative innovative drying process that yields foodstuffs with a very low proportion of oil in comparison to conventional method. Typically, hot-air circulation instead of oil immersion is used to fry food products, and for more effective frying, frying chamber radiates heat from a heat source (Zaghi et al., 2019). According to Andrés et al. (2013), this method involves direct contact between food item in frying chamber and an exterior emulsion of oil droplets in heated air. High-quality fried products are produced as a result of the food item's remarkable high and even distribution of heat transfer rate. Additionally, compared to the normal frying technique, this frying procedure requires a lot less oil, producing food with a reduced oil content. Giovanelli et al. (2017) employed air fryer, deep fat fryer, convection oven, and microwave oven to evaluate the quality of French fries. It was established that air frying enabled a 45%–50% decrease in the usage of oil while the color, texture, flavor, and aroma were considerably distinct when compared to traditional frying. The Maillard reaction and lipid oxidation are primary chemical processes responsible for imparting a deep-fried flavor; thus, decrease in oil content could result in several variations in the product's sensory characteristics (Chang et al., 2020). When compared to deep fat frying, the development of acrylamide reduced due to the fact that this procedure can be carried out with less oil, and problems of oil absorption and degradation of oil are less critical. According to Sansano et al. (2015), air frying reduces the level of acrylamide by roughly 90% when compared to conventional frying. The processes and dynamics of mass transfer events and volume changes during the air frying process are, however, only partially understood by investigations. There is very minimal research on the sensory qualities of fried food, and air frying technology can make fried food with little oil. The knowledge and comprehension of this technology must thus be developed via further study in order to broaden its use to homeowners and industries.

9.4.6 Radiant Frying

In the food-processing sector, radiant heating has been advocated as an alternative to conventional heating, particularly for roasting, baking, freezing, drying, pasteurizing, sterilizing, and blanching. By using high-temperature radiant emission, radiant frying improves heat flux pattern which is created during oil immersion frying of foodstuffs. This technique showed a heat transmission to food items at the frequency between 0.78 and 1,000 mm from the infrared source (Pankaj & Keener, 2017). Microwave heating likewise uses electromagnetic waves; however, heating through infrared showed more thermal effectiveness. Maintaining constant heating flow in radiant frying is considered one of most frequent difficulties. To solve this issue, controlled dynamic radiant (CDR) frying is suggested by Lloyd et al. (2004) for efficient heat transmission. The authors looked at the effects of heating par-fried French fries in the oven, immersion frying, and CDR frying. During their investigation, they found that radiant frying–treated French fries contained lower fat content and had a similar sensory score in term of overall sensory acceptability to the fried food with immersion treated. Nelson et al. (2013) conducted an assessment to determine the efficacy of radiant frying and observed that food cooked with this method had a noticeably paler hue, with an average of 16% less oil content and 19% higher moisture content when compared to the food cooked via immersion frying.

References

Ajo, R. Y. (2017). Application of hydrocolloids as coating films to reduce oil absorption in fried potato chip-based pellets. *Pakistan Journal of Nutrition, 16*(10), 805–812.

Al Faruq, A., Khatun, M. H. A., Azam, S. R., Sarker, M. S. H., Mahomud, M. S., & Jin, X. (2022). Recent advances in frying processes for plant-based foods. *Food Chemistry Advances, 1*, 100086.

Albert, S., & Mittal, G. S. (2002). Comparative evaluation of edible coatings to reduce fat uptake in a deep-fried cereal product. *Food Research International, 35*(5), 445–458.

Albert, A., Perez-Munuera, I., Quiles, A., Salvador, A., Fiszman, S. M., & Hernando, I. (2009). Adhesion in fried battered nuggets: Performance of different hydrocolloids as predusts using three cooking procedures. *Food Hydrocolloids, 23*(5), 1443–1448.

Amboon, W., Tulyathan, V., & Tattiyakul, J. (2012). Effect of hydroxypropyl methylcellulose on rheological properties, coating pickup, and oil content of rice flour-based batters. *Food and Bioprocess Technology, 5*, 601–608.

Andrés, A., Arguelles, Á., Castelló, M. L., & Heredia, A. (2013). Mass transfer and volume changes in French fries during air frying. *Food and Bioprocess Technology, 6*, 1917–1924.

Annapure, U. S., Singhal, R. S., & Kulkarni, P. R. (1999). Screening of hydrocolloids for reduction in oil uptake of a model deep fat fried product. *Lipid/Fett, 101*(6), 217–221.

Archana, G., Azhagu Saravana Babu, P., Sudharsan, K., Sabina, K., Palpandi Raja, R., Sivarajan, M., & Sukumar, M. (2016). Evaluation of fat uptake of polysaccharide coatings on deep-fat fried potato chips by confocal laser scanning microscopy. *International Journal of Food Properties, 19*(7), 1583–1592.

Aukkanit, N., Naowakul, J., & Jamsang, U. (2019). Reducing of Oil Uptake in Rice Cracker During Deep Fat Frying. In Advances in Interdisciplinary Practice in Industrial Design: Proceedings of the AHFE 2018 International Conference on Interdisciplinary Practice in Industrial Design, July 21–25, 2018, Loews Sapphire Falls Resort at Universal Studios, Orlando, Florida, USA 9 (pp. 122–129). Springer International Publishing.

Aydinkaptan, E., Mazi, B. G., & Barutçu Mazi, I. (2017). Microwave heating of sunflower oil at frying temperatures: Effect of power levels on physicochemical properties. *Journal of Food Process Engineering, 40*(2), e12402.

Baeghbali, V., Niakousari, M., & Ngadi, M. (2019). An update on applications of power ultrasound in drying food: A review. *Journal of Food Engineering and Technology*, *8*(1), 29–38.

Belkova, B., Hradecky, J., Hurkova, K., Forstova, V., Vaclavik, L., & Hajslova, J. (2018). Impact of vacuum frying on quality of potato crisps and frying oil. *Food Chemistry*, *241*, 51–59.

Chang, C., Wu, G., Zhang, H., Jin, Q., & Wang, X. (2020). Deep-fried flavor: Characteristics, formation mechanisms, and influencing factors. *Critical Reviews in Food Science and Nutrition*, *60*(9), 1496–1514.

Chen, S., Kuo, Y., Peng, G., & Chen, H. (2009). Study of pretreatments of low fat batter fish nuggets. *Taiwanese Journal of Agricultural Chemistry and Food Science*, *47*(2), 96–106.

Chu, Y. H., & Hsu, H. F. (2001). Comparative studies of different heat treatments on quality of fried shallots and their frying oils. *Food Chemistry*, *75*(1), 37–42.

Da Silva, P. F., & Moreira, R. G. (2008). Vacuum frying of high-quality fruit and vegetable-based snacks. *LWT-Food Science and Technology*, *41*(10), 1758–1767.

Daraei Garmakhany, A., Mirzaei, H. O., Maghsudlo, Y., Kashaninejad, M., & Jafari, S. M. (2014). Production of low fat french-fries with single and multi-layer hydrocolloid coatings. *Journal of Food Science and Technology*, *51*, 1334–1341.

Dehghannya, J., Naghavi, E. A., & Ghanbarzadeh, B. (2016). Frying of potato strips pretreated by ultrasound-assisted air-drying. *Journal of Food Processing and Preservation*, *40*(4), 583–592.

Devi, S., Zhang, M., & Law, C. L. (2018). Effect of ultrasound and microwave assisted vacuum frying on mushroom (Agaricus bisporus) chips quality. *Food Bioscience*, *25*, 111–117.

Dueik, V., & Bouchon, P. (2011). Vacuum frying as a route to produce novel snacks with desired quality attributes according to new health trends. *Journal of Food Science*, *76*(2), E188–E195.

Dueik, V., Moreno, M. C., & Bouchon, P. (2012). Microstructural approach to understand oil absorption during vacuum and atmospheric frying. *Journal of Food Engineering*, *111*(3), 528–536.

Dueik, V., Robert, P., & Bouchon, P. (2010). Vacuum frying reduces oil uptake and improves the quality parameters of carrot crisps. *Food Chemistry*, 119(3), 1143–1149.

Dueik, V., Sobukola, O., & Bouchon, P. (2014). Development of low-fat gluten and starch fried matrices with high fiber content. *LWT-Food Science and Technology*, *59*(1), 6–11.

Farkas, B. E., Singh, R. P., & Rumsey, T. R. (1996). Modeling heat and mass transfer in immersion frying. I, model development. *Journal of food Engineering*, *29*(2), 211–226.

Funami, T., Funami, M., Tawada, T., & Nakao, Y. (1999). Decreasing oil uptake of doughnuts during deep-fat frying using curdlan. *Journal of Food Science*, *64*(5), 883–888.

Gallo, M., Ferrara, L., & Naviglio, D. (2018). Application of ultrasound in food science and technology: A perspective. *Journal of Food*, 7(10), 164.

Galus, S., & Kadzińska, J. (2015). Food applications of emulsion-based edible films and coatings. *Trends in Food Science & Technology*, 45(2), 273–283.

García-Segovia, P., Urbano-Ramos, A. M., Fiszman, S., & Martínez-Monzó, J. (2016). Effects of processing conditions on the quality of vacuum fried cassava chips (Manihot esculenta Crantz). *LWT-Food Science and Technology*, *69*, 515–521.

Garmakhany, A. D., Mirzaei, H. O., Nejad, M. K., & Maghsudlo, Y. (2008). Study of oil uptake and some quality attributes of potato chips affected by hydrocolloids. European journal of lipid science and technology, 110(11), 1045–1049.

Gharachorloo, M., Ghavami, M., Mahdiani, M., & Azizinezhad, R. (2010). The effects of microwave frying on physicochemical properties of frying and sunflower oils. *Journal of the American Oil Chemists' Society*, 87, 355–360.

Giovanelli, G., Torri, L., Sinelli, N., & Buratti, S. (2017). Comparative study of physico-chemical and sensory characteristics of French fries prepared from frozen potatoes using different cooking systems. *European Food Research and Technology*, *243*, 1619–1631.

Granda, C., Moreira, R. G., & Tichy, S. E. (2004). Reduction of acrylamide formation in potato chips by low-temperature vacuum frying. *Journal of Food Science*, *69*(8), E405–E411.

Guilbert, S., Gontard, N., & Gorris, L. G. (1996). Prolongation of the shelf-life of perishable food products using biodegradable films and coatings. *LWT-Food Science and Technology*, *29*(1–2), 10–17.

Guo, Q., Sun, D. W., Cheng, J. H., & Han, Z. (2017). Microwave processing techniques and their recent applications in the food industry. *Trends in Food Science & Technology*, *67*, 236–247.

Huber, K. C. and Embuscado, M. E. (Eds.). (2009). *Edible Films and Coatings for Food Applications*. Springer.

Huse, H. L., Mallikarjunan, P., Chinnan, M. S., Hung, Y. C., & Phillips, R. D. (1998). Edible coatings for reducing oil uptake in production of akara (deep-fat frying of cowpea paste). *Journal of Food Processing and Preservation*, *22*(2), 155–165.

Islam, M., Zhang, M., & Fan, D. (2019). Ultrasonically enhanced low-temperature microwave-assisted vacuum frying of edamame: Effects on dehydration kinetics and improved quality attributes. *Drying Technology*, 37(16), 2087–2104.

Jumras, B., Inprasit, C., & Suwannapum, N. (2020). Effect of microwave-assisted vacuum frying on the quality of banana chips. *Songklanakarin Journal of Science & Technology*, *42*(1), 203–212.

Karimi, N., & Kenari, R. E. (2016). Functionality of coatings with salep and basil seed gum for deep fried potato strips. *Journal of the American Oil Chemists' Society*, *93*, 243–250.

Klangmuang, P., & Sothornvit, R. (2018). Active hydroxypropyl methylcellulose-based composite coating powder to maintain the quality of fresh mango. *LWT*, *91*, 541–548.

Krokida, M. K., Oreopoulou, V., Maroulis, Z. B., & Marinos-Kouris, D. (2001). Effect of pre-drying on quality of French fries. *Journal of Food Engineering*, *49*(4), 347–354.

Kurek, M., & Ščetar, M. (2017). Edible coatings minimize fat uptake in deep fat fried products: A review. *Food Hydrocolloids*, 71, 225–235.

Lalam, S., Sandhu, J. S., Takhar, P. S., Thompson, L. D., & Alvarado, C. (2013). Experimental study on transport mechanisms during deep fat frying of chicken nuggets. *LWT-Food Science and Technology*, *50*(1), 110–119.

Lloyd, B. J., Farkas, B. E., & Keener, K. M. (2004). Quality comparison of French fry style potatoes produced by oven heating, immersion frying and controlled dynamic radiant heating. *Journal of Food Processing and Preservation*, 28(6), 460–472.

Maity, T., Bawa, A. S., & Raju, P. S. (2014). Effect of vacuum frying on changes in quality attributes of jackfruit (Artocarpus heterophyllus) bulb slices. *International Journal of Food Science*, *2014*, 1–8.

Mallikarjunan, P., Chinnan, M. S., Balasubramaniam, V. M., & Phillips, R. D. (1997). Edible coatings for deep-fat frying of starchy products. *LWT-Food Science and Technology*, *30*(7), 709–714.

Mariscal, M., & Bouchon, P. (2008). Comparison between atmospheric and vacuum frying of apple slices. *Food Chemistry*, 107(4), 1561–1569.

Mason, T. J., Chemat, F., & Vinatoru, M. (2011). The extraction of natural products using ultrasound or microwaves. *Current Organic Chemistry*, *15*(2), 237–247.

Mellema, M. (2003). Mechanism and reduction of fat uptake in deep-fat fried foods. *Trends in food science & technology*, *14*(9), 364–373.

Moreira, R. G., Sun, X., & Chen, Y. (1997). Factors affecting oil uptake in tortilla chips in deep-fat frying. *Journal of Food Engineering*, *31*(4), 485–498.

Neetoo, H. C. Haiqiang. (2014). Alternative food processing technologies. In S. Jung, B. Lamsal, & S. Clark (Eds.), *Food Processing: Principles Applications* (2nd ed., pp. 137–169). John Wiley & Sons, Ltd.

Nelson III, L. V., Keener, K. M., Kaczay, K. R., Banerjee, P., Jensen, J. L., & Liceaga, A. (2013). Comparison of the FryLess 100 K Radiant Fryer to oil immersion frying. *LWT-Food Science and Technology*, *53*(2), 473–479.

Norizzah, A. R., Junaida, A. R., & Afifah, A. M. (2016). Effects of repeated frying and hydrocolloids on the oil absorption and acceptability of banana (Musa acuminate) fritters. *International Food Research Journal*, *23*(2), 694.

Oztop, M. H., Sahin, S., & Sumnu, G. (2007). Optimization of microwave frying of potato slices by using Taguchi technique. *Journal of Food Engineering*, *79*(1), 83–91.

Pankaj, S. K., & Keener, K. M. (2017). A review and research trends in alternate frying technologies. *Current Opinion in Food Science*, *16*, 74–79.

Park, S. H., Lamsal, B. P., & Balasubramaniam, V. M. (2014). Principles of food processing. In *Food Processing: Principles and Applications* (pp. 1–15).

Quan, X., Zhang, M., Zhang, W., & Adhikari, B. (2014). Effect of microwave-assisted vacuum frying on the quality of potato chips. *Drying Technology, 32*(15), 1812–1819.

Ramaswamy, H., & Tang, J. (2008). Microwave and radio frequency heating. *Food Science and Technology International, 14*(5), 423–427.

Sansano, M., Juan-Borrás, M., Escriche, I., Andrés, A., & Heredia, A. (2015). Effect of pretreatments and air-frying, a novel technology, on acrylamide generation in fried potatoes. *Journal of Food Science, 80*(5), T1120–T1128.

Sansano, M., De los Reyes, R., Andrés, A., & Heredia, A. (2018). Effect of microwave frying on acrylamide generation, mass transfer, color, and texture in French fries. *Food and Bioprocess Technology, 11*, 1934–1939.

Sham, P. W. Y., Scaman, C. H., & Durance, T. D. (2001). Texture of vacuum microwave dehydrated apple chips as affected by calcium pretreatment, vacuum level, and apple variety. *Journal of Food Science, 66*(9), 1341–1347.

Sharanabasava, M. R. R. (2018). Vacuum processing of food–a Mini Review. *MOJ Food Process Technology, 6*(3), 283–290.

Singh, R. P., & Heldman, D. R. (2001). *Introduction to Food Engineering*. Gulf Professional Publishing.

Su, Y., Zhang, M., Fang, Z., & Zhang, W. (2017). Analysis of dehydration kinetics, status of water and oil distribution of microwave-assisted vacuum frying potato chips combined with NMR and confocal laser scanning microscopy. *Food Research International, 101*, 188–197.

Su, Y., Zhang, M., & Zhang, W. (2016). Effect of low temperature on the microwave-assisted vacuum frying of potato chips. *Drying Technology, 34*(2), 227–234.

Su, Y., Zhang, M., Zhang, W., Liu, C., & Adhikari, B. (2018). Ultrasonic microwave-assisted vacuum frying technique as a novel frying method for potato chips at low frying temperature. *Food and Bioproducts Processing, 108*, 95–104.

Suárez, R. B., Campañone, L. A., Garcia, M. A., & Zaritzky, N. E. (2008). Comparison of the deep frying process in coated and uncoated dough systems. *Journal of Food Engineering, 84*(3), 383–393.

Sumnu, G., & Sahin, S. (2005). Recent developments in microwave heating. In *Emerging Technologies for Food Processing* (pp. 419–444).

Sun, Y., Zhang, M., & Fan, D. (2019). Effect of ultrasonic on deterioration of oil in microwave vacuum frying and prediction of frying oil quality based on low field nuclear magnetic resonance (LF-NMR). *Ultrasonics Sonochemistry, 51*, 77–89.

Tao, Y., & Sun, D.-W. (2015). Enhancement of food processes by ultrasound: A review. *Critical Reviews in Food Science and Nutrition, 55*, 570–594.

Tavera-Quiroz, M. J., Urriza, M., Pinotti, A., & Bertola, N. (2012). Plasticized methylcellulose coating for reducing oil uptake in potato chips. *Journal of the Science of Food and Agriculture, 92*(7), 1346–1353.

Vieira, M. G. A., Da Silva, M. A., Dos Santos, L. O., & Beppu, M. M. (2011). Natural-based plasticizers and biopolymer films: A review. *European Polymer Journal, 47*(3), 254–263.

Yagua, C. V., & Moreira, R. G. (2011). Physical and thermal properties of potato chips during vacuum frying. *Journal of Food Engineering, 104*(2), 272–283.

Yao, Y. (2016). Enhancement of mass transfer by ultrasound: Application to adsorbent regeneration and food drying/dehydration. *Ultrasonics Sonochemistry, 31*, 512–531.

Zaghi, A. N., Barbalho, S. M., Guiguer, E. L., & Otoboni, A. M. (2019). Frying process: From conventional to air frying technology. *Food Reviews International, 35*(8), 763–777.

Zeng, H., Chen, J., Zhai, J., Wang, H., Xia, W., & Xiong, Y. L. (2016). Reduction of the fat content of battered and breaded fish balls during deep-fat frying using fermented bamboo shoot dietary fiber. *LWT, 73*, 425–431.

Zhang, Y., Zhang, T., Fan, D., Li, J. and Fan, L. (2018). The description of oil absorption behavior of potato chips during the frying. *Lwt, 96*, 119–126.

Ziaiifar, A. M., Achir, N., Courtois, F., Trezzani, I., & Trystram, G. (2008). Review of mechanisms, conditions, and factors involved in the oil uptake phenomenon during the deep-fat frying process. *International Journal of Food Science & Technology, 43*(8), 1410–1423.

Ziaiifar, A. M., Courtois, F., & Trystram, G. (2010). Porosity development and its effect on oil uptake during frying process. *Journal of Food Process Engineering, 33*(2), 191–212.

Chapter 10

Alternate Frying Technologies for Healthier and Better Fried Foods

Indira Dey Paul
Haldia Institute of Technology

DOI: 10.1201/9781003329244-10

10.1 Introduction

Frying is one of the oldest and most popular cooking methods. The high organoleptic properties of the fried foods and the simplicity in their preparation make frying process widely desirable at domestic and industrial levels. Frying is a simultaneous heat and mass transfer process which alters the nutritional and sensory attributes of the food through complex food–oil interactions. Fats and oils impart unique flavor and mouthfeel to the food, which enhances its overall palatability. The amalgamation of high temperature (180°C–220°C) and rapid heat transfer makes frying an efficient cooking process. In this process, oil acts as the heat-transferring medium. Frying is basically a dehydration technique where oil replaces the moisture present in the food, thereby exerting a preservative effect by reduction of water activity of food and thermal destruction of microorganisms and enzymes.

Food is a complex matrix of protein, carbohydrates, lipid, water, vitamins, and minerals. As a result, the frying process leads to an infinite number of physical and chemical reactions including lipid oxidation, hydrolysis, polymerization, browning reactions, starch gelatinization, and protein denaturation. All these reactions are majorly dependent upon the quality of oil, frying time, and most importantly frying process. Uncontrolled or improper exposure to oil at high temperatures and atmospheric air can result in highly oxidized and potentially toxic products (Del Re & Jorge, 2006). Apart from that, excessive consumption of high-calorie fried foods leads to obesity, high blood pressure, and diabetes. The extremely busy and instant lifestyle of consumers left them with little time for preparing and cooking fruit- and vegetable-based foods on daily basis. As a result, the demand for ready-to-eat (RTE) foods has increased by many folds, and among the RTE foods, fried foods such as French fries, potato chips, and chicken nuggets have been found to occupy a good place in the snack food industry. However, the increasing concern about obesity-related health complicacies due to high oil intake have encouraged food industries and food researchers to develop alternative frying technologies which can produce delicious as well as nutritious fried foods using the least oil content.

10.2 Types of Traditional Frying Methods and Associated Quality Changes in Food

The interaction between food and oil during frying brings about physical, chemical, and organoleptic changes in the fried food. Frying leads to moisture loss, oil absorption, and acrylamide formation. Here heat and mass transfer occurs due to combined effect of continuous convection (through the oil bulk) and conduction. Since oil is the only medium for heat and mass transfer, the quality of the final product depends upon the type of frying method adopted.

Traditionally, fried foods can be prepared by the following frying methods:

Pan frying – A very small quantity of oil (15 cm^3) is used in this method for just lubricating the frying pan or a flat surface.

Sauteing/stir frying – Sauteing is a basic cooking method, where very little amount of oil (15 cm^3) is heated at medium flame for browning and enhancing flavor of food ingredients. It mainly forms the basis for dishes such as soups and stews. Stir frying is a quick and hotter version of sauteing. It generally involves throwing food ingredients together in a wok and frying them using very hot oil (230°C) with constant stirring. Stir frying is popular as a Chinese cooking technique. Vegetables, fish, eggs, and small meat pieces are prepared by stir frying.

Shallow frying – It is a high-heat frying process, where the food item is partly/half submerged into oil (500 cm^3) for browning and cooking. The food should be flipped to another side for complete frying.

Deep frying – In this method, foods are fully dipped into the heated oil (1,000 cm^3) under atmospheric pressure and fried till the attainment of proper texture and moisture content. Complete immersion in hot oil (≥2 cm) enables all the flavors and juices to get sealed inside the solid crust. Deep frying can be divided into four stages, viz.

- *Stage of primary heating* – It is the shortest period of heating lasting for only few seconds and involves negligible moisture loss.
- *Stage of outward boiling* – This stage involves heat exchange through both convection and conduction. Convective heat transfer takes place from the hot oil to the food, while conductive heat transfer occurs between food and heating source. Bubble formation is also observed in this phase.

- *Falling rate period* – It is the longest and most important stage of deep frying. During this period, the core of the food is heated up to the frying temperature. Various physical and chemical changes occur during falling rate period including evaporation of moisture, oil absorption, thickening of crust, browning reactions (Maillard reaction and caramelization), denaturation of protein, gelatinization of starch, and lipid oxidation.
- *Bubble end point* – Here the moisture evaporation is completed, and the food is completely fried.

The time and temperature of frying depend upon the type of food to be fried and the type of frying oil used. Different food items have variable initial moisture content, size, shape, and thickness, which play significant roles in determining the optimum frying conditions. The various pre-treatments applied on the food before frying also affect the frying parameters.

Various physicochemical alterations of major food constituents and microstructural changes in food are induced by frying, which decides the final quality of both the fried product and frying oil (Dueik & Bouchon, 2011). The atmospheric deep frying–induced changes mainly occur due to exposure to oxygen and high temperature. The fried food can be distinctly divided into outer hard crust and inner soft core (Figure 10.1). The core of the food undergoes mild microstructural changes, viz. swelling of starch granules by intracellular water due to gelatinization at around 60°C–70°C, protein denaturation and disintegration of middle lamellae between cells leading to separation of cells, and thereby development of mealy texture.

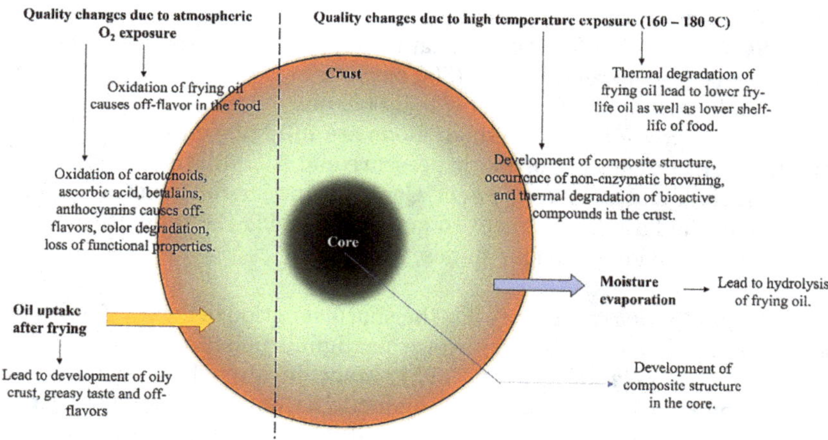

Figure 10.1 Changes in food quality and frying oil during deep fat frying.

The rough crust is formed at a temperature greater than 100°C due to cellular and subcellular alterations in the outermost layers of the food. The outermost layer goes through several physical and chemical changes including release of intracellular materials, starch gelatinization, dehydration, protein denaturation, damage of cellular adhesion, water evaporation, and oil absorption. While studying the characteristics of deep fat-fried potatoes, Reeve and Neel (1960) observed that majority of the inner cells of fried food shrink during frying, with no extended rupture, and retain their unique property. Around dehydrated gelled starch, the cell walls become wrinkled and convoluted (Van Marle et al., 1992; Reeve & Neel, 1960). Sometimes during ordinary cooking, the cell walls break due to slow dehydration. However, rapid dehydration during frying restricts swelling of starch, which helps in maintaining integrity of cell walls. The above-explained microstructural changes result in the formation of a composite structure comprising of a crispy, porous, and oily outer crust and a moist well-cooked inner core. The textural properties of the final fried product are significantly affected by these microstructural changes. The firmness of the fried products (especially potato chips) is often attributed to gelatinization and swelling of starch, as well as to the stability of the pectic substances of the middle lamellae and cell wall (Krokida et al., 2001).

The amount of oil absorbed by the fried food is of prime importance as it significantly affects consumer health. As shown in Figure 10.1, there are mainly two mass transfers in fried food material: water escaping out of the food and oil seeping inside the food. Therefore, it can be inferred that oil absorption mainly depends upon the moisture content of the raw food material (Rossi Marquez et al., 2014). As frying proceeds after initial heating, the product microstructure increasingly becomes weak due to formation of cracks, open capillaries, and channels due to evaporation of moisture (Dana & Saguy, 2006). Moreover, the trapped vapor within the pores, formed because of restrictive intercellular diffusion, becomes superheated and expand during frying, thereby resulting in distorted pore walls and increased product porosity (Moreira et al., 1995). Hence, earlier it was assumed that owing to water evaporation, oil absorption into food takes place during frying process. However, recent studies observed that maximum oil uptake occurs when the fried foods go through cooling phase. During cooling, the water vapor condenses in the pores creating pressure, which results in oil inhalation from the food surfaces into the pores (Zhang et al., 2020). According to Ouchon et al. (2003), the oil uptake by fried foods is basically a surface-related phenomenon arising from the competition between drainage of oil from the food and suction of oil into the porous crust as soon as the food is taken out from the oil mass (Figure 10.2).

While studying oil absorption in French fries, Van Koerten et al. (2015) divided oil uptake process into three parts as follows:

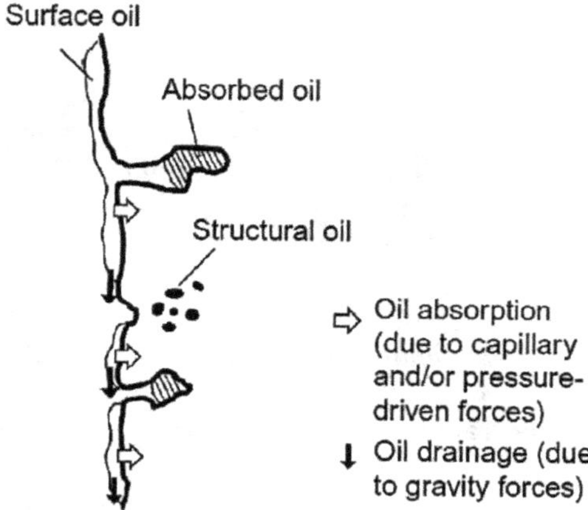

Figure 10.2 Types of oil in the microstructure of fried product and the principle of oil drainage and suction after frying process.

(*Source:* Eichenlaub & Koh, 2015; Ouchon et al., 2003.)

i. Assimilation of structural/internal oil into the chips during frying
ii. Inhalation of absorbed oil by the chips once they are removed from the frying pan
iii. Adhesion of surface oil on the frying surface during cooling

In order to develop alternative frying technologies, which will help reducing oil uptake and in overcoming other limitations of traditional frying methods, it is necessary to have a clear idea about the process of oil penetration during frying operation. This information will also help avoid oil deterioration during frying, improve oil turnover rates, and produce relatively healthier fried products.

10.2.1 Deterioration of Valuable Compounds and Development of Toxic Compounds in Food during Traditional Frying Processes

Frying variables, viz. temperature, oxygen, moisture content, light, and frying time, have remarkable effects on the deterioration of valuable or beneficial components present in the food to be fried (Lešková et al., 2006). Among these, high temperature and exposure to oxygen have major detrimental effects on the nutritional value of the fried foods. Moreover, the solubility of these also plays a significant role as water-soluble compounds (e.g., ascorbic acid) may get lost with water while fat-soluble compounds (e.g., carotenoids) can leach into the oil (Rojas-Gonzalez et al., 2006).

Deep fat frying of carbohydrate-rich foods results in the Maillard reaction between reducing sugars and amino acids. The Maillard reaction is responsible for the development of characteristic color, flavor, and aroma in fried food products, but it is also the principal reason behind the generation of toxic compounds. According to Mottram et al. (2002) and Rosén and Hellenäs (2002), the Maillard reaction at a temperature greater than 120°C and under low humidity may lead to formation of a potential carcinogenic substance known as acrylamide. Zyzak et al. (2003) suggested that the abundant presence of asparagine in potatoes is responsible for the generation of high concentration of acrylamide in French fries and chips. Another important product of the Maillard reaction is hydroxy methyl furfural, which occurs in products such as bread, biscuits, breakfast cereals, honey, and few processed fruits. Hydroxy methyl furfural has been found out to be cytotoxic, genotoxic, mutagenic, and carcinogenic (Teixidó et al., 2006). Heterocyclic amines, another carcinogenic compound, may be generated during deep fat frying and barbecuing meat and fish products, which is again linked to the Maillard reaction in protein-rich food products of animal origin (Skog et al., 1998). Heterocyclic amines are generally formed at a frying temperature above 150°C, and their concentration is directly proportional to frying time and temperature (Ngadi & Hwang, 2007).

10.3 Pre-Frying Treatments for Reduction of Oil Absorption

As the extent of oil absorption into the fried food is dependent upon the composition of the raw material, various pre-treatments can be applied before actual frying for modifying frying parameters as well as preserving frying oil quality. The need of pre-treatments and post-treatments of frying primarily arose from two significant research findings, viz. the effect of crust microstructure on oil absorption, and competition between drainage and suction of oil once the fried food starts to cool (Dueik & Bouchon, 2011).

The pre-frying treatments such as blanching, osmotic pre-treatment, pre-drying, and freezing significantly affect product yield, moisture removal kinetics, final moisture and fat content, and oil distribution in the final product. Few of the effective pre-frying methods have been discussed in subsequent subsections.

10.3.1 Pre-Drying

As discussed in Section 10.2, the oil absorption depends upon the initial moisture content of the food material. Hence, pre-drying is widely used for reducing oil content in the final fried products. The lower moisture content of the pre-dried food results in generation of lower vacuum pressure inside the fried food material after removal from the frying container, thereby allowing less oil penetration (Gupta et al., 2000). The amount of free water removal before frying depends upon the duration of pre-drying operation. According to Rahimi and Ngadi (2014), pre-drying-induced shrinkage decreases volume of open pores of the crust creating compact matrix, which helps in reduction of oil absorption into fried foods. Here, it is important to understand that oil reduction potential of pre-drying is not just because of the removal of food moisture, but it is actually due to the structural alteration of the food surface which reduces its permeability. For example, gelatinization of starch in starchy materials leads to formation of barrier on the food surface which hinders oil uptake during frying (Kawas & Moreira, 2001). On the other hand, although freeze drying reduced moisture content of the food material, it increased oil absorption considerably (Moreno & Bouchon, 2008). Pre-drying is not considered as the best pre-treatment, as it may lead to color changes and heat-induced nutrient degradation.

10.3.2 Freezing

The initial moisture content of the food can be reduced by another preconditioning method, freezing, which also helps in maintaining the original properties of the food material. The pre-frozen fried products exhibit a porous, sponge-like texture (Ren et al., 2018). Maity et al. (2012) observed that freezing prior to frying increased activation energy of food molecules during frying and produced end products having sensory acceptability comparable with the ones fried without pre-freezing. While investigating the influence of pre-freezing on vacuum-fried jackfruit chips, Maity et al. (2018) noticed increase in redness (*a* value increase) and decrease in yellowness (fall in *b* value) during frying operation. The pre-frozen jackfruit chips exhibited the highest crispy texture and the lowest instrumental breaking force. Freezing carrot snacks prior to frying resulted in the final product with improved organoleptic value and antioxidant activity (Albertos et al., 2016a). Owing to its several advantages, freezing might be considered as an efficient and feasible pre-frying method.

10.3.3 Blanching

The process of boiling or steaming food materials for short span in order to inactivate enzymes is known as blanching. The enzymes are responsible for enzymatic browning reactions and leaching of soluble sugars during potato frying (García-Segovia et al., 2016). It has been reported that blanching as pre-frying treatment can induce color change and alter textural properties by producing crisper crust and lesser surface shrinkage. Blanching helps in lowering oil uptake through surface starch gelatinization (Xin et al., 2015). The weakened cell structure caused by blanching increases resistance to oil absorption during frying. While researching on vacuum-fried shiitake mushroom slices, Ren et al. (2018) found out that pre-treatment of blanching combined with osmotic dehydration and freezing resulted in the highest drying rate. He again observed that combined pre-treatment of blanching, osmotic dehydration, and coating produced fried products having the lowest oil content and the best organoleptic properties.

10.3.4 Osmotic Dehydration

The process of osmotic dehydration has also been attempted by few researchers as a pre-treatment for lowering oil uptake. According to Moreno and Bouchon (2008), the fried food materials pre-treated with osmotic dehydration exhibit lower oil content owing to the increased solid content instead of reduced oil absorption. As a matter of fact, the oil inhalation of osmotically dehydrated food samples was found to be higher than those of control food samples. Osmotic dehydration has mainly been observed to prevent discoloration. In order to enhance oil-lowering ability of osmotic dehydration, this pre-treatment method has been combined with other pre-frying treatments such as high hydrostatic pressure, vacuum, high-intensity electric field, and ultrasound and microwave processing (Karizaki et al., 2013; Mujumdar et al., 2010). Ultrasound treatment, a non-thermal technology, has been applied to strengthen the activity of osmotic dehydration. Based on the findings of Karizaki et al. (2013), ultrasound-assisted osmotic dehydration reduces oil absorption into the fried products by lowering their initial water vapor content.

10.3.5 High-Pressure Processing

High-pressure processing (HPP) is a well-established non-thermal technology, which has been considered as a potential pre-frying treatment. Al Khusaibi and Niranjan (2012) observed that HPP could reduce frying period producing fried products having ≤2% final moisture content, but increased oil absorption marginally. This result was attributed to higher cell membrane permeability due to application of high pressure (200–800 MPa), which allowed higher water removal and thereby made way for more oil penetration. Albertos et al. (2016a) reported that HPP in combination with freezing prior to vacuum frying produced crispy carrot chips having better flavor and retaining phenolic compounds. The said products were also found to be less susceptible to oxidation. However, the adverse effects of HPP on product color as well as enzymatic and non-enzymatic reactions during frying should be taken into account before its application.

10.3.6 Edible Coating

Another promising route to reduced oil content after frying is application of edible films and coatings. Edible films are thin layers (maximum thickness of 0.3 mm) which are applied as coatings on food surfaces. These coatings have the ability to restrict or control moisture, oxygen, UV light, and solute movement. Sothornvit (2011) reported that edible coatings can prevent food from absorbing oil and water evaporation in the first frying period. García et al. (2004) attributed this hindrance in moisture and fat transfer between the food and frying medium to thermally induced gelation above 60°C. However, it should be kept in mind that frying is basically a dehydration process; hence, inhibition of water removal leads to longer frying duration, which may affect nutritional quality of the final product. Now edible coatings can be of numerous varieties, and their classification is essentially based on molecular structure. The types and characteristics of edible coatings are tabulated in Table 10.1. Among different edible coating ingredients, cellulose and its derivatives have been known widely for their film-forming properties and unique thermal gelation capacity (Balasubramaniam et al., 1997).

10.3.7 Pulsed Electric Field Method

Pulsed electric field (PEF), a well-known non-thermal technology, has been considered an effective pre-frying treatment. PEF uses short-voltage pulses with high electric field strength to soften the texture of fruits and vegetables such as apples, potatoes, beet, and carrot (Liu et al., 2017a; Lebovka et al., 2004). Pre-treating food materials with PEF prior to frying cause remarkable reduction in oil content (≥18%) (Janositz et al., 2011; Liu et al., 2017a, 2018a). Botero-Uribe et al. (2017) reported that PEF-pretreated fried products have a flat and smooth surface which helps in lower oil uptake. The formation of smooth surface allows less oil absorption without application of coating. Liu et al. (2017a) observed that the oil content of PEF-pre-processed fried sweet potatoes was decreased by 18% compared to the non-PEF-treated products. Moreover, the non-thermal nature of PEF results in less degradation of color, vitamin, and flavor compounds. Ostermeier et al. (2020), while working on PEF-pretreated onions, found PEF to enhance internal diffusion. Liu et al. (2018b) noticed a significant decrease in frying time for PEF-pre-treated,

Table 10.1 Types of Edible Coating and Their Applications in Frying of Different Food Materials

Type of Coating	*Raw Materials*	*Characteristics*	*Applications and Salient Findings*
Protein coating	Corn, milk, soy, whey, and wheat	Reduce water vapor permeability and loss of water, thereby helping in prevention of food conversion from glassy to rubbery state and reduction in oil absorption.	• Applied fried food like 'teralli' biscuits with added transglutaminase (Rossi Marquez et al., 2014). • Corn zein coating on fried mashed potato balls reduced oil content by 59% (Mallikarjunan et al., 1997). • Whey protein coating on potato chips reduced oil content by 5%. Potato chips coated with sodium caseinate coating contained 14% less oil than non-coated chips (Aminlari et al., 2005). • Soy protein isolate coatings on potato pellet chips (Angor, 2016).
Batter coating	Water, flour, starch, and seasoners	Higher cross-linking degree makes starch more resistive to gelatinization, resulting in enhanced crispiness.	• Addition of 10% barley flour in wheat chips remarkably decreased oil content (Yuksel et al., 2015). • Addition of wheat bran in starch food matrix resulted in 70% reduction in oil absorption (Dueik et al., 2014).

(Continued)

Table 10.1 (*Continued*) Types of Edible Coating and Their Applications in Frying of Different Food Materials

Type of Coating	*Raw Materials*	*Characteristics*	*Applications and Salient Findings*
Hydrocolloid coating	Gums, cellulose derivatives, guar, xanthan, methyl cellulose, hydroxypropyl cellulose, carboxy methyl cellulose (CMC), etc.	• Bind carbon dioxide, oxygen, and lipids enabling reduction in oil uptake and acrylamide formation in fried products. • Delay dehydration • Inhibit respiration • Improve texture • Retain flavor • Reduce microorganisms	• Complex coating of polysaccharide (hydroxypropyl methylcellulose, HPMC) and protein decreased oil suction in fried poultry products by 33.7% (Balasubramaniam et al., 1997). • Combined coating of CMC and HPMC reduced oil uptake and enhanced sensory properties of chickpea and gram splits (Phule & Annapure, 2013). • Use of egg white coating on chips led to lower moisture and fat content. The lightness of chips was found to be directly proportional to the concentration of hydrocolloid, whereas browning index and fracturability were inversely proportional to hydrocolloid concentration (Alimi et al., 2013).

vacuum-fried potato tissue. Frying of apples pre-processed with PEF exhibited increased drying rate (by 27%), decrease in energy consumption by 21%, and reduction in drying time by 23% (Wu et al., 2011).

Apart from the above-mentioned pre-frying treatments, mild heat processing (45°C–55°C) alone or in combination with other pre-treatment methods, such as infrared radiation and pre-treatment with asparaginase, has also been successfully attempted to fruits and vegetables as a pre-processing step before actual frying process.

10.4 Alternative Novel Frying Technologies

In light of the adverse effects of traditional frying technologies, the food technologists and engineers felt the need for development of alternative frying methods and fabrication of innovative fryers, which can produce healthy fried foods by reducing oil uptake and generation of toxic compounds. Avoiding degradation of frying oil is another major objective of novel frying technologies.

10.4.1 Vacuum Frying

Vacuum technology has gained popularity because of its protective effect on heat-sensitive food materials. The concept of vacuum frying is not new as vacuum fryers have been first fabricated in the 1960s for production of superior-quality chips. However, later its use decreased due to popularity of blanching and other pre-treatments of raw materials (Moreira et al., 1999). The increasing concern about the health implications of conventionally fried products brought this technology back in demand.

In vacuum frying (also termed as pressure frying) operation, the food materials are fried in vacuum chamber (<50 Torr or 6.65 kPa) in the absence of air (Banerjee & Sahu, 2017). Here, the boiling points of the food moisture and oil are substantially lowered in a low-oxygen environment. As a result, the frying can be accomplished at a lower temperature (<90°C) compared to the temperature used in atmospheric frying leading to better retention of heat-labile nutrients (Liu et al., 2017b; Albertos et al., 2016b; Tharasena & Lawan, 2014). Natural antioxidants such as ascorbic acid, carotenoids, betalains, and anthocyanins are highly sensitive to normal atmospheric frying conditions. Vacuum frying has proved to be a suitable frying option for such delicate bioactive compounds. Application of vacuum also has other advantages including preservation of flavor, less tendency to lipid oxidation, and reduction in browning reactions, which in turn helps in reducing discoloration and acrylamide formation (Belkova et al., 2018; Karimi et al., 2017). This method is an amalgamation of vacuum drying and organic frying.

Commercially both batch and continuous vacuum fryers are available. A batch vacuum fryer (Figure 10.3) is basically comprised of the following components:

a. Stainless steel vacuum chamber
b. Frying basket with lift rod
c. Refrigerated condenser
d. Vacuum pump.

The working principle of vacuum fryer is quite simple. At the start of the frying operation, the vacuum chamber is filled with oil and the chamber is hermetically sealed by closing the lid. Next the pressure inside the chamber is brought down to the required value and oil heating starts. The frying basket is located inside the vacuum chamber as shown in Figure 10.3. During oil heating, the basket is placed in the headspace. Once the oil reaches desired frying temperature, the basket is dipped into the hot oil with the help of the basket rod and kept there for a stipulated duration. After adequate frying, the basket is lifted up from the oil, the chamber is brought back to the

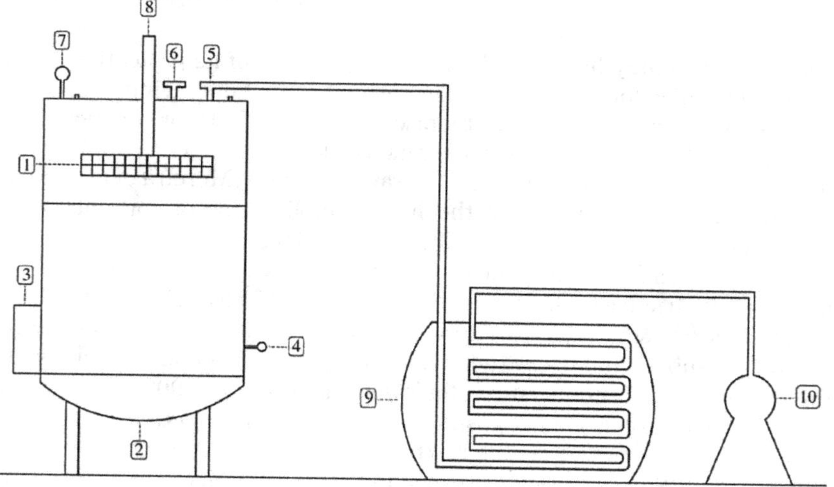

Figure 10.3 Schematic diagram of batch vacuum fryer, which consists of (1) frying basket, (2) heater, (3) temperature controller, (4) thermocouple, (5) vacuum link, (6) pressure valve, (7) pressure gauge, (8) lift rod for basket, (9) condenser, and (10) oil vacuum pump.

(*Source:* Dueik & Bouchon, 2011.)

atmospheric pressure by releasing pressure valve, and the lid is opened. The release of pressure should be carefully done as it has critical influence on oil absorption.

Several researchers have successfully applied vacuum frying for processing fruits and vegetables as well as fishes and crustaceans. Few of the studies are tabulated in Table 10.2.

10.4.2 Microwave Frying

Frying using microwave heating has been considered as an efficient alternative frying method owing to requirement of lesser frying time and exposure to lower temperature. Microwave heating uses electromagnetic waves in the range of 300 MHz to 300 GHz frequency and wavelengths from 1 mm to 1 m.

The direct interaction of alternating microwaves with dipolar water molecules and salt ions present in the food causes friction among the water molecules and ions, which generates heat inside the food. This volumetric mode of heating in microwave oven significantly reduces heating time because whole body of the food material is heated up simultaneously. On the other hand, surface heating mode of conventional frying takes longer duration to heat up the whole food piece. Actually, microwave frying encompasses two energy sources; while the microwave penetration into food generates heat from within, the surface of the food is cooked by the frying oil (Zhang et al., 2020). During microwave frying of food materials, higher water evaporation is observed because of pressure-driven flow (Feng & Tang, 1998). This higher rate of evaporation leads to greater moisture loss, which in turn results in more oil absorption compared to atmospheric deep frying. However, shorter duration of frying compensates for the higher moisture loss and helps in lowering total oil content of the final product. In fact, microwave frying is the quickest heating process as observed by various researchers (Schiffmann, 2017; Su et al., 2016a; Oztop et al., 2007). The other advantages of microwave frying include less oxidation of frying oil, better color and crispier texture of fried foods, and lower acrylamide formation (Chen et al., 2009; Sahin et al., 2007). Lower acrylamide formation during microwave frying is attributed to use of lower heating temperatures (<120°C) and relatively higher humidity than conventional frying processes (Zhang et al., 2020). Microwave frying also proved to have better economic feasibility in terms of reduced manpower, floor space, increased yield, and cost savings.

Microwave frying can be accomplished using domestic microwave oven. In this method, at first, the frying oil is heated at the desired temperature and power level. After the oil gets heated up, the raw food pieces are immersed in the hot oil and put inside the oven for frying for stipulated duration at pre-determined microwave power levels. At industrial level,

Table 10.2 Application of Vacuum Frying on Different Food Materials

Food Products	*Important Salient Findings*	*References*
Squid (ring, tentacle and fin) fry	• Vacuum-fried (VF) products had lower moisture, fat, and looser microstructures than atmospheric-fried (AF) products. • Proteins were well preserved in VF products; however, their secondary structures became modified (increase in β-sheet configurations). • Carbonyl content in the VF products was found to be more than the fresh samples, but less than that of AF products.	Pei et al. (2021)
Red-fleshed papaya chips	• Lycopene and β-carotene contents increased. • The VF chips had low a_w (0.1–0.3) and less color degradation. • The VF chips suffered lower loss of β-cryptoxanthin (≤30%).	Soto et al. (2021)
Potato crisps	• Vacuum frying decreased acrylamide formation by 98%. It also reduced alkylpyrazines. • Lesser extent of oxidative changes occurred in the VF oil. • The VF crisps were declared tasty and appreciated by sensory panelists for their potato-like fresh flavor. However, most of the panelists preferred the taste of AF crisps over their VF counterparts.	Belkova et al. (2018)
Breaded shrimps	• Vacuum treatment significantly decreased oil uptake of the final product. • The reduction in moisture loss was attributed to reduction of moisture loss during vacuum frying. • The VF shrimps showed lesser acrylamide content. • Better color retention and softer texture observed in VF shrimps.	Pan et al. (2015)
Chicken nuggets	• Higher rate of water removal was noticed in VF nuggets during the first 2 minutes of frying. • The VF and traditionally fried nuggets showed comparable sensory properties. • VF nuggets did not produce lower fat content than the conventionally fried nuggets.	Teruel et al. (2014)

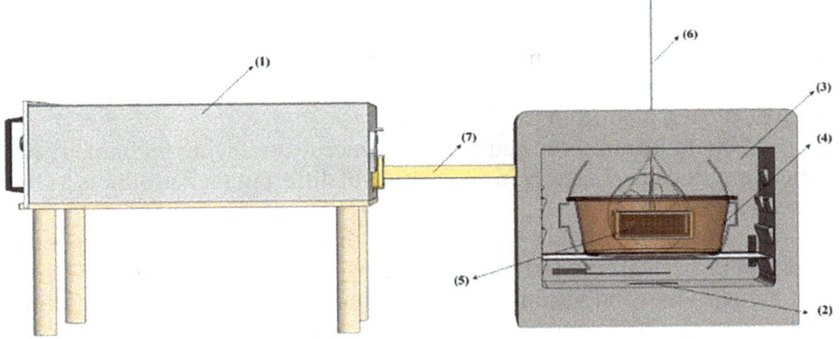

Schematic diagram of microwave frying unit

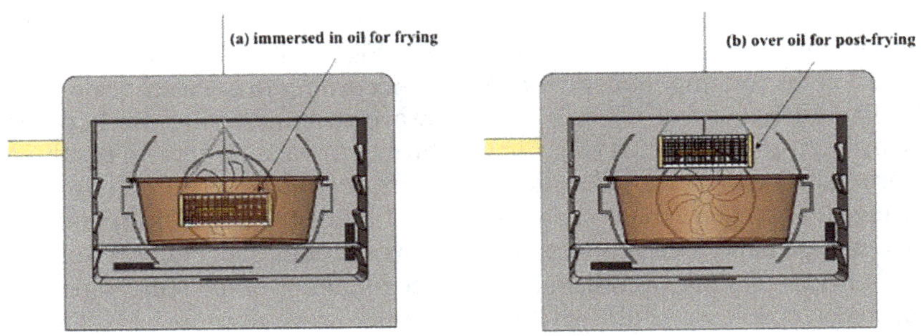

Two working locations of sample holder during (a) frying and (b) post-frying

Figure 10.4 Dual-frequency microwave fryer designed and fabricated by Zhou et al. (2022) for French fries. The microwave fryer consists of seven components, viz. (1) 5.85 GHz solid-state microwave generator, (2) magnetron of 2.45 GHz capacity, (3) frying chamber or microwave cavity, (4) oil container made of glass, (5) sample holder made of nylon, (6) nylon wire for lowering and lifting sample holder, and (7) waveguide.

microwave frying is still at its developing stage. Till date, the only successful industrial application of microwave frying is microwave-assisted donut frying (Schiffmann, 2017).

Zhou et al. (2022) designed and fabricated a dual-frequency (2.45 and 5.85 GHz) microwave fryer to study its post-frying effect on oil absorption in French fries. The schematic diagram of the microwave fryer fabricated by Zhou et al. (2022) is shown in Figure 10.4. This frying unit was capable of operating both at single frequency (2.45 or 5.85 GHz) and two frequencies as mentioned above. During frying, the food material in the sample holder is dipped into the glass oil container containing heated oil. After completion of frying, the food is lifted out of the oil and suspended just above the oil container with the help of nylon wire for post-frying microwave treatment. The frying unit has the provision for turning on microwave power after the removal of fried food from the frying oil.

Owing to its numerous advantages, microwave frying technology has been attempted successfully in production of different food products as discussed in Table 10.3.

In order to enhance the yield of higher-quality fried products, researchers have attempted to develop a hybrid frying technique by combining microwave and vacuum frying methods (Su et al., 2017). According to Ekezie et al. (2017) and Su et al. (2016b), microwave-vacuum frying can increase rate of heating and rate of dehydration, and decrease frying time and final oil content. The integration of the advantages of microwave and vacuum frying can lead to improvement in both energy efficiency and product quality (Wang et al., 2013). The schematic diagram of a laboratory-scale microwave-vacuum frying unit is shown in Figure 10.5. The fryer has the provision for rotating foodstuffs at 5 rpm. According to Gharachorloo et al. (2010) and Sensoy et al. (2013), microwave-vacuum frying method can enable frying at a lower frying temperature than vacuum frying, resulting in products having lower oil content and more desirable texture. While comparing dewatering dynamics, moisture and oil distribution of microwave-vacuum-fried and vacuum-fried potato chips, Su et al. (2017) observed that the hybrid technique significantly enhanced kinetics of moisture vaporization and available water diffusivity. Higher microwave power level (1,000 W) directly influenced rate of water vaporization and available water diffusivity. The same study also found that the microwave-vacuum frying time was prolonged at 100°C, which resulted in the best color preservation.

10.4.3 Ultrasound-Assisted Frying

Ultrasound treatment has been widely accepted as an effective non-thermal method of food processing. Ultrasound technique is comprised of acoustic waves having frequency in the range of 20 kHz to 10 MHz, which can propagate through gas, liquid, and solid. Application of ultrasound creates sponge effect and microscopic channels, which helps in increasing mass transfer and inducing chemical and physical changes in food through cell wall ruptures (Cárcel et al., 2007; Mason et al., 1996). The acoustic waves of high intensity create cavitation, resulting in heating of the food material. The micro-vibration and heating effect of ultrasound have made it suitable for various unit operations of food processing including drying (Duan et al., 2008; Szadzińska et al., 2016) and frying (Al Faruq et al., 2019; Devi et al., 2018; Su et al., 2018a).

Table 10.3 Application of Microwave Frying on Different Food Materials

Food Products	*Important Salient Findings*	*References*
French fries	• The frying time was reduced by 30%–40% as compared to deep oil frying. • Although oil absorption increased with increasing moisture loss during microwave frying, the shorter frying period nullified the adverse effect. • Microwave heating for 60 s after frying in oil reduced the oil content further by 18%–23%. • The oil content of the microwave-fried products obtained at 5.85 GHz was statistically lower than those obtained at 2.45 GHz; the product color and texture were also found to be superior at 5.85 GHz.	Zhou et al. (2022)
Donut	• Microwave-assisted frying of donuts resulted in significantly greater volume (15%–25%) than conventionally fried ones. • Elimination of donut core during microwave frying resulted in lesser fat absorption and longer shelf-life than conventionally fried ones. • Sugar-based coating on microwave-fried donuts had longer shelf-life due to lower fat uptake. • Frying time was reduced by about 20% during microwave-assisted frying.	Schiffmann (2017)
Potato slices	• Used frying oil (a blend of palm olein and soybean oil) had lower polar compounds, higher induction period, and more saturated fatty acids compared to sunflower oil. • Lower temperature and shorter duration of frying helped in avoiding hydrolysis of peroxides, formed in the frying oil as a result of oxidation.	Gharachorloo et al. (2010)

(*Continued*)

Table 10.3 (*Continued*) Application of Microwave Frying on Different Food Materials

Food Products	*Important Salient Findings*	*References*
Chicken fry	• Microwave-fried (MF) products fried having three types of flour (soy, chickpea, and rice flour) had lighter color and lower acrylamide content as compared to the products prepared by conventional frying for 5 minutes. • The acrylamide content of MF chicken fry using rice flour batter was observed to be the lowest among all other samples.	Barutcu et al. (2009a)
Potato slices	• It was observed that the moisture content of MF products decreased with increasing frying time and microwave power level. • The oil content, hardness, and color of the MF products increased with increasing frying time and microwave power level. • Frying at 550 W for 2.5 minutes using sunflower oil produced fried potatoes containing lower oil content than those fried using conventional method.	Oztop et al. (2007)

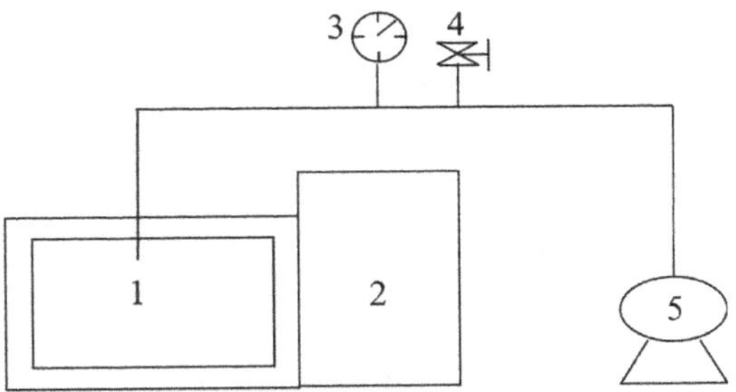

Figure 10.5 Schematic diagram of laboratory-scale microwave-vacuum frying system, comprising of (1) microwave cavity, (2) magnetron, (3) vacuum gauge, (4) valve, and (5) vacuum pump.

(***Source:*** **Song et al., 2007.)**

Ultrasound-assisted frying is a relatively new frying technology. Some studies proved that when coupled with other frying methods, ultrasound treatment resulted in superior quality of final products. Oladejo et al. (2017) found that rate of oil suction of ultrasound-pretreated deep-fried sweet potato chips was lower than the untreated sweet potato chips. Ultrasound has been associated with microwave-assisted vacuum drying by several researchers. The products obtained through this hybrid frying technique were crispier and low in oil content. During ultrasound-assisted microwave-vacuum frying process, the acoustic waves spread through the oil (frying medium) and food, which leads to the alternation of boundary layer, thereby affecting the pressure and oil viscosity significantly (Su et al., 2018b). Figure 10.6 shows the schematic diagram of a laboratory-scale ultrasound microwave-assisted vacuum frying equipment. The fryer is comprised of an ultrasound source, a microwave source, a heating system, and a vacuum fryer. The frying vessel made of polytetrafluorethylene (PTFE) has a capacity of 15 L. The vessel material, PTFE, does not absorb microwave and can sustain temperature as high as 150°C. The sole heating source of the equipment is a microwave source, containing three microwave devices (each with 0–1,200 W capacity; 2,450 MHz); it uniformly surrounds the vacuum chamber. The ultrasound system, made up of five ultrasound devices (each with 0–600 W capacity; 28 kHz), is located at the bottom of the frying chamber/vessel. A fryer is also equipped with a centrifugation system, which removes oil from sample surface after deep frying at 370 rpm. The list of products prepared using ultrasound-assisted frying along with their major inferences is given in Table 10.4.

Despite many positive effects of ultrasound-assisted frying, the widespread application of this novel method is hindered by its production of

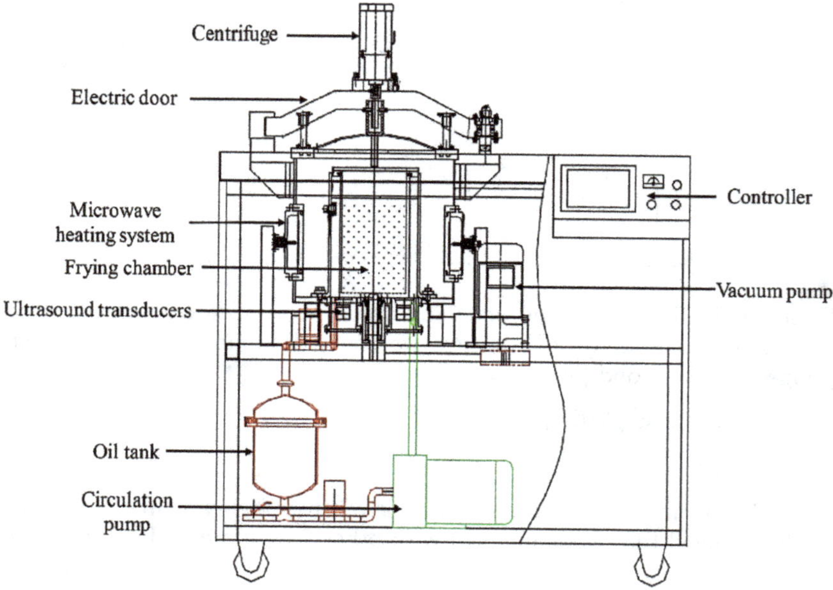

Figure 10.6 Schematic diagram of a laboratory-scale ultrasonic microwave-assisted vacuum fryer.

(*Source:* Devi et al., 2020.)

intolerable sound (Azam et al., 2020). The ultrasound-induced noise can become hazardous to the health of operators. Hence, design improvement is needed for eliminating undesirable noise. However, such modification can lead to increase in capital cost of the equipment.

10.4.4 High-Pressure Deep Frying

A relatively new alternative frying method is high-pressure or super atmospheric deep frying. In this method, the food is fried under controlled pressure application in a hermetically sealed vessel. Pressure is generated by frying pan through the release of water from the food product (Pankaj & Keener, 2017). Application of high pressure facilitates adjustment of boiling point temperatures of different solvents present in the food based on the desired characteristics of the final product. Juicier and tenderer texture is exhibited by high-pressure fried products due to better water retention and lower oil uptake (Pawar et al., 2013). This method of frying also helps in

Table 10.4 Application of Ultrasound-Assisted Frying on Different Food Materials

Food Products	*Important Salient Findings*	*References*
Button mushroom chips	• The moisture removal rate was higher and oil content was lower for ultrasound-assisted-microwave-vacuum-fried (UMVF) products compared to vacuum-fried, microwave-vacuum-fried, and ultrasound-assisted vacuum-fried products. • The higher oil partitioning coefficient of UMVF products indicated a negative effect of sonication on oil penetration into the mushroom chips. • The UMVF chips had most acceptable flavor compounds.	Devi et al. (2020)
Pork-beef meatballs	• Ultrasound-assisted frying decreased frying time (by 1–3 minutes) and increased cooking yield compared to conventional frying. • Ultrasound-treated meatballs retained high levels of moisture due to formation of fibril network cross-linking. • Ultrasound treatment increased whiteness index (L^* value) of the meatballs, while their redness index (a^* value) was decrease. • Ultrasound-treated meatballs had better sensory quality than the untreated meatballs.	Wang et al. (2019)
Apple slices	• The rate of moisture loss was higher in the case of UMVF products. • Color retention and crispiness of UMVF products were observed to be better than microwave-vacuum-fried products. • UMVF also helped in reducing frying time. • Analysis through E-nose showed production of lesser Maillard reaction products in the case of UMVF products.	Al Faruq et al. (2019)

(*Continued*)

Table 10.4 (*Continued*) Application of Ultrasound-Assisted Frying on Different Food Materials

Food Products	*Important Salient Findings*	*References*
Pumpkin chips	• UMVF resulted in lower oil absorption and crispier products. • UMVF helped in better preservation of product cellular structure. • UMVF products had significantly lower moisture content than microwave-vacuum-fried products and vacuum-fried products.	Huang et al. (2018)
Sweet potato fries	• The UMVF process resulted in increased rate of water evaporation and effective moisture diffusivity. • The UMVF products absorbed less oil, and the final texture was crispier compared to other products of the study. • UMVF samples were observed to retain higher total anthocyanin content. • The microstructure of the UMVF products were more porous and disrupted.	Su et al. (2018b)

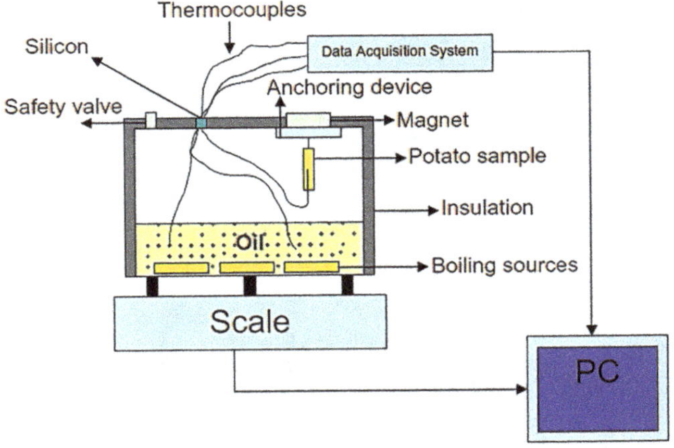

Figure 10.7 High-pressure frying system, designed by Erdogdu and Dejmek (2010), for potato slabs.

reducing cooking time and extent of oil deterioration. Erdogdu and Dejmek (2010), while investigating on convective heat transfer coefficient during conventional frying and high-pressure frying, found out that high-pressure frying almost doubles the convective heat transfer coefficient compared to conventional deep frying.

Figure 10.7 shows the high-pressure frying unit, designed by Erdogdu and Dejmek (2010). It consists of a modified pressure cooker (made of cast aluminum), thermocouple, magnet, anchoring device, and a computer to display result. After the oil (inside the cooker) reached desired temperature, the potato slab (with centrally positioned thermocouple) was held in the vapor space of the cooker by the help of a metal-attaching device. Neodym magnet was fitted outside the cooker, which held the sample in required positions. Pressure inside the cooker was generated by adding a certain quantity of potatoes into the oil (served as the boiling source for pressure build-up), and the lid was closed. After complete stabilization of the pressure system, the potato slabs were dropped into the oil by removing the magnet.

Application of high-pressure deep frying is not yet widely spread as the technology and has been mostly used for meat and poultry frying. The researchers are trying to combine high-pressure deep frying with different types of edible coatings for further reduction in oil absorption by fried foods. Few of the works done on high-pressure deep frying are tabulated in Table 10.5.

Table 10.5 Application of High-Pressure Deep Frying on Different Food Materials

Food Products	*Important Salient Findings*	*References*
Fried chicken leg and breast	• Frying-induced shrinkage was significantly lesser in the pressure-fried products than conventionally fried products. • Pressure-fried chicken was juicier and contained less fat/oil than the conventionally fried ones. • Pressure-fried products suffered lesser lipid oxidation compared to conventionally fried products. • Higher shear force was required by the conventionally fried chicken as compared to the pressure-fried ones.	Das et al. (2013)
Fried chicken chunks	• Neutral lipid content of pressure-fried chicken was lesser than the conventionally fried chicken. • Glycolipid content of pressure-fried chicken was higher than the conventionally fried chicken. • Phospholipid was observed the least among all the lipid classes for both frying methods. • Linoleic acid and oleic acid contents of frying oil remained high for both the frying methods. • Pressure-fried chicken retained more moisture (56%–58%) and lesser fat (14%) compared to the conventionally fried ones (49%–52%; 18%). • Thiobarbituric acid and FFA values of pressure-fried chicken and oil were statistically lesser than conventionally fried chicken and oil.	Pawar et al. (2013)
Fried potato slabs	• Under 2 bar pressure, heat transfer coefficient vs. frying time study showed that the heat transfer coefficient almost doubled compared to the atmospheric pressure frying.	Erodogdu and Dejmek (2010)

(Continued)

Table 10.5 (*Continued*) Application of High-Pressure Deep Frying on Different Food Materials

Food Products	*Important Salient Findings*	*References*
Chicken nuggets	• Nitrogen gas was used as a pressurizing medium in a deep fat fryer. • Application of high pressure during frying reduced moisture loss resulting in juicier and tender nugget production. • Oil uptake was also reduced during high-pressure frying. • Use of nitrogen gas instead of steam for pressure generation produced similar or better quality fried products in terms of water retention and texture of the final product.	Innawong et al. (2006)
Breaded chicken nuggets	• Effects of two edible coatings [methylcellulose (MC) and whey protein isolate (WPI)] and two pressure sources [nitrogen gas and steam] were studied. • Both the edible coatings and pressure sources had significant effects on crispiness, moisture and fat content, juiciness, color, and texture of the nuggets. • The products coated with MC and fried with nitrogen gas were crispier than other products. • Nitrogen-fried products were tenderer than steam fried ones.	Ballard and Mallikarjunan (2006)

10.4.5 Radiant Frying

It is well known that there are three mechanisms of heat transfer, viz. conductive, convective, and radiative. Radiant heating is based on radiative heat transfer. The potential of radiant heating as an alternative frying technique has been investigated through application-based studies. Majority of radiant frying uses the principle of infrared heating. In this case, food materials are exposed to heat generated from an infrared source at the wavelength range of 0.78–1,000 µm. Electromagnetic waves are also used during microwave heating; however, higher thermal efficiency and faster heating ability have been observed in infrared radiant frying technique.

Initially, radiant heating was attempted primarily for maintaining a constant air temperature. Several researchers have developed and patented different types of radiant frying equipment based on the above-mentioned requirement (Alden, 1993; Dauliach, 1999; Mestnik, 1998). These instruments generate a constant heat flux from an emission source for maintaining product environment at a fixed temperature. However, the use of constant flux radiant fryer alone resulted in products with charred surface and under-cooked core; the fried foods had tougher texture and fried mouthfeel.

The limitations of constant flux fryer led to the development of controlled dynamic radiant (CDR) frying technology. It uses high heat flux radiant emitters (capacity > 100 kW/m^2) with precise output control. Farkas et al. (2007) patented continuous CDR frying equipment having multi-zone radiant frying system, each zone comprising of two emitters positioned on each side of a conveyer (Figure 10.8). Each of the emitters can be controlled independently by adjusting power settings, geometry of sample or

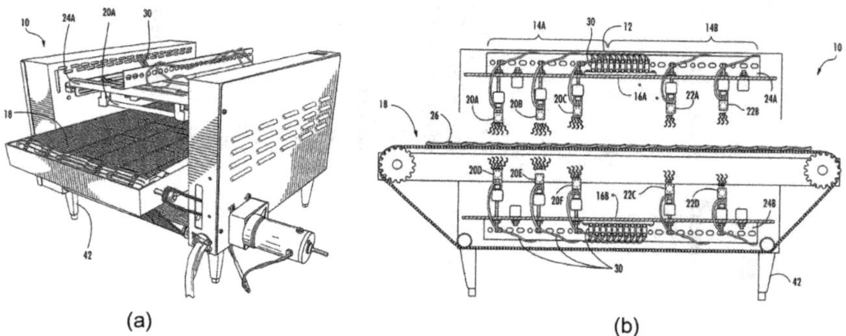

Figure 10.8 (a) Perspective and (b) cross-sectional views of controlled radiant frying equipment with multi-zone facility.

(*Source:* Farkas et al., 2007.)

emitter, and duration of product exposure. CDR-fried products exhibited significantly lesser oil content and organoleptic properties compared to deep-fried products (Lloyd et al., 2004). CDR frying has proved to be an effective alternative frying technology as it has the ability to imitate the heat flux generated during deep fat frying for moisture removal, formation of crust, and browning. Despite its huge potential, very few studies have been done on its application on different food products (Table 10.6).

10.4.6 Air Frying

Air frying is a highly popular and established alternative frying method. It is a rapid cooking process, where raw food products are fried by blowing hot-air over them at high velocity. Instead of immersing into hot oil, in this method, the food material is brought in direct contact with a fine mist consisting of oil droplets and hot air inside the frying chamber.

Most air fryers allow intimate interaction between food and oil-laden air, which increases heat transfer area resulting in an extremely high heat transfer rate and uniform distribution of heat throughout the product. High heat transfer rates could be achieved either by a built-in air blower or by coupling convective and radiative heat transfer. According to Erickson and Minn (1989), the frying chamber of air fryers is shaped in such a way that air can blow at significantly higher velocities than in typical convective hot-air ovens. The conflicting, colliding turbulent hot air flow toward the food facilitates accelerated cooking. Since the food item is positioned close to the heating element (used for heating air), it gets cooked effectively without much energy loss. The hot-air circulation of air fryer mimics the movement and flow of heat currents of boiling oil in a pot; this fashion of hot-air movement cooks the food's interior while developing crust on its surface (responsible for crispiness) (Joshy et al., 2020). The schematic diagram of a typical air fryer is shown in Figure 10.9.

Even distribution of heat helps in enhancing texture and organoleptic properties of fried product. According to Usman and Vanhaverbeke (2017), air drying can reduce oil absorption by about 80%. Andrés et al. (2013) observed that deep oil-fried French fries absorbed almost ten times more oil than air-fried French fries. However, the air-fried fries had less consistent crust which led to lesser resistance to moisture loss, and ultimately adversely affected sensory quality. They also studied the kinetics of mass transfer and volume changes in hot-air frying and immersion frying at the same operating temperature of 180°C, and observed that both the methods are affected by type of heating medium. The heat transfer was slower in the case of air medium than in the case of oil medium because the heat transfer coefficient of air is lower than oil. Based on few studies, the taste,

Table 10.6 Application of Radiant Frying on Different Food Materials

Food Products	*Important Salient Findings*	*References*
Chicken patties	• FryLess 100 K radiant fryer was used for frying chicken patties. • Radiant-fried patties were slightly lighter in color than immersion-fried patties. • Radiant-fried patties had about 16% lesser oil content than immersion-fried patties. • Radiant-fried patties had about 19% more moisture than immersion-fried patties. • According to sensory analysis, the flavor and mouthfeel of radiant-fried patties were better than those of the immersion-fried patties, whereas the crispiness and appearance were better for immersion-fried patties. Both radiant- and immersion-fried patties had comparable overall acceptability.	Nelson III et al. (2013)
Wheat donuts	• Infrared radiation has been used to finish fry partially fried wheat donuts. • All infrared-treated donuts had a significantly lower fat content than the untreated ones. • The overall acceptability of infrared-finished donuts was comparable with the untreated ones.	Melito and Farkas (2012)
French fries	• CDR-fried products had the highest peak breaking force. • Color analysis showed a similar yellow index (*b* value) of CDR- and immersion-fried fries. • Sensory analysis showed a similar overall acceptability of CDR- and immersion-fried fries.	Lloyd et al. (2004)

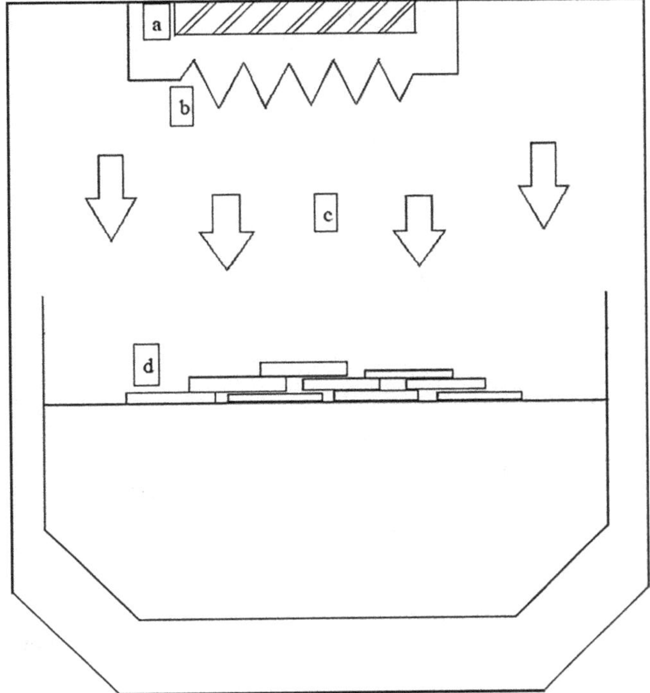

Figure 10.9 Schematic diagram of a typical air fryer consisting of (a) fan, (b) electrical resistance heater, (c) hot air directed toward food surface, and (d) food samples.

(*Source:* Teruel et al., 2015.)

color, odor, appearance, hardness, and crispiness of air-fried food products are quite different from those of deep-fried ones. However, the investigation of Shaker (2014) found no significant difference between the sensory characteristics of deep- and air-fried products. Moreover, the air-fried products were preferred over deep-fried products because of their crispiness, hardness, less oily surface, and color. The scanning electron microscopy (SEM) and differential scanning calorimetry (DSC) studies conducted by Teruel et al. (2015) showed that the immersion-fried products undergo a greater extent of starch gelatinization, characterized by formation of thick and dry crust. This is a result of intense evaporation of the surface water. In the case of air-fried products, the water evaporation is relatively slower, and the crust is thinner, homogenous, and smooth, resulting in the development of a visibly different texture. The lower degree of gelatinization (under low temperature) in air-fried products leads to formation of a harder core compared to that of immersion-fried products. Table 10.7 presents some of the studies conducted on air frying of different food materials. From the studies conducted on air frying, it has been observed that air frying is a costly and least efficient method among all the alternative frying technologies discussed. Combination of air frying with other treatments such as microwave heating may help in compensating its limitations.

10.5 Post-Frying Treatments for Reduction of Oil Absorption

The primary purpose behind the development of different pre-frying treatments, alternative frying technologies, and post-frying treatments is reduction of final oil content without adversely affecting the flavor and taste of the fried product. Majority of oil absorption occurs after frying, especially when the fried products are left for cooling. During cooling, the oil adhered on the surface penetrates into the fried food through the porous opening formed by the expelled moisture. As post-frying moisture adsorption is a surface phenomenon, the modification of the surface texture by making it less irregular and/or coating it with anti-adsorbents can significantly reduce excessive oil uptake (Moreno et al., 2010; Ananey-Obiri et al., 2018; Kurek et al., 2017; Barutcu et al., 2009b). According to Debnath et al. (2009), the oil uptake by the fried product can be reduced by holding the fried product at high temperatures and using a high-absorbent paper after frying. In the case of vacuum frying, the fried products can be effectively deoiled through centrifugation (Quan et al., 2014; Yamsaengsung et al., 2011; Moreira et al., 2009). Sothornvit (2011) observed that the combined effect of edible film coating and centrifugation lowered oil absorption of vacuum-fried banana chips by 33.71%. The fried products, after defatting, have the tendency to become rancid. In this regard, Banerjee and Sahu (2017) suggested the use of polyethylene tetraphthalate (PET) or aluminum film laminate with nitrogen gas filling as packaging material for the fried products. Ahmad Tarmizi and Niranjan (2013) successfully reduced oil absorption of vacuum-fried potato chips up to 48% by conducting two-stage vacuum frying process. This process involved frying under moderate vacuum condition followed by application of higher vacuum for post-frying oil drainage.

10.6 Conclusion

Harmful health implications due to consumption of highly oily and toxic fried food products prepared using traditional deep fat frying necessitated

Table 10.7 Application of Air Frying on Different Food Materials

Food Products	*Important Salient Findings*	*References*
Falafel	• Frying temperature and time were inversely proportional to the moisture content, fat content, and whiteness index (L*) of the air-fried falafels. • Hardness and overall color difference (ΔE^*) of air-fried falafels were directly proportional to frying temperature and time. • Air frying led to reduction in the fat content of fried products by 45% compared to deep-fried ones. • The appearance and crispiness of the air-fried falafels scored over deep-fried falafels.	Fikry et al. (2021)
Eggplant	• Eggplant air fried for 25 minutes and extracted with 50% ethanol had the highest flavonoid content and a significant reserve of polyphenols. • Air drying for 25 minutes with 50% ethanolic extraction also resulted in the highest DPPH scavenging activity. • Overall, air frying method proved to be beneficial in terms of retaining valuable bioactive compounds of eggplant.	Salamatullah et al. (2021)
Fish skin	• Fried fish skin was prepared using different frying techniques, viz. air frying, vacuum frying, electrostatic frying, and conventional deep fat frying, to study their effect on oil uptake, moisture loss, breaking force, color, and microstructure. • Air-fried fish skins (180°C for 2–12 minutes) exhibited the lowest fat content with a smooth and uniform microstructure. • Air frying resulted in slower moisture evaporation rate and higher skin puffing ratio.	Fang et al. (2021)

(Continued)

Table 10.7 (*Continued*) Application of Air Frying on Different Food Materials

Food Products	*Important Salient Findings*	*References*
Surimi	• Increase in time and temperature of air frying led to decrease in surface moisture content of surimi while keeping the core moisture content intact. • Air frying enhanced hardness, gumminess, and chewiness of the fried products. • Hot-air circulation at a temperature greater than 160°C led to a significant change in thiobarbituric acid reactive substance (TBARS). • Lipid peroxidation and degradation of air-fried products led to development of new flavor compounds. • Under optimum conditions, air frying produced surimi with crispy texture, desirable flavor, and lower oil content.	Yu et al. (2020)
French fries	• The effect of frying technique (deep fat frying and air frying) and pre-treatment type (blanching and freezing) was investigated on the mechanical and optical properties of French fries. • Air-fried products underwent lesser increase in a^* value (red-green index) compared to the deep fat-fried ones indicating lower generation of acrylamide. • Texture analysis reports revealed that rate of both the initial softening stage and the following crust formation stage were slower in the case of air frying than in the case of deep fat frying.	Heredia et al. (2014)

development of innovative and healthier frying technologies. This chapter shed light on the working principles of some popular and potential alternative frying technologies along with few examples of their food applications. Some simple pre- and post-frying treatments were also discussed, which can be practiced with traditional deep fat frying as well as alternative frying methods for reducing oil absorption and preserving the characteristic sensory quality of the fried foods. The novel technologies such as vacuum frying, microwave frying, microwave-vacuum frying, ultrasound-assisted frying, high-pressure deep frying, radiant frying, and air frying successfully reduced oil absorption rate and retained moisture. However, some problems were still encountered, viz. non-uniformity of microwave heating, high capital cost of vacuum frying, unpleasant and hazardous noise generation during ultrasound-assisted frying, limited applications of high-pressure deep frying and radiant frying, and less preferable texture of air-fried products. Hence, further investigations are needed for overcoming the limitations of the above-discussed alternate frying technologies and producing healthy as well as decently palatable fried products.

References

Ahmad Tarmizi, Azmil Haizam, and Keshavan Niranjan. "Combination of moderate vacuum frying with high vacuum drainage—Relationship between process conditions and oil uptake." *Food and Bioprocess Technology* 6, no. 10 (2013): 2600–2608.

Al Faruq, Abdulla, Min Zhang, and Benu Adhikari. "A novel vacuum frying technology of apple slices combined with ultrasound and microwave." *Ultrasonics Sonochemistry* 52 (2019): 522–529.

Albertos, I., A. B. Martin-Diana, M. A. Sanz, J. M. Barat, A. M. Diez, I. Jaime, and D. Rico. "Effect of high pressure processing or freezing technologies as pretreatment in vacuum fried carrot snacks." *Innovative Food Science & Emerging Technologies 33* (2016a): 115–122.

Albertos, I., A. B. Martin-Diana, I. Jaime, A. M. Diez, and D. Rico. "Protective role of vacuum vs. atmospheric frying on PUFA balance and lipid oxidation." *Innovative Food Science & Emerging Technologies* 36 (2016b): 336–342.

Alden, Lorne B. "Method for cooking food in an infra-red conveyor oven." U.S. Patent 5,223,290, issued June 29, 1993.

Al-Khusaibi, Mohammed K., and Keshavan Niranjan. "The impact of blanching and high-pressure pretreatments on oil uptake of fried potato slices." *Food and Bioprocess Technology* 5, no. 6 (2012): 2392–2400.

Alimi, Buliyaminu Adegbemiro, Taofik Akinyemi Shittu, Lateef Oladimeji Sanni, and A. Toyin Arowolo. "Effect of some hydrocolloids as adjuncts on the quality of whole egg or egg white coated fried yam chips." *Journal of Food Agriculture and Environment* 11 (2013): 19–24.

Aminlari, M., R. Ramezani, and M. H. Khalili. "Production of protein-coated low-fat potato chips." *Food Science and Technology International* 11, no. 3 (2005): 177–181.

Ananey-Obiri, Daniel, Lovie Matthews, Malak H. Azahrani, Salam A. Ibrahim, Charis M. Galanakis, and Reza Tahergorabi. "Application of protein-based edible coatings for fat uptake reduction in deep-fat fried foods with an emphasis on muscle food proteins." *Trends in Food Science & Technology* 80 (2018): 167–174.

Andrés, Ana, Ángel Arguelles, Maria Luisa Castelló, and Ana Heredia. "Mass transfer and volume changes in French fries during air frying." *Food and Bioprocess Technology* 6, no. 8 (2013): 1917–1924.

Angor, Malak M. "Reducing fat content of fried potato pellet chips using carboxymethyl cellulose and soy protein isolate solutions as coating films." *Journal of Agricultural Science* 8, no. 3 (2016): 162–168.

Azam, S. M. Roknul, Haile Ma, Baoguo Xu, Shoma Devi, Md Abu Bakar Siddique, Sarah L. Stanley, Bhesh Bhandari, and Junsong Zhu. "Efficacy of ultrasound treatment in the removal of pesticide residues from fresh vegetables: A review." *Trends in Food Science & Technology* 97 (2020): 417–432.

Balasubramaniam, V. M., M. S. Chinnan, P. Mallikarjunan, and R. D. Phillips. "The effect of edible film on oil uptake and moisture retention of a deep-fat fried poultry product." *Journal of Food Process Engineering* 20, no. 1 (1997): 17–29.

Ballard, Tameshia S., and Parameswarakumar Mallikarjunan. "The effect of edible coatings and pressure frying using nitrogen gas on the quality of breaded fried chicken nuggets." *Journal of Food Science* 71, no. 3 (2006): S259–S264.

Banerjee, Soumitra, and Chandan Kumar Sahu. "A short review on vacuum frying-a promising technology for healthier and better fried foods." *International Journal of Nutrition and Health Sciences* 2 (2017): 68–71.

Barutcu, I., S. Sahin, and G. Sumnu. Acrylamide formation in different batter formulations during microwave frying. *LWT-Food Science and Technology* 42, no. 1 (2009a): 17–22.

Barutcu, Isil, Serpil Sahin, and Gulum Sumnu. "Effects of microwave frying and different flour types addition on the microstructure of batter coatings." *Journal of Food Engineering* 95, no. 4 (2009b): 684–692.

Belkova, Beverly, Jaromir Hradecky, Kamila Hurkova, Veronika Forstova, Lukas Vaclavik, and Jana Hajslova. "Impact of vacuum frying on quality of potato crisps and frying oil." *Food Chemistry* 241 (2018): 51–59.

Botero-Uribe, Maria, Melissa Fitzgerald, Robert G. Gilbert, and Jocelyn Midgley. "Effect of pulsed electrical fields on the structural properties that affect french fry texture during processing." *Trends in Food Science and Technology* 67 (2017): 1–11.

Cárcel, J. A., J. Benedito, C. Rosselló, and A. Mulet. "Influence of ultrasound intensity on mass transfer in apple immersed in a sucrose solution." *Journal of Food Engineering* 78, no. 2 (2007): 472–479.

Chen, Su-Der, Hui-Huang Chen, Yu-Chien Chao, and Rong-Shinn Lin. "Effect of batter formula on qualities of deep-fat and microwave fried fish nuggets." *Journal of Food Engineering* 95, no. 2 (2009): 359–364.

Dana, Dina, and I. Sam Saguy. "Mechanism of oil uptake during deep-fat frying and the surfactant effect-theory and myth." *Advances in Colloid and Interface Science* 128 (2006): 267–272.

Das, Rashmi, Deepthi P. Pawar, and Vinod Kumar Modi. "Quality characteristics of battered and fried chicken: Comparison of pressure frying and conventional frying." *Journal of Food Science and Technology* 50, no. 2 (2013): 284–292.

Dauliach, Michel Henri. "Electrical cooking apparatus for cooking precooked, deep-frozen or fresh food, of the fryer type, with no oil bath." U.S. Patent 5,910,264, issued June 8, 1999.

Debnath, Sukumar, N. K. Rastogi, A. G. Gopala Krishna, and B. R. Lokesh. "Oil partitioning between surface and structure of deep-fat fried potato slices: A kinetic study." *LWT-Food Science and Technology* 42, no. 6 (2009): 1054–1058.

Del Re, P. V., and N. Jorge. "Behavior of vegetable oils for frying discontinuous frozen pre-fried products." *Ciênc. Tecnol. Aliment* 26 (2006): 56–53.

Devi, Shoma, Min Zhang, and Chung Lim Law. "Effect of ultrasound and microwave assisted vacuum frying on mushroom (*Agaricus bisporus*) chips quality." *Food Bioscience* 25 (2018): 111–117.

Devi, Shoma, Min Zhang, Ronghua Ju, and Bhesh Bhandari. "Water loss and partitioning of the oil fraction of mushroom chips using ultrasound-assisted vacuum frying." *Food Bioscience* 38 (2020): 100753.

Duan, Xu, Min Zhang, Xinlin Li, and Arun S. Mujumdar. "Ultrasonically enhanced osmotic pretreatment of sea cucumber prior to microwave freeze drying." *Drying Technology* 26, no. 4 (2008): 420–426.

Dueik, V., and P. Bouchon. "Development of healthy low-fat snacks: Understanding the mechanisms of quality changes during atmospheric and vacuum frying." *Food Reviews International* 27, no. 4 (2011): 408–432.

Dueik, Verónica, Olajide Sobukola, and Pedro Bouchon. "Development of low-fat gluten and starch fried matrices with high fiber content." *LWT-Food Science and Technology* 59, no. 1 (2014): 6–11.

Eichenlaub, S., and C. Koh. "Modeling of food-frying processes." In *Modeling Food Processing Operations*, pp. 163–184. Woodhead Publishing, 2015.

Ekezie, Flora-Glad Chizoba, Da-Wen Sun, Zhang Han, and Jun-Hu Cheng. "Microwave-assisted food processing technologies for enhancing product quality and process efficiency: A review of recent developments." *Trends in Food Science & Technology* 67 (2017): 58–69.

Erdogdu, Ferruh, and Petr Dejmek. "Determination of heat transfer coefficient during high pressure frying of potatoes." *Journal of Food Engineering* 96, no. 4 (2010): 528–532.

Erickson, C. S., and P. Minn. "Air fryer." *Erickson (Ed.)* (1989).

Fang, Mingchih, Guan-Jing Huang, and Wen-Chieh Sung. "Mass transfer and texture characteristics of fish skin during deep-fat frying, electrostatic frying, air frying and vacuum frying." *LWT* 137 (2021): 110494.

Farkas, Brian, Brian Lloyd, and Kevin Keener. "Dynamic radiant food preparation methods and systems." U.S. Patent 7,307,243, issued December 11, 2007.

Feng, H., and J. Tang. "Microwave finish drying of diced apples in a spouted bed." *Journal of Food Science* 63, no. 4 (1998): 679–683.

Fikry, Mohammad, Ibrahim Khalifa, Rokkaya Sami, Ebtihal Khojah, Khadiga Ahmed Ismail, and Mokhtar Dabbour. "Optimization of the frying temperature and time for preparation of healthy falafel using air frying technology." *Foods* 10, no. 11 (2021): 2567.

García, M. A., C. Ferrero, A. Campana, N. Bértola, M. Martino, and N. Zaritzky. "Methylcellulose coatings applied to reduce oil uptake in fried products." *Food Science and Technology International* 10, no. 5 (2004): 339–346.

García-Segovia, Purificación, A. M. Urbano-Ramos, S. Fiszman, and Javier Martínez-Monzó. "Effects of processing conditions on the quality of vacuum fried cassava chips (*Manihot esculenta* Crantz)." *LWT-Food Science and Technology* 69 (2016): 515–521.

Gharachorloo, Maryam, Mehrdad Ghavami, Maryam Mahdiani, and Reza Azizinezhad. "The effects of microwave frying on physicochemical properties of frying and sunflower oils." *Journal of the American Oil Chemists' Society* 87, no. 4 (2010): 355–360.

Gupta, P., U. S. Shivhare, and A. S. Bawa. "Studies on frying kinetics and quality of French fries." *Drying Technology* 18, no. 1–2 (2000): 311–321.

Heredia, A., M. L. Castelló, A. Argüelles, and A. Andrés. "Evolution of mechanical and optical properties of French fries obtained by hot air-frying." *LWT-Food Science and Technology* 57, no. 2 (2014): 755–760.

Huang, Meng-Sha, Min Zhang, and Bhesh Bhandari. "Synergistic effects of ultrasound and microwave on the pumpkin slices qualities during ultrasound-assisted microwave vacuum frying." *Journal of Food Process Engineering* 41, no. 6 (2018): e12835.

Innawong, Bhundit, Parameswarakumar Mallikarjunan, Joseph Marcy, and John Cundiff. "Pressure conditions and quality of chicken nuggets fried under gaseous nitrogen atmosphere." *Journal of Food Processing and Preservation* 30, no. 2 (2006): 231–245.

Janositz, A., A-K. Noack, and D. Knorr. "Pulsed electric fields and their impact on the diffusion characteristics of potato slices." *LWT-Food Science and Technology* 44, no. 9 (2011): 1939–1945.

Joshy, C. G., G. Ratheesh, George Ninan, K. Ashok Kumar, and C. N. Ravishankar. "Optimizing air-frying process conditions for the development of healthy fish snack using response surface methodology under correlated observations." *Journal of Food Science and Technology* 57, no. 7 (2020): 2651–2658.

Karimi, Solomon, Michael Wawire, and Francis M. Mathooko. “Impact of frying practices and frying conditions on the quality and safety of frying oils used by street vendors and restaurants in Nairobi, Kenya.” *Journal of Food Composition and Analysis* 62 (2017): 239–244.

Karizaki, Vahid Mohammadpour, Serpil Sahin, Gulum Sumnu, Mohammad Taghi Hamed Mosavian, and Alexandru Luca. “Effect of ultrasound-assisted osmotic dehydration as a pretreatment on deep fat frying of potatoes.” *Food and Bioprocess Technology* 6, no. 12 (2013): 3554–3563.

Kawas, M. L., and R. G. Moreira. “Effect of degree of starch gelatinization on quality attributes of fried tortilla chips.” *Journal of Food Science* 66, no. 2 (2001): 300–306.

Krokida, M. K., V. Oreopoulou, Z. B. Maroulis, and D. Marinos-Kouris. “Effect of pre-treatment on viscoelastic behaviour of potato strips.” *Journal of Food Engineering* 50, no. 1 (2001): 11–17.

Kurek, Mia, Mario Ščetar, and Kata Galić. “Edible coatings minimize fat uptake in deep fat fried products: A review.” *Food Hydrocolloids* 71 (2017): 225–235.

Lebovka, Nikolai I., Iurie Praporscic, and Eugene Vorobiev. “Effect of moderate thermal and pulsed electric field treatments on textural properties of carrots, potatoes and apples.” *Innovative Food Science & Emerging Technologies* 5, no. 1 (2004): 9–16.

Lešková, Emília, Jana Kubíková, Eva Kováčiková, Martina Košická, Janka Porubská, and Kristína Holčíková. “Vitamin losses: Retention during heat treatment and continual changes expressed by mathematical models.” *Journal of Food Composition and Analysis* 19, no. 4 (2006): 252–276.

Liu, Tingting, Ethan Dodds, Sze Ying Leong, Graham T. Eyres, David John Burritt, and Indrawati Oey. “Effect of pulsed electric fields on the structure and frying quality of “kumara” sweet potato tubers.” *Innovative Food Science & Emerging Technologies* 39 (2017a): 197–208.

Liu, Rui, Yingquan Zhang, Liang Wu, Yanan Xing, Yan Kong, Junmao Sun, and Yimin Wei. “Impact of vacuum mixing on protein composition and secondary structure of noodle dough.” *LWT-Food Science and Technology* 85 (2017b): 197–203.

Liu, Caiyun, Nabil Grimi, Nikolai Lebovka, and Eugene Vorobiev. “Effects of preliminary treatment by pulsed electric fields and convective air-drying on characteristics of fried potato.” *Innovative Food Science & Emerging Technologies* 47 (2018a): 454–460.

Liu, Caiyun, Nabil Grimi, Nikolai Lebovka, and Eugene Vorobiev. “Effects of pulsed electric fields treatment on vacuum drying of potato tissue.” *LWT* 95 (2018b): 289–294.

Lloyd, Brian J., Brian E. Farkas, and Kevin M. Keener. “Quality comparison of French fry style potatoes produced by oven heating, immersion frying and controlled dynamic radiant heating.” *Journal of Food Processing and Preservation* 28, no. 6 (2004): 460–472.

Maity, Tanushree, P. S. Raju, and A. S. Bawa. "Effect of freezing on textural kinetics in snacks during frying." *Food and Bioprocess Technology* 5, no. 1 (2012): 155–165.

Maity, Tanushree, A. S. Bawa, and P. S. Raju. "Effect of preconditioning on physicochemical, microstructural, and sensory quality of vacuum-fried jackfruit chips." *Drying Technology* 36, no. 1 (2018): 63–71.

Mallikarjunan, P., M. S. Chinnan, V. M. Balasubramaniam, and R. D. Phillips. "Edible coatings for deep-fat frying of starchy products." *LWT-Food Science and Technology* 30, no. 7 (1997): 709–714.

Mason, Timothy J., Larysa Paniwnyk, and J. P. Lorimer. "The uses of ultrasound in food technology." *Ultrasonics Sonochemistry* 3, no. 3 (1996): S253–S260.

Melito, Helen S., and Brian E. Farkas. "Impact of infrared finishing on the mechanical and sensorial properties of wheat donuts." *Journal of Food Science* 77, no. 9 (2012): E224–E230.

Mestnik, Frank G., and Billy J. Alexander. "Oil-free fryer, food cooker." U.S. Patent 5,780,815, issued July 14, 1998.

Moreira, Rosana G., Jaime E. Palau, and X. Sun. "Deep-fat frying of tortilla chips: an engineering approach." *Food Technology (Chicago)* 49, no. 4 (1995): 146–150.

Moreira, Rosana G., M. Elena Castell-Perez, and Maria A. Barrufet. "Deep fat frying: Fundamentals and applications." (1999), Springer.

Moreira, Rosana G., Paulo F. Da Silva, and Carmen Gomes. "The effect of a de-oiling mechanism on the production of high quality vacuum fried potato chips." *Journal of Food Engineering* 92, no. 3 (2009): 297–304.

Moreno, M. C., and P. Bouchon. "A different perspective to study the effect of freeze, air, and osmotic drying on oil absorption during potato frying." *Journal of Food Science* 73, no. 3 (2008): E122–E128.

Moreno, María Carolina, Christopher A. Brown, and Pedro Bouchon. "Effect of food surface roughness on oil uptake by deep-fat fried products." *Journal of Food Engineering* 101, no. 2 (2010): 179–186.

Mottram, Donald S., Bronislaw L. Wedzicha, and Andrew T. Dodson. "Acrylamide is formed in the Maillard reaction." *Nature* 419, no. 6906 (2002): 448–449.

Mujumdar, Arun S., and Chung Lim Law. "Drying technology: Trends and applications in postharvest processing." *Food and Bioprocess Technology* 3, no. 6 (2010): 843–852.

Nelson III, Louis V., Kevin M. Keener, Kaitlin R. Kaczay, Preetha Banerjee, Jean L. Jensen, and Andrea Liceaga. "Comparison of the FryLess 100 K Radiant Fryer to oil immersion frying." *LWT-Food Science and Technology* 53, no. 2 (2013): 473–479.

Ngadi, Michael O., and Dae Kun Hwang. "Modelling heat transfer and heterocyclic amines formation in meat patties during frying." Agricultural Engineering International: the CIGR Ejournal Vol. IX, (2007): Manuscript BC 04 004.

Oladejo, Ayobami Olayemi, Haile Ma, Wenjuan Qu, Cunshan Zhou, Bengang Wu, Xue Yang, and Daniel I. Onwude. "Effects of ultrasound pretreatments on the kinetics of moisture loss and oil uptake during deep fat frying of sweet potato (*Ipomea batatas*)." *Innovative Food Science & Emerging Technologies* 43 (2017): 7–17.

Ostermeier, Robin, Oleksii Parniakov, Stefan Töpfl, and Henry Jäger. "Applicability of pulsed electric field (PEF) pre-treatment for a convective two-step drying process." *Foods* 9, no. 4 (2020): 512.

Ouchon, P. B., J. M. Aguilera, and D. L. Pyle. "Structure oil-absorption relationships during deep-fat frying." *Journal of Food Science* 68, no. 9 (2003): 2711–2716.

Oztop, Mecit Halil, Serpil Sahin, and Gulum Sumnu. "Optimization of microwave frying of potato slices by using Taguchi technique." *Journal of Food Engineering* 79, no. 1 (2007): 83–91.

Pan, Guangkun, Hongwu Ji, Shucheng Liu, and Xiaoqing He. "Vacuum frying of breaded shrimps." *LWT-Food Science and Technology* 62, no. 1 (2015): 734–739.

Pankaj, Shashi K., and Kevin M. Keener. "A review and research trends in alternate frying technologies." *Current Opinion in Food Science* 16 (2017): 74–79.

Pawar, Deepthi P., S. Boomathi, Swapna C. Hathwar, Amit Kumar Rai, and Vinod Kumar Modi. "Effect of conventional and pressure frying on lipids and fatty acid composition of fried chicken and oil." *Journal of Food Science and Technology* 50, no. 2 (2013): 381–386.

Pei, Zhisheng, Tingting Ma, Pan Wen, Changfeng Xue, Aiguo Feng, Chuan Li, Yunsheng Xu, and Xuanri Shen. "Effect of vacuum frying and atmospheric frying on the quality and protein oxidation of squid (*Loligo chinensis*)." *Journal of Food Science* 86, no. 10 (2021): 4316–4329.

Phule, Asmita S., and Uday S. Annapure. "Effect of coating of hydrocolloids on chickpea (*Cicer arietinum* L.) and green gram (*Vigna radiata*) splits during deep fat frying." *International Food Research Journal* 20, no. 2 (2013): 565–573.

Quan, Xiaojian, Min Zhang, Weiming Zhang, and Benu Adhikari. "Effect of microwave-assisted vacuum frying on the quality of potato chips." *Drying Technology* 32, no. 15 (2014): 1812–1819.

Rahimi, Jamshid, and Michael O. Ngadi. "Effect of batter formulation and pre-drying time on oil distribution fractions in fried batter." *LWT-Food Science and Technology* 59, no. 2 (2014): 820–826.

Reeve, R. M., and E. M. Neel. "Microscopic structure of potato chips." *American Potato Journal* 37, no. 2 (1960): 45–52.

Ren, Aiqing, Siyi Pan, Weirong Li, Guobao Chen, and Xu Duan. "Effect of various pretreatments on quality attributes of vacuum-fried shiitake mushroom chips." *Journal of Food Quality* 2018 (2018): 1–7.

Rojas-Gonzalez, Juan A., Sylvie Avallone, Pierre Brat, Gilles Trystram, and Philippe Bohuon. "Effect of deep-fat frying on ascorbic acid, carotenoids and potassium contents of plantain cylinders." *International Journal of Food Sciences and Nutrition* 57, no. 1–2 (2006): 123–136.

Rosén, Johan, and Karl-Erik Hellenäs. "Analysis of acrylamide in cooked foods by liquid chromatography tandem mass spectrometry." *Analyst* 127, no. 7 (2002): 880–882.

Rossi Marquez, Giovanna, Prospero Di Pierro, Marilena Esposito, Loredana Mariniello, and Raffaele Porta. "Application of transglutaminase-cross-linked whey protein/pectin films as water barrier coatings in fried and baked foods." *Food and Bioprocess Technology* 7, no. 2 (2014): 447–455.

Sahin, Serpil, Gulum Sumnu, and Mecit Halil Oztop. "Effect of osmotic pre-treatment and microwave frying on acrylamide formation in potato strips." *Journal of the Science of Food and Agriculture* 87, no. 15 (2007): 2830–2836.

Salamatullah, Ahmad Mohammad, Mohammed Asif Ahmed, Mohammed Saeed Alkaltham, Khizar Hayat, Najla Sulaiman Aloumi, Alhanouf Mohammed Al-Dossari, Laila Naif Al-Harbi, and Shaista Arzoo. "Effect of air-frying on the bioactive properties of eggplant (*Solanum melongena* L.)." *Processes* 9, no. 3 (2021): 435.

Schiffmann, R. "Microwave-assisted frying." In *The microwave processing of foods*, pp. 142–151. Woodhead Publishing, 2017.

Sensoy, Ilkay, Serpil Sahin, and Gulum Sumnu. "Microwave frying compared with conventional frying via numerical simulation." *Food and Bioprocess Technology* 6, no. 6 (2013): 1414–1419.

Shaker, M. Arafat. "Air frying a new technique for produce of healthy fried potato strips." *Journal of Food and Nutrition Sciences* 2, no. 4 (2014): 200–206.

Skog, K. I., M. A. E. Johansson, and M. I. Jägerstad. "Carcinogenic heterocyclic amines in model systems and cooked foods: a review on formation, occurrence and intake." *Food and Chemical Toxicology* 36, no. 9–10 (1998): 879–896.

Song, Xian-Ju, Min Zhang, and Arun S. Mujumdar. "Effect of vacuum-microwave predrying on quality of vacuum-fried potato chips." *Drying Technology* 25, no. 12 (2007): 2021–2026.

Sothornvit, Rungsinee. "Edible coating and post-frying centrifuge step effect on quality of vacuum-fried banana chips." *Journal of Food Engineering* 107, no. 3–4 (2011): 319–325.

Soto, Marvin, Ana Mercedes Pérez, Adrien Servent, Fabrice Vaillant, and Nawel Achir. "Monitoring and modelling of physicochemical properties of papaya chips during vacuum frying to control their sensory attributes and nutritional value." *Journal of Food Engineering* 299 (2021): 110514.

Su, Ya, Min Zhang, and Weiming Zhang. "Effect of low temperature on the microwave-assisted vacuum frying of potato chips." *Drying Technology* 34, no. 2 (2016a): 227–234.

Su, Ya, Min Zhang, Weiming Zhang, Benu Adhikari, and Zaixing Yang. "Application of novel microwave-assisted vacuum frying to reduce the oil uptake and improve the quality of potato chips." *LWT* 73 (2016b): 490–497.

Su, Ya, Min Zhang, Weiming Zhang, Chunquan Liu, and Benu Adhikari. "Ultrasonic microwave-assisted vacuum frying technique as a novel frying method for potato chips at low frying temperature." *Food and Bioproducts Processing* 108 (2018a): 95–104.

Su, Ya, Min Zhang, Zhongxiang Fang, and Weiming Zhang. "Analysis of dehydration kinetics, status of water and oil distribution of microwave-assisted vacuum frying potato chips combined with NMR and confocal laser scanning microscopy." *Food Research International* 101 (2017): 188–197.

Su, Ya, Min Zhang, Bhesh Bhandari, and Weiming Zhang. "Enhancement of water removing and the quality of fried purple-fleshed sweet potato in the vacuum frying by combined power ultrasound and microwave technology." *Ultrasonics Sonochemistry* 44 (2018b): 368–379.

Szadzińska, J., S. J. Kowalski, and M. Stasiak. "Microwave and ultrasound enhancement of convective drying of strawberries: Experimental and modeling efficiency." *International Journal of Heat and Mass Transfer* 103 (2016): 1065–1074.

Teixidó, E., F. J. Santos, L. Puignou, and M. T. Galceran. "Analysis of 5-hydroxymethylfurfural in foods by gas chromatography–mass spectrometry." *Journal of Chromatography A* 1135, no. 1 (2006): 85–90.

Teruel, M., M. Gordon, M. B. Linares, M. D. Garrido, A. Ahromrit, and K. Niranjan. "A comparative study of the characteristics of french fries produced by deep fat frying and air frying." *Journal of Food Science* 80, no. 2 (2015): E349–E358.

Teruel, M., P. García-Segovia, J. Martínez-Monzó, M. Belén Linares, and M. Dolores Garrido. "Use of vacuum-frying in chicken nugget processing." *Innovative Food Science & Emerging Technologies* 26 (2014): 482–489.

Tharasena, Busaba, and Siriporn Lawan. "Phenolics, flavonoids and antioxidant activity of vegetables as Thai side dish." *APCBEE Procedia* 8 (2014): 99–104.

Usman, Muhammad, and Wim Vanhaverbeke. "How start-ups successfully organize and manage open innovation with large companies." *European Journal of Innovation Management* 20, no. 1 (2017): 171–186.

Van Koerten, K. N., M. A. I. Schutyser, D. Somsen, and R. M. Boom. "A pore inactivation model for describing oil uptake of French fries during pre-frying." *Journal of Food Engineering* 146 (2015): 92–98.

Van Marle, J. T., A. C. M. Clerkx, and A. Boekestein. "Cryo-scanning electron microscopy investigation of the texture of cooked potatoes." *Food Structure* 11, no. 3 (1992): 3.

Wang, Yuchuan, Min Zhang, Arun S. Mujumdar, Kebitsamang Joseph Mothibe, and S. M. Roknul Azam. “Study of drying uniformity in pulsed spouted microwave–vacuum drying of stem lettuce slices with regard to product quality.” *Drying Technology* 31, no. 1 (2013): 91–101.

Wang, Yan, Wangang Zhang, and Guanghong Zhou. “Effects of ultrasound-assisted frying on the physiochemical properties and microstructure of fried meatballs.” *International Journal of Food Science & Technology* 54, no. 10 (2019): 2915–2926.

Wu, Yali, Yuming Guo, and Dongguang Zhang. “Study of the effect of high-pulsed electric field treatment on vacuum freeze-drying of apples.” *Drying Technology* 29, no. 14 (2011): 1714–1720.

Xin, Ying, Min Zhang, Baoguo Xu, Benu Adhikari, and Jincai Sun. “Research trends in selected blanching pretreatments and quick freezing technologies as applied in fruits and vegetables: A review.” *International Journal of Refrigeration* 57 (2015): 11–25.

Yamsaengsung, Ram, Thaworn Ariyapuchai, and Kulchanat Prasertsit. “Effects of vacuum frying on structural changes of bananas.” *Journal of Food Engineering* 106, no. 4 (2011): 298–305.

Yu, Xina, Linqiu Li, Jing Xue, Jie Wang, Gongshuai Song, Yiqi Zhang, and Qing Shen. “Effect of air-frying conditions on the quality attributes and lipidomic characteristics of surimi during processing.” *Innovative Food Science & Emerging Technologies* 60 (2020): 102305.

Yuksel, F., Safa Karaman, and A. Kayacier. “Barley flour addition decreases the oil uptake of wheat chips during frying.” *Quality Assurance and Safety of Crops & Foods* 7, no. 5 (2015): 621–628.

Zhang, Xiaotian, Min Zhang, and Benu Adhikari. “Recent developments in frying technologies applied to fresh foods.” *Trends in Food Science & Technology* 98 (2020): 68–81.

Zhou, Xu, Shuang Zhang, Zhongwei Tang, Juming Tang, and Pawan S. Takhar. “Microwave frying and post-frying of French fries.” *Food Research International* 159 (2022): 111663.

Zyzak, David V., Robert A. Sanders, Marko Stojanovic, Daniel H. Tallmadge, B. Loye Eberhart, Deborah K. Ewald, David C. Gruber, Thomas R. Morsch, Melissa A. Strothers, George P. Rizzi, and Maria D. Villagran. “Acrylamide formation mechanism in heated foods.” *Journal of Agricultural and Food Chemistry* 51, no. 16 (2003): 4782–4787.

Chapter 11

Techno-Functional Attributes of Frying Additives

Richa Singh and Anusha Kishore
ICAR-National Dairy Research Institute

DOI: 10.1201/9781003329244-11

11.1 Introduction

The frying process entails immersing food in hot oil at temperatures ranging from 150°C to 190°C (Choe and Min, 2007). Frying degrades oil by accumulating polar compounds, resulting in a series of physicochemical reactions such as thermo-oxidation, hydrolysis, and polymerization (Orozco et al., 2011; Tena et al., 2014). Polar compounds include triacylglycerols that have been oxidized, dimerized, or polymerized, as well as diacylglycerols and free fatty acids. They are derived from triacylglycerols and accelerate the degradation of oil. These degradation products eventually accumulate in fried food due to the mass transfer process, lowering the nutritional quality of both the oil and the food (Zhang et al. 2012). Frying fat's quality also influences the finished product's quality, including flavor, texture, shelf life, and nutritional properties (Dunford, 2003), as well as storage stability.

11.2 Frying Oils and Their Physical and Chemical Properties

Frying fats range from non-hydrogenated completely refined fats and oils to especially hydrogenated frying goods. The key factors for selecting frying fat for a given application are the contribution of no off flavors and simplicity of handling (Dunford, 2003). Furthermore, the availability and cost of oil are crucial for the sector. Even the best-performing frying oil is ineffective if it is not accessible in sufficient quantities (Gupta, 2005). Specific product criteria such as appearance, seasoning adhesion, mouth feel, and fat retention are all significant factors to consider when choosing frying oil (Dunford, 2003). Regional preferences also exist for specific oils used in frying items. The sort of frying fats is used to be virtually entirely determined by where people lived (Blumenthal, 1996), i.e., the availability of oil and the flavor of oil they were accustomed to. Cottonseed oil, for example, is used in the United States for chips; groundnut oil on the Indian subcontinent; olive oil in the Mediterranean region; and animal fats in Northern Europe (Blumenthal, 1996; Gupta, 2005). Table 11.1 describes the chemical properties of vegetable oils, which are mostly used for frying food items.

11.3 Changes Occurring in Oil during Frying

The frying process is influenced by the food to be fried, the oil used, the temperature, and the time of frying. Deep fat frying decreases the proportion of unsaturated fatty acids in the oil and increases viscosity, density, specific heat, free fatty acids, foaming, color, polar materials, and polymeric

Table 11.1 Chemical Properties of Vegetable Oils (FSSAI, 2011)

	Coconut Oil	*Cotton Seed Oil*	*Ground Nut Oil*	*Mustard Oil*	*Soybean Oil*	*Virgin Olive Oil*
BR reading at 40°C	34.0–35.5	55.6–60.2	54.0–57.1	58.6–61.7	58.5–68	51.0–55.6
RI at 40°C	1.4481–1.4491	1.4630–1.4660	1.4620–1.4640	1.465–1.467	1.4649–1.4710	1.4600–1.4630
Saponification value (mg KOH/g oil)	Not less than 250	190–198	188–196	182–193	189–195	184–196
Iodine value	7.5–10	98–112	85–99	-	120–141	75–94
Polenske value	Not less than 13	-	-	-	-	-
Unsaponifiable matter	Not more than 1%	Not more than 1.5%	Not more than 1.0%	Not more than 20 g/kg	Not more than 1.5%	Not more than 15 g/kg
Acid value	Not more than 6.0	Not more than 0.5	Not more than 6.0	Not more than 0.6	Not more than 2.5	Not more than 6.0

compounds. Deep fat frying changes the flavor stability and quality of the oil through processes such as hydrolysis, oxidation, and polymerization. Essential amino acids, tocopherols, and fatty acids in foods are degraded during deep fat frying. The factors affecting the processes of hydrolysis, oxidation, polymerization, and eventually frying oil deterioration are unsaturation of fatty acids, type of frying oil, type of food, metals in oils and food, initial oil quality, degradation products of oils, and additives. Frying oil degradation can be managed and even inhibited by controlling these factors. Frying additives (antioxidants, antifoam agents, etc.) are added to food to prevent oxidation and foaming and increase the frying oil's shelf life (Table 11.2).

11.4 Frying Additives

As per codex (2019), whether or not it has nutritional value, a food additive is any substance that is intentionally added to food for a technological (including organoleptic) purpose in the manufacture, processing, preparation, treatment, packing, packaging, transport, or holding of such food and results in, or may be reasonably expected to result in, the production, processing, preparation, treatment, packaging, transport, or holding of the food. The term excludes contaminants or substances added to food to maintain or improve nutritional qualities. A frying additive is any substance added to frying oils to prevent oxidation and foaming, as well as to increase the shelf life of the frying oil. Additives can be of two types in regard to deep frying: (i) those substances added to food products that reduce oil uptake by the fried food and (ii) those substances added to frying oil to reduce oxidation and foam formation. Various Types of Frying Additives with the Permissible Limit in Oil is presented in Table 11.3. Frying additives can be

Table 11.2 Changes in Some Physical and Chemical Parameters of Oils/Fat during Deep Frying

	Changes during Deep Frying	*Reason*
Physical Parameters		
Refractive index	Increases	Accumulation of conjugated fatty acids
Density	Increases	Polymerized triacylglycerols
Dielectric coefficient	Decreases	Polar-oxidized components
Intense color	Maillard browning	Brown-pigmented compounds accumulation
Conductivity	Increases	Polar compounds
Surface tension	Decreases	Polar compounds

(*Continued*)

Table 11.2 (*Continued*) Changes in Some Physical and Chemical Parameters of Oils/Fat during Deep Frying

	Changes during Deep Frying	*Reason*
Smoke point	Decreases	Volatile-oxidized decomposition products
Specific heat	Increases	Polar compounds
Viscosity	Increases	Polymerized triacylglycerols
Chemical Parameters		
Anisidine value	Increases	Secondary oxidation products
Iodine value	Decreases	Formation of oxidized fat products
Peroxide value	Increases but can also decrease	Primary oxidation products
Petrol ether-insoluble oxidized fatty acids	Increase	Oxidized polymerization products
Polar compounds	Increase	Oxidized and polymerized degradation products including unchanged polar fat components
Polymerized triacylglycerols	Increases	Oxidized and not oxidized polymerized triacylglycerols
Acid value	Increases	Formation of oxidation products with free carboxyl groups

Table 11.3 Various Types of Frying Additives with the Permissible Limit in Oil (FSSAI, 2011; Gertz, 2004; Tokuşoğlu, 2014)

Type of Frying Additive	*International Numbering*	*Permissible Limit*
Antioxidants		
Ascorbic acid	E300	GMP
Ascorbic palmitate	E304	500 ppm max
Mixed tocopherols	E306	GMP
Alpha-tocopherol	E307	
Synthetic gamma-tocopherol	E308	
Synthetic delta-tocopherol	E309	
Propyl gallate (PG)	E310	200 ppm max
Octyl gallate	E311	
Dodecyl gallate	E312	

(*Continued*)

Table 11.3 (*Continued*) Various Types of Frying Additives with the Permissible Limit in Oil (FSSAI, 2011; Gertz, 2004; Tokuşoğlu, 2014)

Type of Frying Additive	*International Numbering*	*Permissible Limit*
Butylated hydroxyanisole (BHA)	E320	200 ppm max
Butylated hydroxyltoluene (BHT)	E321	-
Tertiary butylhydroquinone (TBHQ)	E319	200 ppm max
Organic Acids		
Lactic acid	E270	-
Citric acid	E330	100 ppm max
Sodium citrates (sodium dihydrogen citrate, disodium monohydrogen citrate, trisodium citrate)	E331	
Potassium citrates (potassium dihydrogen citrate, tripotassium citrate)	E332	
Calcium citrates	E333	
Anti-Spattering Agents		
Lecithin	E322	-
Emulsifiers		
Mono- and diglycerides of fatty acids	E471	-
Citric ester of mono- and diglycerides	E472c	-
Antifoaming Agents		
Dimethyl polysiloxane (DMPS)	E900	10 ppm max
Adsorbent/Filter Aids		
Minerals		Trade Name
Calcium silicate/magnesium silicate	E552/E553	Hubersorb® 600, Magnesol® XL
Sodium silicate	E550	Britesorb® F100, C201
Perlite/citric acid/water		Frypowder®
Silica		TriSyl®
Bentonite	E558	Tonsil® 314FF
Organic Materials		
Cellulose/citric acid		Maxfry® Filter Aid
Cellulose/charcoal		SuperSorb® CarbonPad

broadly classified into natural and synthetic frying additives. These additives in oils influence the quality of oil during deep fat frying. Tocopherols, butylated hydroxytoluene (BHT), butylated hydroxyanisole (BHA), propyl gallate (PG), and tert-butyl hydroquinone (TBHQ) slow down the oxidation of oil at room temperature. These additives become less effective at frying temperatures due to losses through volatilization or decomposition. An antifoaming agent, dimethyl polysiloxane (DMPS), is added to the long-life frying oil. Frying oil stability can be moderately improved by different kinds of antioxidants.

11.4.1 Antioxidants

Antioxidants are used to enhance the shelf life by delaying lipid oxidation and preserving the quality of edible oils. Synthetic antioxidants are added to the oil to retard oxidation during storage and frying. Antioxidants for frying oils should remain stable when exposed to frying temperatures and also provide protection to the foods fried in these oils, thus increasing the shelf life of the final products. Antioxidants are added to frying oils early in their production to protect the oils during handling, transport, and storage. The ability of antioxidants to delay lipid oxidation in fats and oils is well understood, though most available literature has focused on their effects at room temperature during storage or at the moderate temperatures of accelerated tests. Autoxidation reactions that occur at room or moderate temperatures are relatively slow; the major products formed are hydroperoxides, and their concentration increases until the advanced stages of oxidation. Primary antioxidants' activity under these conditions is due to their ability to donate hydrogen to lipid free radicals, preventing a new lipid molecule from entering the chain to be oxidized. However, oxidation at high temperatures in food processes such as frying is far more complex because it involves both oxidative and thermal reactions. At high temperatures, new compounds form very quickly, oxygen pressure decreases, and hydroperoxides decompose rapidly and are practically absent above 150°C, indicating that hydroperoxide decomposition becomes faster than their formation (Dobarganes et al., 1998). As a result of the combination of two variables: high temperature and low oxygen pressure, dimeric and oligomeric triglycerides form from the very early stages of heating, and a significant portion of the new compounds formed are non-oxygenated compounds (Dobarganes and Márquez-Ruiz, 2007). Stabilizing edible oils and fats under frying conditions can be difficult because the rate of oxidation is quite high under extreme temperature and time conditions, and the poor thermal stability of some antioxidants may cause premature decomposition. The majority of the antioxidants are volatile and evaporate during the frying process. Antioxidants are rapidly consumed during free-radical oxidation and degraded at high temperatures during frying. As a result, the desired level of antioxidants is not maintained during the frying process.

11.4.1.1 The Fate of Natural Antioxidants during Deep Frying

Plant-based antioxidants or natural antioxidants are natural and safer alternatives to synthetic antioxidants used in frying oils, e.g. tocopherols, tocotrienols, phytosterols, sesamol, sesamin, and sesamolin, and various natural extracts (grape seed, rosemary, thyme, oregano, pomegranate, sage, citrus peel, olive, and green tea). Tocopherols, tocotrienols, and carotenoids are naturally present in oils, which act as natural frying additives. With prolonged frying, these natural frying additives will decompose. Antioxidants delay frying oil oxidation, but the effectiveness of antioxidants decreases with an increase in frying temperature. Lignan compounds in sesame oil are also effective antioxidants in deep fat frying. Citric acid plays a synergist with tocopherols. Tocopherols are the primary antioxidants in vegetable oils. Tocopherols, when present in relatively low levels in vegetable oil, exert their utmost antioxidant activity. At very high concentrations, tocopherols behave as pro-oxidants. As free-radical scavengers, tocopherols play a significant role in protecting vegetable oils from oxidative degradation during frying.

Tocopherols in soybean oil, palm oil, and beef tallow decompose at 12.5%, 100%, and 100%, respectively, after 8 hours of frying at 150°C. Tocopherol retention in soybean oil reduced oxidation more than in beef tallow or palm oil that did not retain tocopherols. Palm oil contained tocotrienols in addition to tocopherols, which were all decomposed during the 8-hour frying of steamed noodles at 150°C (Choe and Lee, 1998).

In the absence of other antioxidants, carotenes do not protect the oil from thermal oxidation. Carotenes are important compounds in red palms that react with oil radicals (Schroeder et al., 2006). Carotene radicals are converted into carotenes by tocotrienols. The combination of tocotrienols and carotenes reduced oil oxidation synergistically during the 163°C frying of potato slices.

Sesamol, sesamin, and sesamolin, which are lignan compounds in sesame oil, can survive while in heating hours and thus contribute to the high oxidative stability of roasted sesame oil even at 170°C. When frying at 160°C, a mixture of soybean oil and roasted sesame oil reduced the formation of conjugated dienoic acids. The formation of conjugated diene decreased as the sesame oil content in the blended oil increased, possibly due to the antioxidants in sesame oil (Chung and Choe, 2001).

The addition of rice bran oil and sesame oil improved the oxidative and flavor stability of high oleic sunflower oil, which is due to avenasterol, which is stable at high temperatures (Kochhar, 2000). During deep fat frying, ascorbyl palmitate reduces dimer formation in oil and sterols and their fatty acid esters improved oil oxidative stability (Gordon and Kourimska, 1995).

11.4.1.2 The Fate of Synthetic/Artificial Antioxidants during Deep Frying

Synthetic antioxidants include BHA, BHT, THHQ, ferulic acid, and esterified ferulic acids. TBHQ is a phenolic antioxidant that is synthesized and used to improve the stability of edible oils during frying. At high temperatures, TBHQ is relatively stable and less volatile than BHA or BHT. The protective effects of TBHQ not only improve the stability of oils during frying, but also extend the shelf life of fried foods.

According to the literature, BHT and TBHQ have the highest frying temperature volatility, while BHA has intermediate volatility and PG is less volatile. In terms of thermal oxidation resistance, the order discovered was BHT>PG>BHA>TBHQ (Hamama and Nawar, 1991). However, Allam and Mohamed (2002) reported that among BHA, BHT, TBHQ, and propyl gallate, TBHQ showed the highest thermal stability. They have conducted the most comprehensive study, with the goal of clarifying the effect of the most important synthetic antioxidants, BHT, BHA, PG, and TBHQ, as well as their binary and tertiary mixtures with synergists and tocopherols. The authors measured the loss of the oil stability index of each system in sunflower oil after 1 hour of heating at 180°C. It was discovered that the protection provided by antioxidants at high temperatures was unrelated to the protection provided by antioxidants at low temperatures, and it was also discovered that the interaction of antioxidants at high temperatures could result in a negative or positive synergism.

The major degradation product of TBHQ was discovered to be tertiary butyl benzoquinone (TBBQ). In terms of the remaining antioxidant activity of synthetic antioxidant degradation products, the low effectiveness of BHT and BHA under frying conditions suggests that their oxidation products have no antioxidant activity, despite the fact that some of them have been reported as potent antioxidants at room or moderate temperatures (Kurechi and Kato, 1980). TBBQ was detected as a major degradation compound from TBHQ at frying temperature, and rapid TBHQ degradation is compatible with an important carry-through effect attributed to the TBHQ–TBBQ cycling system. In this regard, novel antioxidants derived from TBHQ but with higher oil solubility, namely lauryl TBHQ and lauryl TBBQ, have been discovered to have stronger antioxidant activity than TBHQ at temperatures above 140°C (Zhang et al., 2004).

11.4.2 Antifoaming Agents

The appearance of bubbles during the heating of the oil is called foaming. When food is placed in frying oil, the moisture in the food is heated, and the resulting steam rises to the surface. This phenomenon is responsible for the characteristic bubbling of frying food. As the oil in the frying vat heats up, it becomes more viscous, trapping an increasing amount of moisture and air. After frequent use, moisture, air, starch, and other contaminants from the food or already existing in the oil accumulate on the surface and form the dreaded foam. "Antifoaming agent" means a substance that retards deteriorative changes and foaming height during heating. As per the Food Safety and Standards Authority of India (FSSAI), DMPS (food grade) may be used as an antifoaming agent in edible oils and fats for deep fat frying up to a maximum limit of 10 parts per million. Defoamers based on silicone emulsions of the DMPS type are commonly used in the food business (food code E-900). These products have a strong and quick antifoam impact, but their effectiveness fades with time. This necessitates the continuous injection of the antifoaming chemical during the manufacturing process in order to maintain its effectiveness. Furthermore, food legislation restricts the use of silicones to specific food groups.

Fats and oils become very vulnerable to break down at high temperatures during frying and tend to create unwanted foam due to oxidative polymerization. Silicones are widely used as antifoaming agents in aqueous environments, and their action is explained by the surface chemical nature of the polymer (Miura, 1993). The addition of silicones, particularly DMPS, to oils at extremely low concentrations has been shown to improve oil stability at high temperatures during the frying process. Under these conditions, the antifoaming effects discovered may be an indirect outcome of the lower occurrence of oxidation (Zwobada, 1979).

11.4.3 Additives to Reduce Oil Uptake by Fried Food

The level of oil content in fried foods has been identified as a crucial element contributing to various health concerns. In fried raw materials, there are two mass transfers: the first is where water and soluble components escape from the inside of foods to the surface crust, and from here water is evaporated and migrates out of the fried items, while the second is oil penetration into the food products. The water vapor leaves voids for the oil to infiltrate; hence, oil uptake is mostly determined by the water content of fried raw materials (Rossi Marquez et al., 2014). To minimize the oil uptake while frying without harming the quality of fried dishes, it is vital to understand the mechanism of oil penetration during the frying process.

Previously, it was assumed that oil uptake occurs mostly during the frying process, as the moisture from the capillaries evaporates out of the meal. However, recent studies have indicated that the majority of oil uptake of fried goods occurs during the cooling phase when they are suctioned out of the oil. Cooling induces condensation of water vapor in the pores to create pressure, resulting in oil inhalation from the surface into the pores of the fried items. Consider the oil absorption in French fries, which can be broken down into three stages (Van Koerten et al., 2015):

1. Internal oil/structural oil absorbed by the chips during the frying process;
2. The absorbed surface oil was taken up by the chips immediately after removing them from the frying pan;
3. During the cooling process, the surface oil adhered to the frying surface;

The oil type may have an effect on oil uptake by food to be fried due to differences in the viscosity of the oil. In the literature, contradictory results have been reported, and some studies noted that oil origin and viscosity had no effect on oil uptake by fried foods (Kim et al., 2010); however, Surojanametakul et al. (2020) reported that frying in coconut oil had higher oil absorption than palm oil. Coconut oil had a viscosity of 73.30 ± 0.14 centipoise, while palm oil had a viscosity of 89.35 ± 0.21 centipoise. Coconut oil exhibited a lower viscosity than palm oil, which is linked negatively with the heat transfer coefficient (Debnath et al., 2012) and reduced batter expansion during frying. According to Krokida et al. (2000), higher oil viscosity results in reduced porosity values of fried items. According to Gertz (2014), also oil viscosity influences oil migration to items.

11.4.4 Protective Gases

According to reports, flushing frying oils with nitrogen or carbon dioxide reduces dissolved oxygen as well as thermal deterioration during oil frying. When compared to ordinary frying conditions, carbon dioxide blanketing and vacuum frying reduce the amount of total polar compounds and unsaturated fatty acids in canola oil. Carbon dioxide blanketing and vacuum frying both reduced the rate of tocopherol breakdown (Aladedunye and Przybylski, 2009). Nitrogen or carbon dioxide flushing minimizes dissolved oxygen in oil and thus oil oxidation during deep fat frying. Because of its higher solubility and density, carbon dioxide protects against oxidation better than nitrogen. Prior to heating, a minimum of 5 minutes of carbon dioxide or 15 minutes of nitrogen flushing reduces oil oxidation during deep fat frying (Przybylski and Eskin, 1988). In containers, the linear flow of nitrogen and carbon dioxide should be 50 cm/min.

11.4.5 Others

Anti-spattering agents, emulsifiers, and filtering aids are used to enhance the quality of frying oils. Anti-spattering agents are compounds added to frying oils and fats to reduce potentially dangerous spattering, such as lecithin, sucrose esters of fatty acids, and derivatives of mono- and diglycerides. The water droplets' surrounding phospholipid membrane prevents coalescence. Severe coalescence would result in an explosive evaporation and the hot fat would spatter. Anti-spattering refers to the movement of the stabilized water droplets to the surface, where the water slowly evaporates (van Nieuwenhuyzen and Szuhaj, 1998). They work by preventing water droplets from clumping together. A number of substances are promoted with the intention of regenerating used frying oils. These substances are known as "active" filter aids. They are said to adsorb polar oil breakdown products such as mono-and diglycerides, oxidized triglycerides, oligomeric triglycerides, and free fatty acids, and to retain them in place for removal by filtration. It is preferable to remove particles by filtration and soluble impurities by adsorption (active filtration) simultaneously, as it will provide superior-quality fried foods and prolong the life of the oil. Yates and Caltwell (1993) investigated the absorptivity per gram for various filter aids, including activated aluminas, acid-activated bleaching earth, diatomaceous earth, silica, carbon, and synthetic magnesium silicate. Aluminas, bleaching earth, magnesium silicate, silica, and synthetic magnesium silicate were found to have higher absorption of polar compounds. Farag and El-Anany (2006) evaluated the quality of frying oil (soybean, sunflower, palm, and cottonseed oils) with and without the presence of filtering aids. To absorb the secondary oxidation products of the oil, they utilized synthetic (Magnesol XL) and natural (diatomaceous earth and kaolin) filter aids at 1%, 2%, and 4% levels. The frying procedure was carried out at 180°C±5°C for 12 hours of continuous heating. The fried oils were treated with synthetic and natural filter aids for 15 minutes at 105°C. Frying of oils resulted in significant increases in refractive index, color, foam height, viscosity, acid value, peroxide value, TBA value, conjugated diene, and polymer contents, and a decrease in iodine value that represents the degradation of frying oil. However, after treatment with Magnesol XL, diatomaceous earth, and kaolin, the quality of fried oil was improved due to adsorption of oil degradation products by these adsorbents.

Organic acids are commonly used in combination with antioxidants, to produce special or synergistic effects. Citric acid does not act as an

Table 11.4 Commercially Available Frying Oil Formulations (Tokuşoğlu, 2014)

Product	*Composition*
RE08	Rosemary extract, citric and fatty acid esters of glycerol (E472c), mono- and diglycerides of fatty acids (E471)
RE09	Rosemary extract, polyoxyethylene, sorbitan monooleate, E472c, E471
G1021	Ascorbyl palmitate, tocopherol extract, E472c, E471
G1029	Ascorbyl palmitate, tocopherol extract, diacetyltartaric and fatty acid ester of glycerol (E472e), E471
Good-Fry Constituents	Rice bran, sesame oil
Good-Fry Plus	Rice bran, sesame oil, tocopherol, E472c, E471
Oilmaster	Ascorbyl palmitate, tocopherol extract, E472c, E471
Miroil Fryliquid	Citric acid, water, rosemary extract, curcuma, lactic and fatty acid esters of glycerol (E472b), lecithin, ascorbic acid
Maxfry Classic nat	Rice bran, sesame oil, E471, E472b, E472c, natural extracts, citric acid
Maxfry Classic	Vegetable oil, tocopherol extract, ascorbic palmitate, E471, E472b, E472c, citric acid

acidity regulator here. On the fryer's steel surface, citric acid is thought to chelate or combine with metal ions. Iron and copper promote the formation of radical degradation products. As a result, adding citric acid to frying oils prevents the deposition of brown degradation products on the surface of the fryer's steel (Gertz, 2004). Commercially available blends of frying additives are summarized in Table 11.4.

11.5 Conclusion

Although fried foods have been on the global menu for a long time, one of the main issues with fried foods is their high oil content, which has been linked to an increase in the prevalence of diseases such as obesity, high cholesterol, and high blood pressure. Thermally exploited fats (heated above 180°C) always result in a fried product with a dull appearance, brownish color, and waxy flavor. The use of frying additives consistently and constantly preserves the product's inherent/typical quality attributes. Frying additives have the potential to enhance food quality and extend oil life, but it is an error to believe that by filtering, treating, or adding special additives, the oil can be used indefinitely.

References

Aladedunye, Felix A., and Roman Przybylski. "Protecting oil during frying: A comparative study." *European Journal of Lipid Science and Technology* 111, no. 9 (2009): 893–901.

Allam, Samah Said Mahmoud, and Hani Moustafa Aly Mohamed. "Thermal stability of some commercial natural and synthetic antioxidants and their mixtures." *Journal of Food Lipids* 9, no. 4 (2002): 277–293.

Blumenthal, M. M. "Frying technology." *Bailey's Industrial Oils and Fats Products* 3 (1996): 429–481.

Choe, Eun-Ok, and Jin-Young Lee. "Thermooxidative stability of soybean oil, beef tallow and palm oil during frying of steamed noodles." *Korean Journal of Food Science and Technology* 30, no. 2 (1998): 288–292.

Choe, E., and D. B. Min. "Chemistry of deep-fat frying oils." *Journal of Food Science* 72, no. 5 (2007): R77–R86.

Chung, Jisang, and Eunok Choe. "Effects of sesame oil on thermooxidative stability of soybean oil." *Food Science and Biotechnology* 10, no. 4 (2001): 446–450.

Codex Alimentarius. "Standard for edible fats and oils not covered by individual standards." 402 CODEX STAN 19-1981. International Food Standards. (2019).

Debnath, Sukumar, Navin K. Rastogi, A. G. Gopala Krishna, and B. R. Lokesh. "Effect of frying cycles on physical, chemical and heat transfer quality of rice bran oil during deep-fat frying of poori: An Indian traditional fried food." *Food and Bioproducts Processing* 90, no. 2 (2012): 249–256.

Dobarganes, M. Carmen, and Gloria Márquez-Ruiz. "Formation and analysis of oxidized monomeric, dimeric, and higher oligomeric triglycerides." In *Deep Frying*. Editor: Michael D. Erickson (2nd Edn.). AOCS Press, 2007, pp. 87–110.

Dobarganes, M. Carmen, María del Carmen Pérez-Camino, and Gloria Márquez-Ruíz. "High performance size exclusion chromatography of polar compounds in heated and non-heated fats." *Lipid/Fett* 90, no. 8 (1988): 308–311.

Dunford, Nurhan. *Deep-Fat Frying Basics for Food Services: Fryer, Oil and Frying Temperature Selection*. Oklahoma Cooperative Extension Service, 2003.

Farag, Radwan S., and Ayman M. El-Anany. "Improving the quality of fried oils by using different filter aids." *Journal of the Science of Food and Agriculture* 86, no. 13 (2006): 2228–2240.

FSSAI. "Food safety and standards (food products standards and food additives) regulations." (2011): 280.

Gertz, Christian. "Optimising the baking and frying process using oil-improving agents." *European Journal of Lipid Science and Technology* 106, no. 11 (2004): 736–745.

Gertz, Christian. "Fundamentals of the frying process." *European Journal of Lipid Science and Technology* 116, no. 6 (2014): 669–674.

Gordon, Michael H., and Lenka Kourkimskå. "The effects of antioxidants on changes in oils during heating and deep frying." *Journal of the Science of Food and Agriculture* 68, no. 3 (1995): 347–353.

Gupta, M. K. "Frying oils." In F. Shahidi (ed.) *Bailey's Industrial Oil & Fat Products* (6th ed., Vol. 4). Wiley-Interscience, 2005, pp. 1–31.

Hamama, Anwar A., and Wassef W. Nawar. "Thermal decomposition of some phenolic antioxidants." *Journal of Agricultural and Food Chemistry* 39, no. 6 (1991): 1063–1069.

Kim, Juyoung, Deok Nyun Kim, Sung Ho Lee, Sang-Ho Yoo, and Suyong Lee. "Correlation of fatty acid composition of vegetable oils with rheological behaviour and oil uptake." *Food Chemistry* 118, no. 2 (2010): 398–402.

Kochhar, S. P. "Stable and healthful frying oil for the 21st century." *Inform* 11 (2000): 642–647.

Krokida, M. K., V. Oreopoulou, and Z. B. Maroulis. "Effect of frying conditions on shrinkage and porosity of fried potatoes." *Journal of Food Engineering* 43, no. 3 (2000): 147–154.

Kurechi, T., and T. Kato. "Studies on the antioxidants: XI. Oxidation products of concomitantly used butylated hydroxyanisole and butylated hydroxytoluene." *Journal of the American Oil Chemists' Society* 57, no. 7 (1980): 220–223.

Miura, Takahiro. "Surface chemistry and antifoaming of polysiloxane polymer." *Journal of Japan Oil Chemists' Society* 42, no. 10 (1993): 762–767.

Orozco-Solano, M. I., F. Priego-Capote, and M. D. Luque de Castro. "Influence of simulated deep frying on the antioxidant fraction of vegetable oils after enrichment with extracts from olive oil pomace." *Journal of Agricultural and Food Chemistry* 59, no. 18 (2011): 9806–9814.

Przybylski, R., and N. A. M. Eskin. "A comparative study on the effectiveness of nitrogen or carbon dioxide flushing in preventing oxidation during the heating of oil." *Journal of the American Oil Chemists' Society* 65, no. 4 (1988): 629.

Rossi Marquez, Giovanna, Prospero Di Pierro, Marilena Esposito, Loredana Mariniello, and Raffaele Porta. "Application of transglutaminase-cross-linked whey protein/pectin films as water barrier coatings in fried and baked foods." *Food and Bioprocess Technology* 7, no. 2 (2014): 447–455.

Schroeder, Maria T., Eleonora Miquel Becker, and Leif H. Skibsted. "Molecular mechanism of antioxidant synergism of tocotrienols and carotenoids in palm oil." *Journal of Agricultural and Food Chemistry* 54, no. 9 (2006): 3445–3453.

Surojanametakul, Vipa, Sujinna Karnasuta, and Prajongwate Satmalee. "Effect of oil type and batter ingredients on the properties of deep-frying flakes." *Food Science and Technology* 40 (2020): 592–596.

Tena Pajuelo, Noelia, Ramón Aparicio Ruiz, and Diego Luis García-González. "Use of polar and nonpolar fractions as additional information sources for studying thermoxidized virgin olive oils by FTIR." *Grasa y Aceites* 65, no. 3 (2014) :1–13.

Tokuşoğlu Ö.2014. The Less Oil Uptake Strategies in Deep-Fat Frying. 3rd International Conference and Exhibition on Food Processing & Technology. Hampton Inn Tropicana, Las Vegas, USA LECTURE PRESENTATION.

Van Koerten, K. N., M. A. I. Schutyser, D. Somsen, and R. M. Boom. "A pore inactivation model for describing oil uptake of French fries during pre-frying." *Journal of Food Engineering* 146 (2015): 92–98.

van Nieuwenhuyzen, Willem, and Beranrd F. Szuhaj. "Effects of lecithins and proteins on the stability of emulsions." *Lipid/Fett* 100, no. 7 (1998): 282–291.

Yates, R. A., and J. D. Caldwell. "Regeneration of oils used for deep frying: A comparison of active filter aids." *Journal of the American Oil Chemists' Society* 70, no. 5 (1993): 507–511.

Zhang, C. X., H. Wu, and X. C. Weng. "Two novel synthetic antioxidants for deep frying oils." *Food Chemistry* 84, no. 2 (2004): 219–222.

Zhang, Qing, Cheng Liu, Zhijian Sun, Xiaosong Hu, Qun Shen, and Jihong Wu. "Authentication of edible vegetable oils adulterated with used frying oil by Fourier Transform Infrared Spectroscopy." *Food Chemistry* 132, no. 3 (2012): 1607–1613.

Zwobada, F. "Additives and edible oils. Influence on quality." 26, no. 11 (1979): 435–453.

Chapter 12

Fried Foods

A Global Perspective on Food Safety Guidelines, and Packaging and Labeling Requirements

Sushama P S, Mukul Sain, Pushpalatha K, and Ganga A

ICAR-National Dairy Research Institute (NDRI)

DOI: 10.1201/9781003329244-12

12.1 Introduction

Food safety is an inescapable global public health issue. According to figures from the World Health Organization, contaminated food is projected to be the cause of approximately 420,000 deaths and 600 million illnesses annually (World Health Organization, 2022). Even while government around the world are working hard to increase the safety of the food supply, foodborne illness is still a serious concern for public health and the economy in both industrialized and developing nations. The prevalence of foodborne illnesses is influenced by recent changes to food production and processing methods as well as consumers' ever-evolving food habits. Today, many known pathogens are of significant concern, and new difficulties have emerged in recent years as a result of pathogen interactions with food matrix leading to life-threatening illnesses.

Some of the major trends in the emergence of microbiological food safety concerns and challenges include changes in consumer tastes, manufacturing and distribution systems, international trade and travel patterns, climatic and environmental changes, and rising anti-microbial resistance. Food safety is a serious public health concern, and governments, regulatory bodies, and huge multinational corporations are increasingly concerned about regaining customer trust in the food industry.

Ensuring global food safety is an integral part of the Sustainable Development Goal (SDG) too. Access to adequate quantities of safe and nourishing food is a key requirement of the 2030 Agenda for the SDG (United Nations, 2015).

12.2 Food Safety Guidelines around the World

The global food trade involves the production, marketing, and transportation of billions of tons of food annually. In addition to ensuring fair trade practices, it is important to guarantee consumer safety internationally. Globally, many international organizations lay down standards, guidelines, and recommendations to ensure the safety of food.

12.2.1 International Food Law

A universal point of reference is the Codex Alimentarius, sometimes known as the Food Code. The best standards now accessible are the Codex standards, which have been scientifically created by international experts. Standards for all foods, whether raw, semi-processed, or processed, are included in the Codex Alimentarius. Additionally, it covers broad areas including food additives, food hygiene, pesticides and veterinary drugs residues, contaminants, labeling and presentation, method of analysis and sampling, and import and export inspection and certification ("Codex Alimentarius | Food Safety and Quality | Food and Agriculture Organization of the United Nations" n.d.). Several committees are under the Codex that develop standards for various elements of food safety.

12.2.2 World Health Organization (WHO)

The WHO is a UN agency that was founded in 1948 and is accountable for international public health. Earlier in the 1990s, the WHO created the Ten Golden Rules for Safe Food Preparation, and these guidelines were widely followed. But it soon became clear that something more fundamental and comprehensive was required. The WHO released the Five Keys to Safer

Food poster in 2001 after nearly a year of deliberation with different stakeholders (World Health Organization, 2006).

It includes the following five keys:

1. **Keep clean**
Maintaining personal hygiene is important to prevent the contamination of food. In addition, proper cleaning of plates and utensils, and keeping away the pests from food preparation areas is the key to food safety. It is recommended to wash hands thoroughly for at least 20 seconds with soap prior to handling food and having food; after going to toilet, handling raw meat/poultry and chemicals, changing diaper, smoking, blowing one's nose, and handling garbage; and often during food preparation.
2. **Separate raw and cooked**
Poultry, seafood, and raw meat should be separated from other foods as they contain potentially hazardous microorganisms that may be transferred during handling or storage of food. Hence, to prevent cross-contamination, different utensils, knives, and containers should be used for raw and cooked foods.
3. **Cook thoroughly**
Studies have shown cooking food to at least 70°C kills most of the dangerous microorganisms. Food should always be thoroughly cooked, especially meat, poultry, eggs, and seafood. It is best to cook poultry and meat until the fluids are clear and the pink color vanishes.
4. **Keep food at safe temperatures**
The temperature between 5°C and 60°C is considered a "Danger Zone" as microorganisms proliferate fast during this stage. Food that has been cooked shouldn't be kept at room temperature for longer than 2 hours. All cooked and perishable food needs to be promptly refrigerated. Avoid thawing frozen foods at room temperature. Leftover food shouldn't be kept in the refrigerator for longer than three days, and it shouldn't be reheated more than once.
5. **Use safe water and raw materials**
Ice and water are examples of raw materials that can be contaminated with harmful chemicals and microorganisms. Spoilt food may contain toxic substances. The risk may be decreased by using caution when choosing raw materials and taking measures like washing and peeling.

12.2.3 Global Food Safety Initiative

To promote food safety and guarantee that everyone has access to safe food, a global collaboration was started in May 2000 ("Overview - MyGFSI,", n.d.).

This industry-driven initiative provides global recommendations on food safety management systems throughout the supply chain.

12.2.4 The International Food Safety Authorities Network (INFOSAN)

INFOSAN, a global network of national food safety authorities, is managed by the FAO and the WHO in collaboration. It aims to encourage the quick transmission of information during events linked to food safety and aids in the capacity building too (The International Food Safety Authorities Network INFOSAN Connecting Food Safety Authorities to Reduce Foodborne Risks Global Food Safety Food and Agriculture Organization of the United Nations, n.d.). This helps prevent the spread of contaminated food across the globe.

12.2.5 International Organization for Standardization

It is a global network of national standards. Its goal is to streamline global coordination and harmonize industry standards. Over 1,600 of the more than 21,500 of its international standards pertain to the food industry covering diverse areas including food products, essential oils, food safety management, fisheries and aquaculture, microbiology, starch and its by-products, traceability, hygiene, and contaminants. Several international standards created by the ISO are intended to guarantee food safety along the supply chain. The ISO 22000 series of standards covers food safety management by offering best practices and guidance for controlling food safety hazards (International Organization for Standardization, 2018). It includes standards concerned with food manufacturing, farming, catering, packaging, and animal foodstuffs and feed production.

12.3 Food-Packaging Requirements

Packaging is "the act of enclosing or protecting the product using a container to aid its distribution, identification, storage, promotion, and usage" (Nguyen et al., 2020).

The major functions of packaging include:

- Product containment and protection during transport, storage, or consumption;
- For easy usage and handling of the product;

- To differentiate the product;
- It is a new form of product marketing strategy;
- To increase customer convenience;
- It provides a communication media between the customer and the producer;
- It is a way to add aesthetic value to the product.

12.3.1 Packaging Requirements in EU

The Regulation (EC) No. 1935/2004 principally governs packaging requirements in the European Union. It outlines the fundamental standards for all food contact materials (FCMs) (European Commission, 2004).

According to the regulation, materials must not:

i. Release their constituents into food at amounts that are hazardous to human health;
ii. Make unacceptable modifications to food's flavor, aroma, and content;
iii. Mislead the consumers in terms of presentation, advertising, and labeling of a material or article.

Besides these, the regulation covers the following requirements under its ambit:

i. Specific requirements for the following groups of materials and articles: active and intelligent materials and articles, paper and board, ceramics, adhesives, rubbers, regenerated cellulose, glass, ion-exchange resins, cork, metals and alloys, plastics, printing inks, silicone, textiles, varnishes and coatings, waxes, and wood;
ii. General requirements and procedures for authorization and safety evaluation of substances used to make FCMs;
iii. Compliance and traceability.

12.3.1.1 Good Manufacturing Practices Regulation 2023/2006

This regulation lays down that the manufacturing process of FCM enlisted in Regulation (EC) No. 1935/2004 should comply with general and detailed rules on good manufacturing practice (GMP) (European Commission, 2006).

It should conform to the legislation in the following terms:

i. choosing appropriate raw materials for the production process with an eye on the finished products' safety and inertness;

ii. premises suited for the purpose, the staff informed of crucial production steps, documentation in place for quality assurance and quality control;
iii. emphasizing the formulation of printing inks, the process involving its application, and subsequently the handling and storage conditions of printed materials in detail.

12.3.1.2 EU Regulation on Specific Materials

i. **Ceramics**: Cadmium and lead migration limitations have been established under Ceramics Directive 84/500/EEC, the two heavy metals that are prone to low-level migration (Council Directive, 1984). At all marketing phases including retail, it must be accompanied by a written declaration as per Regulation 1935/2004, Article 16 (European Commission, 2004).
ii. **Regenerated cellulose film**: A positive list of compounds that can be used in its manufacturing is provided in Directive 2007/42/EC. Additionally, food products should not be in contact with printed surfaces (European Commission, 2007). At marketing phases other than retail, cellulose films meant to contact food must be accompanied by a written declaration.
iii. **Recycled plastics material**: Regulation (EU) No. 10/2011 of the Commission specifies the compositional standards for new plastic materials (European Commission, 2011c). However, after being used, these materials cease to be compliant with the plastic. Consequently, Regulation EC 282/2008 is in place to control the recycling processes (European Commission, 2008).
iv. **Active and intelligent packing materials**: The formation of a Union list of compounds approved for the production of active and intelligent materials is provided in Commission Regulation (EC) No. 450/2009 (European Commission, 2009).
v. **Plastics**: The compositional requirements include the Union list of authorized substances like monomers, polymer production aids, colorants, and microbial-fermented macromolecules for use in plastic manufacturing regulated by Commission Regulation 10/2011 (European Commission, 2011c). They are also subject to be compliant with certain specifications in terms of specific migration and overall migration limits. The overall migration limit (OML) is 60 mg/kg food of the contact material to maintain its overall safety. The regulation lays out specific guidelines for migration testing. Although testing for migration in food is common, migration is often tested using "simulants." These substitutes serve as the standard for a certain food group, such as alcoholic food with an alcohol content of up to 20% or lipophilic food, which is allocated

ethanol at 20% (w/v). The maximum shelf life of packaged foods is covered by migration testing, which is carried out under standardized time/temperature settings that are representative of a specific food application.

12.3.1.3 Specific Regulation in Regard to Certain Chemical Contaminants

i. **Bisphenol A**: Regulation EU 321/2011 restricts the use of Bisphenol A in infant feeding bottles (European Commission, 2011a). Regulation EU 2018/213 established the use of bisphenol A in plastic FCMs concerning its use in varnishes and coatings bottles (European Commission, 2018).
ii. **Epoxy derivatives**: Commission Regulation 1895/2005/EC limits the use of specific epoxy derivatives in FCMs used in bottles (European Commission, 2005).
iii. **N-nitrosamines and N-nitrosatable substances**: While acknowledging that N-nitrosamines and N-nitrosatable substances may pose a threat to human health due to their toxicity, the Scientific Committee for Food has advised that migration of these substances from the aforementioned articles be kept below the detection limit of an appropriate sensitive method. The release of these substances from elastomer or rubber teats and soothers is regulated under Commission Directive 93/11/EEC (European Commission, 1993).
iv. **Melamine and polyamide**: Plastic kitchenware from China and Hong Kong should comply with the import procedures regulated under Regulation EU 284/2011 (European Commission, 2011b). A declaration and a lab report on the examination of primary aromatic amines (for polyamide) and formaldehyde are required for shipments (for melamine) (European Commission, 1993).
v. **Printing inks**: At the EU level, the laws governing printing inks on items that come in touch with food are not entirely unified. Printing inks must adhere to the GMP Regulation (EC) No. 2023/2006 and Regulation (EC) No. 1935/2004, which are both parts of the EU's food contact legislation (European Commission, 2004, 2006). Once this is proven, printing inks may be sold within the Union, subject to applicable Member State laws and based on mutual recognition. However, Switzerland is the only nation in Europe that presently has legislation Swiss Ordinance RS 817.023.21 in place expressly regulating food-packaging printing inks, despite not being an EU member state ("Switzerland Publishes Updated Ordinance on Materials and Articles Intended to Contact Foodstuffs | PackagingLaw.com", n.d.).

12.3.2 Packaging Requirements in the US

The Food, Drug, and Cosmetic Act of 1958, which is the primary law governing FCMs as well as other pertinent laws, are enforced by the FDA. Most of the time, it is governed by Title 21 of the Code of Federal Regulations (C.F.R., Parts 176–186; Title 21 Code of Federal Regulations §. 170–186. n.d.). For indirect additives, which are substances that might reasonably be expected to migrate into the food product, clearance for FCMs is necessary.

FCMs and/or substances used in FCMs in the US should comply with one of the following:

- Regulatory requirements of Title 21 Code of Federal Regulations (Parts 176–186);
- Fulfilling GRAS status criteria including but not limited to a GRAS notice or GRAS regulation;
- A prior sanction letter: Prior-sanctioned substances are those whose use in contact with food was covered by a letter from the FDA or USDA issued before 1958 offering no objection to a specific usage of a specific substance (21 C.F.R. 181);
- A Threshold of Regulation (TOR) exemption request (Title 21, C.F.R. Part 170.39);
- An effective Food Contact Substance Notification (FCN): The Federal Food, Drug, and Cosmetic Act was amended by the Food and Drug Administration Modernization Act of 1997 to allow for the submission of food contact notifications (FCNs).

A manufacturer or supplier of an FCM may submit an FCN to FDA under this system outlining the name and planned application of a novel food contact substance, as well as any data that would support that conclusion. The submitter and its clients may market the substance if FDA does not object in writing to its usage for safety reasons within 120 days. When the notification goes into effect, FDA will include it in its inventory of food contact substance notifications that have gone into force.

12.3.3 Packaging Requirements in Japan

In Japan, both national law and voluntary industry rules govern the use of FCMs. The Food Sanitation Act (1947) governs food, food additives, equipment, and packages/containers. The Food Safety Basic Act (2003) supports it. The Ministry of Health, Labor, and Welfare (MHLW) also specifies specifications and standards for foods, food additives, and other materials under Notification No. 370. The Japanese regulation has an article that holds manufacturers accountable for the harm caused by their substances even if they abide by the law in addition to controlling materials used in food contact (Food Sanitation Law Enforcement Regulations, Chapter 1, Art. 16). Each business in the supply chain for food packaging must communicate information about compliance with the final consumers (Food Sanitation Act, Art. 52–53).

According to Japan's food sanitation law, the term "container/package" means articles in which foods or food additives are contained or packaged and in which foods and food additives are offered when such foods and food additives are delivered.

12.3.3.1 Salient Features of Food Sanitation Law

1. The production, sale, import, or use of any products that pose a threat to human health is prohibited under this law.
2. From the perspective of public health, the Minister of Health, Labor, and Welfare may establish standards for this after consultation with the Pharmaceutical Affairs and Food Sanitation Council.
3. Articles 15–18 under Chapter 1 of the Food Sanitation Act is concerned with Utensils, Containers, and Packaging (UCP).
4. As per the latest amendment to the Food Sanitation Act concerning UCP, 2019, the following major changes are incorporated (Agenda, 2019):
   i. Article 52 of the newly amended act prescribed General Sanitary Management and GMPs to manage production.
   ii. A list of additives that may be used in various FCMs is provided in the food sanitation law. Several countries besides Japan have adopted the Positive List System for FCMs like the US, EU, India, Israel, Vietnam, New Zealand, Australia, China, Indonesia, Saudi Arabia, Brazil, and more. A Positive List System basically prohibits usage and then lists the compounds that are acceptable for use after assuring their safety.
   iii. Only UCP materials that have been approved using the positive list procedure will be allowed under the new standard. However, MHLW will permit UCP use for materials not on the positive list if the following requirements are met:
      a. An unapproved material is used in a UCP component that has no direct interaction with food.
      b. When present in quantities above those deemed safe by MHLW, an unapproved substance does not contaminate food when present in UCP.

c. The maximum safe level of unapproved UCP material in food is defined by MHLW at 0.01 mg/kg under this proposal. Any UCP material will be put through a leakage test using a solvent that simulates food because leaking would result in the UCP material getting into food. In this system, the maximum safe dose of 0.01 mg/kg will be equated to 0.01 mg/L of food-simulating solvent.

12.3.3.2 Major Specifications and Standards of the Positive List (PL) System

i. The following substances shall be managed under the PL system:
   - The principal elements of synthetic resin (base polymers);
   - Ingredients that are anticipated to stay in finished products and alter the physical or chemical characteristics of synthetic resin.
ii. For the polymerization of monomers, catalysts and polymerization aids are employed. They are not anticipated to remain in finished goods and do not mix with base polymers to form them. As a result, they must be controlled using traditional risk management techniques rather than the PL.
iii. Assuming that additives and other substances will be handled in accordance with the amount contributed (content) to each substance, the migration and other required constraints for additives and other substances shall be stipulated as needed.
iv. Colorants will be thoroughly monitored and included in the Positive List.
v. Upon request, manufacturers of raw materials should disseminate information regarding UCP.
vi. According to their characteristics and present use status, synthetic resins must be categorized into a few groups, and each category must specify and manage the amount of additive use as show in Table 12.1 (Agenda, 2019). The term "consumption factor" refers to the factor determined by measuring the percentage of the diet that comes into contact with particular materials used for UCP overall, whereas the factor generated by measuring the percentage of UCP needed for foods of a particular food type (aqueous, acid, oil, alcoholic, dairy, dry foods, etc.) by a material is known as the "food-type factor."

An overview of differences in Positive List System in the US and EU is shown in Table 12.2 (Agenda, 2019; Food Sanitation Act, 1947).

Table 12.1 Classification of Synthetic Resin Group

Synthetic Resin Group	*Examples*	*Consumption Factor*
Group 1	Engineering plastic, thermosetting resin, etc.	0.05 All < 0.001
Group 2	Olefin, etc.	0.07 Polystyrene: 0.06 Other < 0.001
Group 3	Polyester, polyamide, etc.	0.05 All < 0.001
Group 4	Polyvinyl chloride (PVC), polyvinylidene chloride (PVDC)	0.05 PVC: 0.02 PVDC < 0.001
Group 5	Polyethylene	0.25
Group 6	Polypropylene	0.16
Group 7	Polyethylene terephthalate	0.22

12.3.4 Packaging Requirements in China

According to Chinese law, "under intended usage, items that have contact with food or food additives or have the potential to come into contact with food or release their constituents into food" are called FCMs. During the manufacture, packing, transportation, storage, preparation, and serving of food, for instance, packaging materials, containers, utensils, printing ink, and machinery may come into direct contact with food or have the potential to do so. Figure 12.1 illustrates an overview of the food safety law in place in China (World Packaging Organisation, 2013).

The initial two Chinese language characters, "GB," stand for "Chinese National Standard." FCM's regulatory framework consists of general, product, testing, and production standards. By making use of the relevant legislation, one may ascertain the legal requirements for the goods in China.

12.3.4.1 Major Regulations on Food Packaging in Food Safety Law

The Food Safety Law of China forbids the import, usage, or sale of any goods connected to food (such as food additives or food-packaging materials) that do not meet the relevant national Food Safety Standards (Luo and Wu, 2017). Some of the major regulations include:

Table 12.2 Comparison of PL System in the US and the EU

Parameter	*US*	*EU*
Year of inclusion	1958	2010
Coverage	Synthetic resins, paper, and rubber products	Synthetic resins
Regulation in force	The CFR specifies the monomers, additives, and amounts that may be used for each type of polymer in synthetic resins. In order to hasten the process of inclusion in the Positive List, the Food Contact Notice (FCN) system, a premarket notification for compounds used in food contact, was established in 2000. It restricts its use to the notifier for each product	It restricts the amounts of migration, the usage circumstances, and other crucial factors for each monomer and additive. The system also controls the overall migration level of the chemicals used to make the products and their components
Good manufacturing practices (GMP)	Manufacturers, including those who produce raw materials, are required to abide by GMP	Manufacturers, including those who produce raw materials, are required to abide by GMP
Communication	Information sharing among business operators is not specifically regulated, and is instead left to self-management and self-declaration	It is required to disseminate information among business operators through the issuance of a "declaration of compliance," which attests to the compliance of FCMs with the PL

i. General Safety Requirements for FCMs – GB 4806.1-2016

The following is a list of the primary criteria of GB 4806.1-2016 that apply to all sorts of FCMs:

- It must not alter food's composition, taste, or odor or release their contents at amounts that are detrimental to human health.
- The manufacturing of FCM and their products must adhere to the GHP for Production of FCM and its products criteria of GB 31603-2015 National Food Safety Standard.
- By pertinent food safety regulations, FCMs and food contact additives must adhere to applicable limitation restrictions.

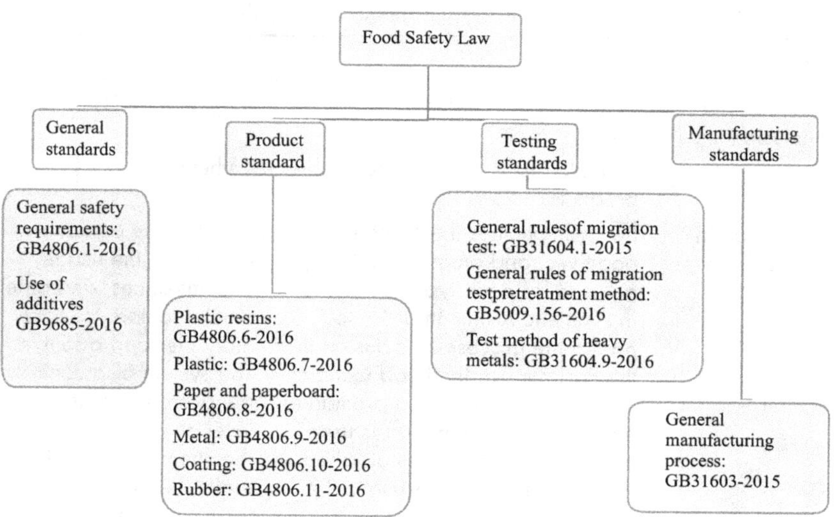

Figure 12.1 Overview of the regulation system for food contact materials in China.

- From manufacture to distribution, producers must build a traceability system for FCMs.
- Product details: They must come with adequate product details. This information includes the name of the product, the material, the declaration of compliance with applicable standards, the name, address, and contact details of the manufacturer and (or) salesman, the date of production, the guarantee period (if applicable), the instructions for use, and the qualification certificate.
- Labeling: They must be marked with "food contact use," "used as food packaging," or be adhered with the spoon- or chopstick-shaped symbol. Articles with obvious food contact do not need to be labeled in this way.

ii. Positive List of Food Contact Additives: GB 9685-2016

A list of additives that may be used in various products and materials intended for interaction with food is provided in GB 9685. Before they may be utilized to create FCM, new food contact additives or novel uses must first have the approval of National Health and Family Planning Commission in China.

For each additive, its name in China, CAS no., the scope for usage, maximum level of usage, specific migration limit or maximum residue, and other requirements should be given.

12.3.5 Packaging Requirements in India

The central authority in charge of overseeing food in India is the Food Safety and Standards Authority of India (FSSAI). Food Safety and Standard (Packaging) Regulations (2018) is the principal regulation governing food packaging in India.

12.3.5.1 Salient Features of the Regulation

i. Every person who runs a food business must make sure that the packaging they use complies with these rules;
ii. When applicable International Standards are not accessible, Indian Standards may be followed;
iii. Any material that will come into direct touch with food or is likely to do so when used to package, prepare, store, wrap, transport, sell, or serve food must be of food-grade quality;
iv. Materials used for packaging should be appropriate for storage conditions offered, the equipment used to fill, seal, and package food, type of product, and the conditions of transportation;
v. Packaging materials should be able to endure chemical, mechanical, or thermal stressors that may be present during routine transportation. Flexible or semi-rigid containers require overwrap packing;
vi. Food goods must be packaged in a hygienic and tamper-evident container or package;
vii. The sealing substance must be appropriate for the product, the containers, and the closing mechanisms employed on the containers;
viii. Once used, tin containers must not be used again to package food;
ix. Reusable plastic containers with a volume of five liters or more and glass bottles used for food packing must be durable and simple to disinfect and clean;
x. Printing ink on food packaging must adhere to IS: 15495;
xi. It is forbidden for printed surfaces on packing materials to encounter food items directly;
xii. For food storage or packaging, newspaper or any other similar material is not permitted to be used;
xiii. Every person who operates a food business must get a conformity certificate from a NABL-accredited laboratory to FCMs;
xiv. Specifications for primary food-packaging material for various food products like canned products, edible oil, milk, milk products, fruits, vegetables, and drinking water are also enlisted. For instance, bottles/containers used for packaged drinking water should be hygienic, clean, colorless, tamperproof, and made of polyethylene;

Table 12.3 Plastic Materials Standards and Migration Limit

Substances	*Maximum Migration Limit (mg/kg)*
Fe	48
Zn	25
Cu	5
Ba	1
Li	0.6
Mn	0.6
Co	0.05

xv. The maximum migration limit permitted under the law for various metals is enlisted in Table 12.3 (Food Safety and Standard (Packaging) Regulations, 2018);

xvi. Recycled polyethylene terephthalate materials can be used to package, carry, store, or dispense food products.

An overview of differences in packaging regulation in India, the US, and the EU is shown in Table 12.4 (Food Safety and Standard (Packaging) Regulations, 2018, C.F.R. Title 21; European Commission, 2004).

12.4 Labeling Requirements

Labeling provides detailed information about the product and helps the customer to recognize the required product easily.

Objectives of labeling are as follows:

- It helps to differentiate a product from one another.
- It describes the details of the product.
- It helps for easier comparison with other similar products.
- It protects the customers from buying the wrong product.
- It provides information to the customer about the product as per law.
- It helps customer to choose a healthier option.

12.4.1 Labeling Requirements in the EU

Food Information to Consumers (FIC) Regulation 1169/2011 is the principal food-labeling regulation of the European Union in force (European Parliament, Council of the European Union, 2011).

Table 12.4 Comparative Analysis of Packaging Regulation

Element	*India*	*US*	*EU*
Authority in control	FSSAI	FDA	Functions are clearly separated between EFSA (European Food Safety Authority) and European Commission
Safety of packaging material	Migration testing is done in India	Regulation of FCMs is based more on dietary exposure rather than on migration, as in the EU	Safety is assessed through the risk of migration, overall using a worst-case scenario in which each citizen is expected to consume 1 kg of food per day
Printing inks, waxes, paper, and paperboard components	Paraffin wax must adhere to Type I of IS 4654 Printing inks for use on food packages should conform to IS: 15495	Printing inks/colorants for polymers (Part 178.3297) paper and paperboard components (Part 176), waxes (Part 174–180)	There is no specific regulation for these other than those covered under regulation EC 1895/2005
Legal enforcement	FSSAI regulates only food packaging	The provisions are only primarily enforced for food packaging	In Europe, the legal enforcement covers both food packaging and containers for transporting food, machinery to process food, and housewares articles equally

(*Continued*)

Table 12.4 (*Continued*) Comparative Analysis of Packaging Regulation

Element	*India*	*US*	*EU*
Exemptions for testing of packaging material	In India, there are no choices to waive testing Every owner or operator of a food business must get a certificate of conformity for the packaging material from a NABL-recognized laboratory	Exemptions for testing applies as per: • Threshold rule • GRAS status • Prior sanction letter	Testing cannot be exempted
Overall migration limit for plastics	Overall migration limit of 60 mg/kg when tested as per IS 9845 with no visible color migration Plastic having specific migration limit for heavy metals	As per Title 21 part §176.170©, the extractives shall not extend 0.5 mg/sq. inch of food contact surface nor exceed 50 ppm of the water capacity of the container in general or otherwise limits specified for specific material when tested as per the prescribed method	The overall migration limit is 60 mg/kg food of the contact material Some free monomers have specific limit
Recycled plastics	The use of recycled plastics for food purpose is banned in India	Only on issuance of favorable opinion on the suitability of material by FDA, post-consumer-recycled plastics can be used as FCM	Recycling process is controlled by Regulation EC 282/2008

The key principles of this regulation are as follows:

i. This regulation establishes the framework for the assurance of a high degree of consumer protection concerning food information, taking into account the variations in consumer perception and their information needs.
ii. The general specifications governing food information, specifically food labeling, are established by this regulation. It lays down the methods and ensures that consumers have access to information

on food, while also taking into account the need for sufficient flexibility to respond to upcoming developments and new information requirements.

iii. Food business operators who engage in activities involving the dissemination of food information to customers at any point in the food chain are subject to this regulation. It will apply to all foods destined for the final consumer, including food served by mass caterers and food supplied to them.

iv. The regulation lays down the following 12 elements which must appear on the labeling:
   - Name of the food;
   - Ingredients list;
   - Allergens;
   - Quantitative ingredient declaration;
   - Net quantity;
   - Minimum durability;
   - Country of origin;
   - The actual alcohol strength by volume with beverages having an alcohol content of more than 1.2%;
   - Special storage conditions;
   - Nutrient declaration;
   - Instructions for use;
   - Business name along with address.

12.4.1.1 Allergen Information

This regulation mandates the declaration of 12 allergens if present, i.e., mustard, celery, sulfites (>10 ppm), sesame seeds, lupin, Molluscs, milk, peanuts, egg, crustacean shellfish, fish, tree nuts, cereals containing gluten, and soybeans.

12.4.1.2 Country of Origin Labeling (COOL)

It is compulsory to have this labeling for certain products like fruit and vegetables, honey, olive oil, fishery and aquaculture products, beef, fresh, frozen, and chilled meat from swine, sheep, goats, and poultry.

As per the new labeling rule in force from April 1, 2020, it is mandatory to mention the origin of the primary ingredient (>50%) if it is not similar to the product origin as "EU," "non-EU," or "EU and non-EU" or other different possibilities as mentioned in the regulation or else must mention as

> (name of the primary ingredient) does/does not originate from (the country of origin or the place of provenance of the food)

12.4.1.3 Nutrition and Health Claims

Regulation 1924/2006 governs nutrition and health claims in EU (European Parliament, Council of the European Union, 2006).

A "nutrition claim" is any assertion, implication, or suggestion that food possesses specific, beneficial nutritional properties attributed to either its low or high or negligible calorific value or other nutrients.

"Health claim" refers to any assertion, implication, or suggestion that there is a link between a food category, a food, or one of its constituents, and health.

At the EU, claims for various nutrient components are declared as shown in Table 12.5 (European Parliament, Council of the European Union, 2011).

12.4.2 Labeling Requirements in the US

Title 21 Code of Federal Regulations, §. 101 (2006) under the Federal Food, Drug & Cosmetic Act defines the major labeling requirements in the US.

Table 12.5 Overview of Claims Declaration in the EU

Component	*Claim*	*EU Conditions*
(Not More Than)		
Energy	Low	Solids: 40 kcal (170 kJ)/100 g Liquids: 20 kcal (80 kJ)/100 mL
	Reduced	Reduced by at least 30%
Fat	Low	Solids: 3 g of fat/100 g Liquids: 1.5 g of fat/100 mL
	Free	0.5 g of fat per 100 g or 100 mL
Sugar	Low	Solids: 5 g of sugars per 100 g Liquids: 2.5 g of sugars per 100 mL
	Free	0.5 g of sugars per 100 g or 100 mL
Saturated fat	Low	1.5 g per 100 g for solids or 0.75 g/100 mL for liquids
	Free	0.1 g of saturated fat per 100 g or 100 mL
(Not Less Than)		
Protein	High	20% of the energy value of the food
Fiber	High	6 g of fiber per 100 g or at least 3 g of fiber per 100 (kcal)

Both the US and the EU have well-established systems and laws in place, but differ in their approaches and requirements. The major dissimilarities concerning labeling requirements are shown in Table 12.6 (Study on Food Safety Policy, 2005, C.F.R. Title 21; European Parliament, Council of the European Union, 2011).

12.4.2.1 Food Allergen Labeling Requirements

Specific labeling requirements including the name of the allergen source are required in foods or ingredients that contain a major food allergen (Food Allergen Labeling and Consumer Protection Act, 2004). Earlier, eight foods were considered as the major food allergens by this law which includes milk, fish, peanuts, wheat, shellfish, eggs, tree nuts, and soybeans. Sesame is recently added to the list as of January 1, 2023 (Scott, 2021).

12.4.3 Labeling Requirements in Japan

The Food Labeling Act was initiated in the year 2013 to ensure the safety of food. It ensures the suitability of food labeling by establishing standards and defining other essential information regarding the labeling of food intended for sale, thereby promoting consumer interest and contributing to the protection and promotion of the human rights.

The following are some of the most important modifications to current food-labeling practices:

i. **Systematic classifications for processed and fresh foods**: Earlier, the same product was classified differently under the Food Sanitation Law and the Japan Agricultural Standard (JAS) Law. By classifying all items into three groups – fresh food, processed food, or food additive – the Standard unified these differences. For instance, the Food Sanitation Law traditionally classified foods that had been gently salted, blanched, or dried overnight as fresh foods; nevertheless, the Standard now classifies such goods as "processed foods," necessitating suitable labeling.
ii. **Manufacturer identification codes and contact details**: The previous legislation allowed an alpha-numeric manufacturer's identity (ID) number which is unique to a certain production facility to take the place of the address and name of a food-manufacturing facility. However, according to the Standard, ID numbers are only accepted when a product is produced at more than two sites. The name and location of the manufacturing plant must thus be listed

Table 12.6 Overview of the Differences in Labeling Regulation in the US and the EU

Parameters	*US*	*EU*
Regulation	FD&C act	FIC Regulation 1169/2011
Mandatory information	Five specific elements are required: i. Statement of identity; ii. Net quantity of ingredients; iii. Nutrition facts; iv. Ingredient statement; v. Manufacturer's statement	Twelve specific elements required: i. Name of the food; ii. List of ingredients; iii. Allergens; iv. QUID; v. Net quantity; vi. Date of minimum durability; vii. Special storage conditions; viii. Business name & address; ix. Country of origin; x. Instructions for use; xi. The actual alcohol strength by volume in relation to beverages having an alcohol content more than 1.2%; xii. Nutrient declaration
Allergen information	There are nine major allergens which must be declared if present: i. Milk, ii. Egg, iii. Fish, iv. Crustacean shellfish, v. Tree nuts, vi. Wheat, vii. Peanuts,	There are 12 major allergens which must be declared if present: i. Celery, ii. Mustard, iii. Sesame seeds, iv. Sulfites (>10 ppm), v. Lupin, vi. Molluscs, vii. Milk, viii. Egg,

(*Continued*)

Table 12.6 (*Continued*) Overview of the Differences in Labeling Regulation in the US and the EU

Parameters	*US*	*EU*
	viii. Soybeans, ix. Sesame	ix. Fish, x. Crustacean shellfish, xi. Tree nuts, xii. Cereals containing gluten, xiii. Peanuts, xiv. Soybeans
Expiry date	Mandatory for infant formula	A mandatory inclusion; on product labels, a date of first freezing will also be displayed when suitable, such as for meat and fish. "Frozen on day/month/year" should be used to denote the freezing date
Trans fat	• To be mentioned in nutritional facts as "trans fat" in Latin on food products with >**0.5 g of trans fats per serving** • Trans fat free: food products <0.5 g of trans fats per serving • Partially hydrogenated oil (PHO) has been banned for use in food products (2015)	• No mandatory labeling in EU Legislation regarding trans fat; for foods meant for consumer, and retail supply, maximum level of trans fat is **2 g per 100 g of fat** • It mandates that "fully" or "partially" hydrogenated oils or fats, together with the specific vegetable origin, be stated in the ingredient list
Gluten labeling	**Gluten free**: contain <20 ppm of gluten Terms like "low gluten" and "very low gluten" are not defined in the rule	**Gluten free**: this criteria is similar to the US **Very low gluten**: only foods derived from or containing one or more ingredients from wheat, rye, barley, oats, or their derivatives may be labeled as having "extremely low gluten"

(*Continued*)

Table 12.6 (*Continued*) Overview of the Differences in Labeling Regulation in the US and the EU

Parameters	*US*	*EU*
Genetically modified organism (GMO)	As per the National Bioengineered Food Disclosure Standard, it is required that by 2022, foods that are bioengineered or have bioengineered ingredients must have information on their packaging using one of the approved methods, including text on the package that says "bioengineered food," the bioengineered food symbol, or directions for using gadgets to find the disclosure	In the case of pre-packaged GM food/feed products constituting >0.9% of the food/feed ingredients, ingredient list must indicate "genetically modified" or "produced from genetically modified [name of the organism]" Voluntary "GM-free labels" exists to avoid misleading the consumer
Beef labeling scheme	As long as the beef is subjected to further processing in the US; **beef derived** from livestock born, raised, and slaughtered in **third** countries is **labeled** "Product of the U.S.A."	Beef labeling scheme mandates that all bovine meat labels must mention the following information: • "Born in: name of third country" • "Reared in: name of third country or third countries" • For beef obtained from livestock born, raised, and slaughtered in the same third country, the above indications may be combined as "Origin: name of third country" • The link between the meat and the animal or animals is ensured by a reference number • "Slaughtered in: third country/approval number of slaughter house" • "Cutting in: third country/approval number of cutting plant" • A traceability code

(*Continued*)

Table 12.6 (*Continued*) Overview of the Differences in Labeling Regulation in the US and the EU

Parameters	*US*	*EU*
Nutritional information (mandatory and voluntary information)	• Mandatory nutrients (total calories, total fat, saturated fat, trans fat, cholesterol, sodium, total carbohydrate, dietary fiber, total sugars, added sugars, protein, vitamin D, calcium, iron, potassium) • It must provide Serving per container & serving size, total calories; carbohydrates with added sugars in grams and as % daily value • It requires "total fat," "saturated fat," and "trans fat" on the label • It must declare the **actual amount**, in addition to percent **Daily Value** of vitamin D, minerals – Ca, Fe, and K • Quantity for other vitamins and minerals can be declared on voluntary basis	• Mandatory information: it must provide the energy value and the amounts of fat, saturates, carbohydrate, sugars, protein, and **salt** of the food • Energy value and amount of only total fat and saturated fat are mandatory on label • Amounts of mono-unsaturates and polyunsaturates is voluntary • Declaration of vitamins and minerals is voluntary • Other voluntary information: nutrition declaration may be supplemented voluntarily with the indication of the amounts of polyols, starch, and fiber • On voluntary basis, % Reference Intake (RI) shall be indicated
Engineered nano materials (ENMs)	No labeling requirements in place	In the ingredients list, such ingredients should be clearly mentioned followed by the word "nano" in brackets

(*Continued*)

Table 12.6 (*Continued*) Overview of the Differences in Labeling Regulation in the US and the EU

Parameters	*US*	*EU*
Alcoholic strength declaration	It is not necessary at the federal level, but it can be at the state level. Any alcoholic beverage with a volume alcohol content more than 0.5% must have the following labeling: GOVERNMENT WARNING: (1) "According to the Surgeon General, women should not drink alcoholic beverages during pregnancy because of the risk of birth defects" (2) "Consumption of alcoholic beverages impairs your ability to drive a car or operate machinery, and may cause health problems"	It must be declared on beverages containing more than **1.2% alcohol** by volume. On voluntary basis, advisory notice on consumption of alcoholic beverages during pregnancy may be included

on all product labels by smaller Japanese enterprises. A product label must also include one of the following when employing an ID code: (1) names, locations, and ID numbers for all manufacturing facilities, and (2) customer service and (3) website information of the company.

iii. **Allergen labeling**: Under the former food-labeling legislation, producers were permitted to exclude potentially allergenic substances from package labeling where it was reasonable to conclude that the ingredients were allergic. For instance, it would be reasonable to presume that mayonnaise-containing items have eggs. According to the Standard, each allergen must have its label, such as dried egg yolk (containing an egg). The law requires the declaration of 7 allergens on mandatory basis and 24 allergens on voluntary basis. The use of "may contain" wording is not permitted under the Standard, and thus the Consumer Affairs Agency's (CAA) present guidelines the term "Product contains". Mandatory and recommended requirements for various allergens are shown in Table 12.7 (Labelling, Marking & Packaging, 2022).

iv. **Labeling requirements for nutrients**: The Standard mandates labeling of nutrition on all processed, pre-packaged food. Products made by factories with less than 20 staff members or those imported by businesses with less than five workers are exempted from the dietary labeling requirement. The proper serving size for nutritional labeling may be chosen by manufacturers or importers.

v. **Content claims**: Content claim requirements under Japanese food law aligns with the current Codex standards. The percentage difference between the prior or standard formulation and the present formulation must be 25% or more to make the claim "Reduced X" or "Less X." Vitamins and minerals can be labeled as "Enhanced X," and there must be an absolute difference of 10% or more between the present formulation and the nutritional reference value. Under the Standard, the phrases "No added sugar" and "No added salt" were first used.

Table 12.7 Labeling Requirements for Allergens in Japan

Labeling Requirements	*List of Allergens*
Mandatory requirements	Egg, milk, buckwheat, wheat, peanuts, crab, shrimp/prawn
Recommended requirements	Abalone, mackerel, squid, salmon, salmon roe, cashew nut, walnut, matsutake mushroom, sesame, soybean, yam, apple, banana, kiwifruit, orange, peach, beef, chicken, gelatin, pork

vi. **Labeling of food additives for sale**: When a product is a food additive intended for sale, the Standard mandates that it be labeled with extra information. For instance, a product like vanilla essence for flavoring would be subject to the labeling requirement (i.e., the product is a food additive in and of itself). In addition, it mandates the declaration of net composition of the food additive along with the name and address of the manufacturer.
vii. **Implementation of CAA notices**: The Standard includes several food-labeling requirements published as CAA "notices," which were previously exempted from all three laws governing food labeling. A prior article on food-labeling policies to prevent unintended poisoning by blowfish toxin serves as an example of one such notification.
viii. **Improving the layout of food labels**: Regardless of package size, the Standard mandates that a product's name, expiry date, correct storage directions, manufacturer/seller contact details, allergen, and L-phenylalanine constituents (if present) be included on the label.

12.4.4 Labeling Requirements in China

i. Guidelines for Food-Labeling Administration

The provisions came into effect on September 1, 2008 with the main aim of standardizing food labeling and enhancing relevant supervision (USDA Foreign Agricultural Service, 2021). It provides information on mandatory requirements and proper labeling techniques. In 2009, the competent authority revised it, and the amendment came into effect on October 22, 2009. It includes:

- General Standard for the Labeling of Pre-packaged Foods: GB 7718-2011.
- General Standard for the Nutrition Labeling of Pre-packaged Foods: GB 28050-201.

  Pre-packaged foods should comply with this standard; however, foods meant for special dietary use are exempted from such requirements.
- National Food Safety Standard Labeling of Pre-packaged Foods for Special Dietary Uses: GB 13432-2013.

ii. Label filing for import of pre-packaged foods

Before exporting, food that has already been pre-packaged must submit a label-filing application. But as of October 1, 2019, there is no longer a need to file labels for prepared imported foods. A typical sampling inspection item for customs release will be a label inspection for items imported for the first time. The importer needs to present the certificate regarding product quality, origin label and its translation, label sample of China, and other requirements if a pre-packaged food item is chosen to undergo on-site inspection and laboratory testing.

iii. Standards for documentation when labeling food with nutritional information: It should comply with the following requirements if nutrition information is labeled:
   - **For conventional food**: The labeling of nutrition information for conventional food should follow GB 28050-2011, which mandates the declaration of nutritional information like energy, protein, fat, carbohydrate, etc. Certain nutritional components that are added according to GB 14880-2012 or National Health Commission (NHC) notification should be properly labeled. Label should be verified by appropriate testing.
   - **For foods with particular dietary requirements**: GB 13432-2013 should be followed when labeling foods with special dietary requirements, such as baby formula. Energy, protein, fat, carbohydrates, salt, and other key nutrients that are specified in the relevant product standards must be included in mandatory nutrition information. Following the relevant product standards, GB 14880-2012, or NHC notifications, certain nutritional components added to food must be appropriately labeled. Nutritional evaluations should be conducted on each nutrient listed on the label.

12.4.4.1 Recent Revisions to Food-Labeling Requirements in China

Both the NHC and the State Administration of Market Regulation (SAMR) submitted major revisions to food-labeling drafts after public feedback at the end of 2019 (The National Law Review, 2020). Major highlights of this proposal include the following:

i. Minimal revisions to the labeling information

   The proposed measures mandate the declaration of the food's shelf life, production date, and warnings in a conspicuous way on the label in addition to the food name, which must be presented in a distinguishing manner under GB 7718. According to GB 7718-2011, the order of year, month, and day must be followed when declaring the manufacturing date; and for shelf life, it can be done in any sequence.

ii. The introduction of obligatory allergen labeling

   The new proposal makes allergen labeling mandatory, which is now merely advised under GB 7718. Disclosure requirements also apply to allergens that might be brought on by cross-contamination.

The draft GB 7718's Annex E lists the formats that must be used for labeling allergens.

iii. Increased strength of voluntary on-label claims

The proposed measures underline that the China Advertising Law and Anti-Competition Law shall be taken into consideration when deciding whether a voluntary product claim is compliant. The draft GB 7718 recommends more rigorous inspection in this area. Claims like "free..." and "not contain..." are not permitted for things like food additives, pollutants, and un-authorized compounds. In addition, unless specifically allowed by law, statements like "...not added" and "...not utilized" are forbidden.

12.4.5 Labeling Requirements in India

Labeling requirements are specified in Food Safety and Standards (Labeling and Display) Regulations 2020 and Food Safety and Standards (Advertising and Claims) Regulations 2018. These rules specify the information that must be shown on the premises where food is processed, produced, stored, and served as well as the labeling requirements for prepackaged meals.

12.4.5.1 Salient Features of the Regulation

1. Every food packaging must include a name for the food that describes the actual type of food inside the box on the front of the package.
2. Ingredients list: Except for products containing just one ingredient, a list of ingredients must be declared on the label.

   Information about nutrients: It should be per 100 g, 100 mL, or per single consumption pack of the product, and per serve percentage (%) contribution to RDA shall be provided on the label containing the following information:
   - Energy value (kcal).
   - Amounts of
     - carbohydrate (g), protein (g), and total sugars (g) and added sugars (g);
     - saturated fat (g), trans fat and total fat (g), cholesterol (mg), and sodium (mg).
3. Declaration of non-veg or veg

   i. Every non-veg food packaging containing elements from animals, such as food additives and processing aids, must include a declaration to that effect in the form of the symbol and color code described below. The sign must consist of a triangle filled with brown color inside of a square with a brown outline, with sides that are at least the minimum size required.

ii. If an item of food exclusively contains eggs as a non-veg component, the packer or seller can add a statement to this effect to the product along with the above-mentioned sign.
iii. Vegan food that contains ingredients, such as food additives and processing aids derived from plants, must include the symbol and color code described below.

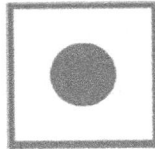

The symbol must have a circle filled in the green inside of a square with a green outline that has a diameter that is at least the minimum size required.

4. Food additive declarations

   As per the Food Safety and Standards (Food Products Standards and Food Additives) Regulations 2011, in the ingredients list, functional classes for food additives must be disclosed together with any names or recognized International Numbering Systems (INS). Whether or not the brand owner is also the producer, marketer, packer, or bottler, as the case may be, his full name and address must be given on the label. The qualifying terms like "Manufactured by (Mfg by, Mfd by)" or "Marketed by (Mkt by)" or "Manufactured & Marketed by" or "Packed & Marketed by," as applicable, must appear before the name and address. Imported food should carry the name and complete address of the importer in India.
5. Name and complete address declaration.
6. License number and FSSAI logo.
7. Fortified food should mark with the logo given below:

8. Retail sale price, net quantity, and consumer care details.
9. Lot/batch identification/code.
10. Date marking.

    "Date of manufacture or packaging" and "Expiry/Use by" should be declared on the label. As additional information, "Best before use" can also be used. In case of special storage conditions, it should be mentioned on the label.
11. Food imported from a foreign country.
12. Usage instruction.
13. Allergen information should be declared for the following eight allergens, i.e., cereals containing gluten, milk and milk products, crustacean and their products, eggs and egg products, peanuts, tree nuts, fish and fish products, soybeans and their products, and sulfite in concentrations of 10 mg/kg or more.
14. Food material not for human consumption should mark with the following symbol:

15. The logo outlined below must be shown on every package of food that has been certified organic as per the FSSR, 2017.

16. The Radura emblem in green and the following statement on the label must appear on any item that has undergone the irradiation procedure mandated by Rule 2.13 of the FSSR 2011, which states:

PROCESSED BY RADIATION

Name of the Product:

Purpose of Radiation Processing:

Operating License No. :

Batch Identification No. (BIN) (as provided by facility):

Date of Processing............................"

17. Gluten free labeling requirements

    As per the Food Safety and Standards (Food Products Standards and Food Additives) Regulations, 2011, the term "Gluten Free" shall be printed next to the product name if the levels are below 20 mg/kg. In addition, food labeled as Low Gluten should bear a warning that it may pose a risk to a celiac person. If any gluten-free product is produced at a facility that also produces gluten-containing products, it must be disclosed on the label as "Processed in a facility that produces gluten-containing products."

12.4.5.2 Recent Amendments

Some significant changes are included in the second amendment on labeling and display regulations in 2022 which are:

- The rules listed under Specialty Ingredient and Minimum Quantity of Specialty Ingredient as% of Flour must be followed when labeling different types of bread.
- In the first year of enforcement, specialty ingredients for multigrain bread must be at least 10%; after that, they must be 20%.
- Bread with added protein must have a minimum of 15% edible protein.
- The health-warning statement on the label for Pan Masala must take up 50% of the front of the package.

In addition, the use of genetically modified organisms (GMOs) in food has long been in question. On November 15, 2021, FSSAI issued the draft regulation FSSR (Genetically Modified or Engineered Foods) 2021 ("India: India Proposes Draft Regulations for Genetically Modified Food for a Second Time" 2022). The regulation covers the following aspects as prior approval for all kinds of activities with GMOs including manufacture, storage, distribution, sale, and import. As per this, if the food contains genetically modified ingredients of 1% or more, then it should be labeled as

"Contains GMO/Ingredients derived from GMO"

References

"Agenda: Overview of Amendments to the Food Sanitation Act (2019)." Ministry of Health, Labor and Welfare. Accessed January 01, 2023. https://www.mhlw.go.jp/content/11130500/000537821.pdf.

"Are You Ready for Chinas New Food Labelling requirement." The National Law Review, 2020. Accessed January 22, 2023. https://www.natlawreview.com/organization/national-law-review-national-law-forum-llc.

Center for Food Safety and Applied Nutrition. "Food Allergen Labelling and Consumer Protection Act of 2004." U.S. Food and Drug Administration (FDA), 2004. Accessed January 02, 2023. https://www.fda.gov/food/food-allergensgluten-free-guidance-documents-regulatory-information/food-allergen-labelling-and-consumer-protection-act-2004-falcpa.

"China: Interpretative Guidance on Imported Food Labelling Requirements." USDA Foreign Agricultural Service, January 19, 2023. https://www.fas.usda.gov/data/china-interpretative-guidance-imported-food-labelling-requirements-decree-249.

"Codex Alimentarius | Food Safety and Quality | Food and Agriculture Organization of the United Nations." n.d. Accessed January 8, 2023. https://www.fao.org/food-safety/food-control-systems/policy-and-legal-frameworks/codex-alimentarius/en/#:~:text=The%20Codex%20Alimentarius%2C%20or%20%E2%80%9CFood, practices%20in%20the%20food%20trade.

Council Directive 84/500/EEC of 15 October 1984 on the Approximation of the Laws of the Member States Relating to Ceramic Articles Intended to Come into Contact with Foodstuffs §. L277, 1984.

European Commission. Commission Directive 93/11/EEC Concerning the Release of the N-Nitrosamines and N-Nitrosatable Substances from Elastomer Or Rubber Teats and Soothers §. L 93, 1993. https://eur-lex.europa.eu/legal-content/EN/TXT/?uri=CELEX%3A31993L0011&qid=1674322983461.

European Commission. Commission Regulation (EU) No 1935/2004 on Materials and Articles Intended to Come into Contact with Food and Repealing Directives 80/590/EEC and 89/109/EEC §, 2004.

European Commission. Commission Regulation (EC) No 1895/2005 on the Restriction of Use of Certain Epoxy Derivatives in Materials and

Articles Intended to Come into Contact with Food §. L 302/28, 2005. https://eur-lex.europa.eu/legal-content/EN/TXT/?uri=CELEX%3A32005R1895&qid=1674322589163.

European Commission. Commission Regulation (EC) No 2023/2006 on Good Manufacturing Practice for Materials and Articles Intended to Come into Contact with Food §. L 384/75, 2006.

European Commission. Commission Directive 2007/42/EC Relating to Materials and Articles Made of Regenerated Cellulose Film Intended to Come into Contact with Foodstuffs §. L 172/71, 2007.

European Commission. Commission Regulation (EU) 2022/1616 on Recycled Plastic Materials and Articles Intended to Come into Contact with Foods, and Repealing Regulation (EC) No 282/2008 §. L 243/3, 2008.

European Commission. Commission Regulation (EC) No 450/2009 on Active and Intelligent Materials and Articles Intended to Come into Contact with Food §. L 135/3, 2009.

European Commission. Commission Implementing Regulation (EU) No 321/2011 Amending Regulation (EU) No 10/2011 as Regards the Restriction of Use of Bisphenol A in Plastic Infant Feeding Bottles §. L 87/1, 2011a.

European Commission. Commission Regulation (EU) No 284/2011 Laying Down Specific Conditions and Detailed Procedures for the Import of Polyamide and Melamine Plastic Kitchenware Originating in Or Consigned from the People's Republic of China and Hong Kong Special Administrative Region, China §. L 77/25, 2011b. https://eur-lex.europa.eu/legal-content/EN/TXT/?uri=CELEX%3A32011R0284&qid=1674323157958.

European Commission. Commission Regulation (EU) No 10/2011 on Plastic Materials and Articles Intended to Come into Contact with Food Text with EEA Relevance §. L 12/1, 2011c.

European Commission. Commission Regulation (EU) No 2018/213 on the Use of Bisphenol A in Varnishes and Coatings Intended to Come into Contact with Food and Amending Regulation (EU) No 10/2011 as Regards the Use of That Substance in Plastic Food Contact Materials, 2018. https://eur-lex.europa.eu/eli/reg/2018/213/oj.

European Parliament, Council of the European Union. Regulation (EC) No 1924/2006 on Nutrition and Health Claims Made on Foods §. L 404/9, 2006. https://eur-lex.europa.eu/legal-content/EN/TXT/?uri=CELEX%3A32006R1924&qid=1674367650125.

European Parliament, Council of the European Union. Regulation (EU) No 1169/2011 on the Provision of Food Information to Consumers, Amending Regulations (EC) No 1924/2006 and (EC) No 1925/2006 of the European Parliament and of the Council, and Repealing Commission Directive 87/250/EEC, Council Directive 90/496/EEC, Commission Directive 1999/10/EC, Directive 2000/13/EC of the European Parliament and of the Council, Commission Directives 2002/67/EC and 2008/5/EC and Commission Regulation (EC) No 608/2004 §. L 304/18, 2011.

"FoodSafetyandStandards (AdvertisingandClaims) Regulations,2018."FSSAI, December 14, 2022. https://www.fssai.gov.in/upload/uploadfiles/files/Compendium_Advertising_Claims_Reg ulations_14_12_2022.pdf.
"Food Safety and Standards (Food Products Standards and Food Additives) Regulations, 2011". FSSAI. https://www.fssai.gov.in/upload/uploadfiles/files/Compendium_Food_Additives_Regulations_20_12_2022.pdf.
"Food Safety and Standards (Labelling and Display) Regulations 2020-FSSAI." FSSAI, September 23, 2021. https://www.fssai.gov.in/upload/uploadfiles/files/Compendium_Labelling_Display_23_0 9_2021.pdf.
"FoodSafetyandStandards (Packaging) Regulations,2018."FSSAI,September9, 2022.https://www.fssai.gov.in/upload/uploadfiles/files/Compendium_Packaging_Regulations_ 09_09_2022.pdf.
"Food Safety Basic Act - Fsc.go.jp." May 23, 2003. https://www.fsc.go.jp/english/basic_act/fs_basic_act.pdf.
"Food Sanitation Act (Act No. 233 of February 24, 1947)." The Cabinet Secretariat. https://www.cas.go.jp/jp/seisaku/hourei/data/fsa.pdf.
"India: India Proposes Draft Regulations for Genetically Modified Food for a Second Time." 2022. USDA Foreign Agricultural Service. Accessed January 11, 2023. https://www.fas.usda.gov/data/india-india-proposes-draft-regulations-genetically-modified-food-second-time.
"ISO and Food En 2018-International Organization for Standardization." ISO, 2018. https://www.iso.org/files/live/sites/isoorg/files/store/en/PUB100297.pdf.
"Labelling, Marking & Packaging." Labelling, Marking & Packaging | EU Business in Japan, 2022. Accessed January 22, 2023. https://www.eubusinessinjapan.eu/procedures/regulatory-legal-issues/labelling-marking-packaging.
Luo, Yunbo, and Guangfeng Wu. "Food Safety Laws and Regulations." *Food Safety in China*, (2017): 345–631. https://doi.org/10.1002/9781119238102.ch21.
Nguyen, Anh Thu, Lukas Parker, Linda Brennan, and Simon Lockrey. "A Consumer Definition of Eco-Friendly Packaging." *Journal of Cleaner Production* 252 (2020): 119792. https://doi.org/10.1016/j.jclepro.2019.119792.
""Overview - MyGFSI." n.d. Mygfsi.com." MyGFSI. Accessed January 7, 2023. https://mygfsi.com/who-we-are/overview/.
Scott, Tim. 2021. "Text - S.578 - 117th Congress (2021-2022): FASTER Act of 2021." Www.congress.gov. April 23, 2021. https://www.congress.gov/bill/117th-congress/senate-bill/578/text.
"Study on Food Safety Policy and Regulation in the United States." 2005. Directorate General for Internal Policies, Policy Department A: Economic and Scientific Policy. https://www.europarl.europa.eu/RegData/etudes/STUD/2015/536324/IPOL_STU(2015)536324_EN.pdf. Accessed January 3, 2023.
"Switzerland Publishes Updated Ordinance on Materials and Articles Intended to Contact Foodstuffs." PackagingLaw.com. n.d. Accessed May 30, 2023.

https://www.packaginglaw.com/news/switzerland-publishes-updated-ordinance-materials-and-articles-intended-contact-foodstuffs.

The International Food Safety Authorities Network INFOSAN Connecting Food Safety Authorities to Reduce Foodborne Risks Global food safety Food and Agriculture Organization of the United Nations. (n.d.). Retrieved May 29, 2023, from https://www.fao.org/3/AU632E/au632e.pdf

Title 21 Code of Federal Regulations §. 101, 2006. https://www.ecfr.gov/current/title-21/chapter-I/subchapter-B/part-101.

Title 21 Code of Federal Regulations §. 170–186. n.d. Accessed January 12, 2023. https://www.ecfr.gov/current/title-21/chapter-I/subchapter-B.

"Transforming Our World: The 2030 Agenda for Sustainable Development | Department of Economic and Social Affairs." United Nations, 2015. https://sdgs.un.org/2030agenda.

World Health Organization. *Five Keys to Safer Food Manual.* France: World Health Organization, 2006.

World Health Organization. "Food Safety." World Health Organization, May 19, 2022. https://www.who.int/news-room/fact-sheets/detail/food-safety#:~:text=An%20estimated%20600%20million%20%E2%80%93%20almost, health y%20life%20years%20(DALYs).

"World Packaging Organisation." 2013. Accessed January 11, 2023. https://www.worldpackaging.org/Uploads/SaveTheFood/China.pdf.

Chapter 13

Safety Aspects of Fried Foods with Reference to Acrylamide and Other Polar Compounds

Ira Tripathi
Kashi Naresh Government PG College

Sandeep Kumar Singh
Mata Gujri College

DOI: 10.1201/9781003329244-13

13.1 Introduction

Evolution developed us to *Homo sapiens*, but fruition of human beings was stimulated by food and its cooking methods. Mortals have always relished foods, due to its sensory characteristics (Stier 2000). Among the various culinary methods from all over the world, frying method is the most prevalent one. Frying method is comprised of heat and mass transfer in conventionally used media such as fat and oil. The thermal exposure makes food primarily more palatable and apposite for the human being, and additionally, it also has a preservative effect against microorganisms and certain enzymes. In the frying method, food is submerged partially or fully in hot fat/oils which then promotes the replacement of moisture and penetrates into the food, and inevitably the food becomes digestible, and crusty and gives delicious mouth feel (Cuesta et al. 1993; Krokida et al. 2000; Budžaki and Seruga 2004). The temperature varies according to food texture and material, and generally the range of temperature of oil or fat falls between 120°C and 180°C (Costa and Oliveira 1999). In the process of deep fat frying, heat shifts by the convection method from oil toward the peripheral portion of food and further leads to core by conduction transfer method (Figure 13.1). The moistness of food outflows via cracks, and this manner of cooking then plows pores which is resultant of pressurized seal-packed water molecules. Subsequently as the moisture is lost, the exterior surface temperature increases almost equivalent to the medium, i.e., oil. This

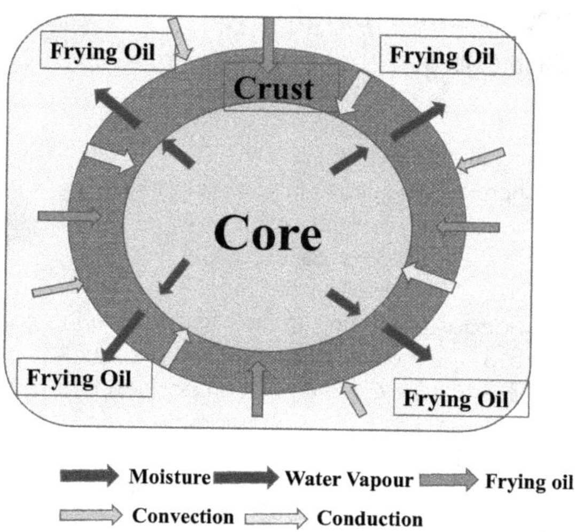

Figure 13.1 Heat and mass transfer process during deep fat frying.

process tends to swap away moisture from food, and through overpressure expansion during frying, a significant amount of oil absorption is inhibited (Mellema 2003; Oke 2017).

The process of frying can be carried out in small batches for domestic use or continuously for industrial applications. Fryers can perform in air pressure, low, high, or even vacuum conditions. However, the majority of industrial activities take place in atmospheric environments (Mallikarjunan et al. 2010). Deep fat frying method involves five steps to complete this process (Figure 13.2). The first stage is initial heating, in this time period the food product's external surface is raised from its original temperature to the water's boiling point. This phase is completed early and normally brief, resulting in a negligible water loss from the food. Surface boiling occurs in the second step. It is prominent due to quick loss of free moisture of exterior surface, and there is also rise in external surface heat transfer coefficient and as a result crust formation starts. The third stage, which can also be referred to as the declining rate period, is the moment when the preponderance of the humidity is lost. When the core region's temperature approaches the water's boiling point, the next and longest stage begins. The final stage of frying is the bubble finale point, which signals the loss of moisture and apparent completion of the frying process (Farkas et al. 1996).

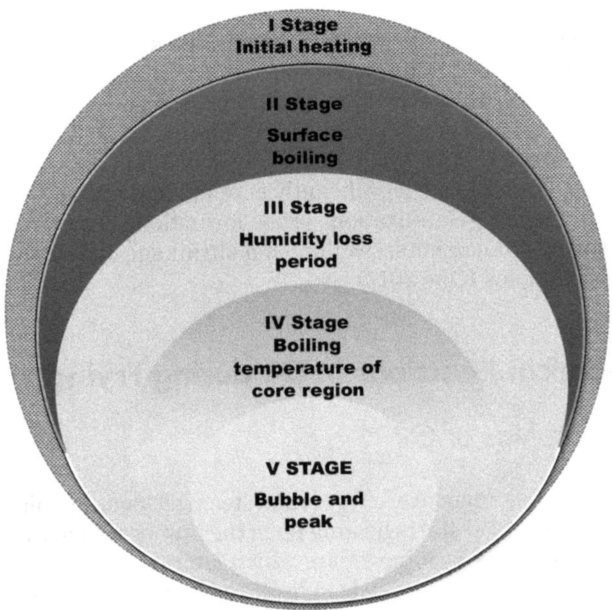

Figure 13.2 Steps of deep fat frying.

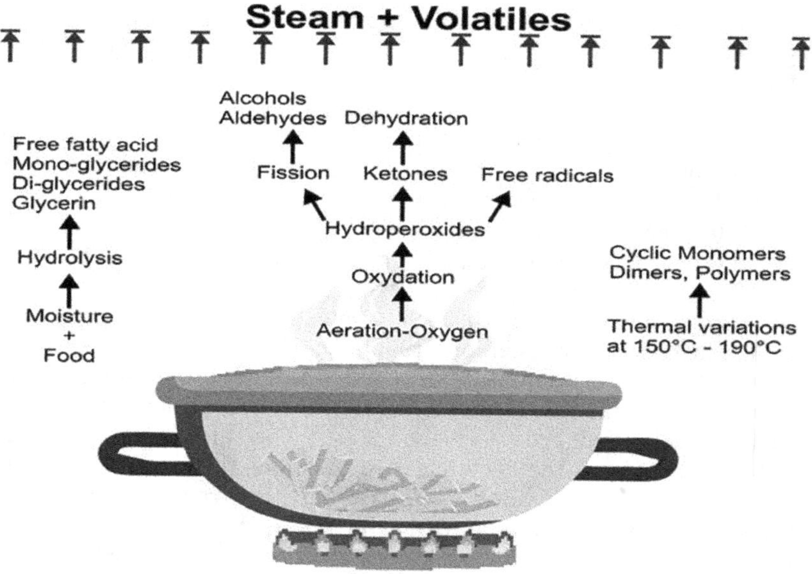

Figure 13.3 Reactions in oil during frying.

When food undergoes the frying process, various kinds of reactions occur. As the temperature rises, significant changes occur as a result of numerous physical and chemical interactions, including cyclization, hydrolysis, oxidation, isomerization, and polymerization (Figure 13.3) (Zribi et al. 2015, 2016; Ben Hammouda et al. 2016). These physicochemical changes alter the quality of food as well as oil, significantly at micro and macro levels. Protein denaturation, starch gelatinization, crust formation, browning, enhancement of flavoring components, swelling and shrinkage of food materials are some of the examples (Oke 2017).

13.2 Chemical Reactions in Oil during Frying

13.2.1 Hydrolysis of Oil

In the frying process, moisture, oxygen, and steam all together initiate a chemical reaction. When food starts heating up in the immersed fatty medium, the packed moisture starts forming steam which later on vaporizes with crackling and bubbling action and slowly subsides from fried food. Di- and monoacylglycerols, free fatty acids, and glycerol are produced when water, a weak nucleophile, breaks the ester link of triacylglycerols. As the quantity of frying increases, free fatty acids accumulate in the frying medium (Chung et al. 2006). Free fatty acids ratio is used to assess the quality of frying oil. Compared to oil with long and saturated fatty acids, the hydrolysis reaction occurs more frequently in fat with short and unsaturated fatty acids. Comparatively, short and unsaturated fatty acids are much more hydrophilic than long and saturated fatty acids. Short-chain fats and oils are effortlessly accessible to moisture of meals for hydrolysis (Nawar 1969). Volume of moisture affects the hydrolysis process positively (Dana et al. 2003). Steam is less efficient in comparison to water for hydrolysis process (Pokorny 1989). However, longer intervals between meals in the oily phase and food in the damp phase can speed up the hydrolysis process (Houhoula et al. 2003). Regularly replacing used oil with new frying oil slows down the hydrolysis process of frying oil (Romero et al. 1998). The amount of time spent on frying had no impact on the oil's hydrolysis (Naz et al. 2005). Free fatty acids and associated oxidized constituents cause off-flavor to be created during frying, thus making the oil lesser appropriate for deep fat frying. Glycerol, free fatty acids, di- and monoacylglycerols all speed up the hydrolysis reaction of oil even more (Frega et al. 1999). Glycerol evaporates around 150°C, and any leftover glycerol in oil later on encourages the breakdown of fatty acids to produce free fatty acids (Naz et al. 2005). According to Stevenson and colleagues (1984), frying oil should have a maximal free fatty acid level between 0.05% and 0.08%.

13.2.2 Polymerization of Oil

There are certain volatile compounds in food products and oils which produce flavor though their concentration are in ppm level (Nawar 1984). Nonvolatile polar compounds, triacylglycerol dimers, and polymers are the key breakdown yields of frying oil. Dimers and polymers are big molecules with molecular weights ranging between 692 to 1,600 grams per mole. Polymers and dimers both have carbonyl groups, hydroperoxy, hydroxyl, epoxy, and –C–O–C– and –C–O–O–C– bonds (Stevenson et al. 1984; Kim et al. 1999). Dimers are either cyclic or acyclic based on the reaction course, and a variety of fatty acids are present in fats and oils (Cuesta et al. 1993; Sánchez-Muniz et al. 1993; Takeoka et al. 1997; Tompkins and Perkins 2000). In deep fat frying, dimerization and polymerization are radical processes. Production of polymers and dimers is determined by the nature of oil, number of frying, and frying temperature. The sum of polymers surges as the frying number and temperature increase (Cuesta et al. 1993; Sánchez-Muniz et al. 1993; Takeoka et al. 1997). The linoleic acid–rich oil is more easily polymerized in the course of deep fat frying in comparison to the oleic acid–rich oil (Takeoka et al. 1997; Tompkins and Perkins 2000; Bastida and Sánchez-Muniz 2001). Radical reactions as well as the Diels–Alder reaction results in the development of cyclic polymers inside or around triacylglycerols. The level of unsaturation and the cooking temperature affect the assemblage of cyclic compounds in cooking oil (Meltzer et al. 1981). Formation of cyclic compounds does not begin in earnest till the oil temperature rises to 200°C–300°C. Deep fat frying produces oxygen-rich polymers. According to Yoon and colleagues (1988), oxidized polymer components speed up the oxidation of oil. Polymers enhance oil viscosity, reduce heat transfer, encourage oil breakdown, produce foam throughout deep fat frying, and alter the color of food in an unfavorable way (Tseng et al. 1996). Additionally, polymers enhance oil absorption of food.

13.2.3 Oxidation of Oil

The deep-frying method makes advantage of responsive oxygen (Peers and Swoboda 1982; Cuesta et al. 1993; Sánchez-Muniz et al. 1993; Houhoula et al. 2003). The chemical process driving the thermal oxidation and autoxidation mechanism is largely the same. Although thermal oxidation reaction happens at a rapid frequency in comparison to the autoxidation, it is precise and complete though systematic evidence and assessments of oxidation rates are still unknown between autoxidation and thermal oxidation. Normal oxygen is a diradical molecule that exists in air, which doesn't react with nonradical singlet state oil. Oxidation of oil takes place when radical oxygen entails radical oil. To interact with radical oxygen for an oil oxidation reaction, oil must be in a radical state. Oil's carbon will first undergo radical oxidation when the hydrogen with the weakest link is eliminated. The different assets of hydrogen–carbon bond of fatty acids clarify the alterations of oxidation degrees, among stearic, linoleic, oleic, and linoleic acids during autoxidation. Temperature, light, metals, and responsive oxygen types accelerate the radical formation of oil. Some polyvalent metals like Fe^{3+} and Cu^{2+}, wipe out hydrogen protons, even at low temperature, from oil to create alkyl radicals through oxidation–reduction process of metals. In contrast to unsaturated fatty acids, the location of radical formation in saturated fatty acids is at the position of carboxyl group with the electron-pulling property. This is known as initiation or beginning phase in the oxidation reaction of fats and oils, when the alkyl radical is formed in oil molecule by eliminating hydrogen. The alkyl radical might also respond with alkyl radicals, alkoxy radicals, and peroxy radicals to create dimers and polymers. Alkyl radicals react rapidly with diradical triplet oxygen and yield peroxy radicals. By taking the hydrogen out of oleic and linoleic acids, this peroxy radical further generates hydroperoxide. When this reaction is repeated with an additional oil molecule in food, a new hydroperoxide and an extra alkyl radical are produced which is called propagation step. This subsequent reaction is known as a free-radical chain reactions. Peroxy radicals can also produce dimers or polymers by responding with some other radicals at a rate of around 1.1 10^6/M/s (Choe and Min 2006). Chain reactions between free alkyl and peroxy radicals expedite the thermal oxidization of oil. Usually, hydroperoxides are unstable during the deep fat frying (Nawar 1984). By homolyzing the peroxide link, hydroperoxides break down into alkoxy radicals and hydroxy radicals. Oxy- and hydroxy radicals are created as a result of the hydroperoxide's breakdown. The alkoxy radical interacts with certain other alkoxy radicals or breaks down to produce compounds that are not radicals. The last phase is termination stage, which occurs at the end of oxidation, and it is the creation of nonradical volatile and nonvolatile molecules. Although the majority of volatiles are eliminated from frying oil by evaporation and decomposition as well as interactions between volatile compounds and other food components, deep fat frying retains certain volatiles (May et al. 1983; Nawar 1984). However, adding water to a frying system causes the amount of volatile chemicals in oil to decrease (Wu and Chen 1992). Further processes involving dimerization, polymerization, and oxidation are carried out on the volatile components in frying oil. Despite all these, volatile compounds significantly contribute to flavor formation in the fried foods and frying oil. On the other hand, as oxygen and free-radical concentrations increase, oxidative degradation reactions become more rapid (Paul et al. 1997). At the beginning of frying, mono- and diacylglycerols are not present in great quantities. High interfacial tension of the frying system causes steam bubbles to burst and creates a steam blanket on the oil's surface. The steam blanket lessens the oil's exposure to oxygen and declines the rate of oil oxidation (Blumenthal 1991). Alkaline substances like potassium or sodium salts of fatty acids are typically present in refined oil in concentrations of less than 1 ppm. Alkaline elements should be below 10 ppm in fresh frying oils (Moreira et al. 1999).

During immersion or deep frying process, oil also decays into a diversity of monomeric, polymeric, and volatile compounds, which influences the sensory quality, healthiness, as well as the shelf life of cooked products (Warner 1999; Juárez et al. 2011). Certain compounds are even accountable for the pleasurable flavor, taste, distinctive crispiness, and unique color when foodstuff is fried under suitable environments. But there are certain other compounds too such as conjugated linoleic acids, free radicals, trans-fatty acids, and several oxidized volatile products (acrolein and other α,β-unsaturated aldehydes) which are also unanimously produced during edible oil degradation. These are actually blamed for undesirable flavors, decreasing the shelf life of edible oils and possibly promoting health ailments (Warner 1999; Fullana et al. 2004a, b).

13.3 Changes Occurring in Fats and Oil during Frying

Due to its ability to add desirable flavor, color, and crisp texture to food, deep fat frying is one of the oldest (Varela 1998) and most widespread food preparation techniques in the culinary world to fry food, thus making them widely acceptable and enjoyable (Boskou et al. 2006). Cooking oil is a widely utilized food commodity in the frying process among various culinary cultures, and hence its excellence check has been a matter of debate nowadays among many customers. Regarding the quality and suitability of cooking oils, heating and other processing conditions have a significant impact on food items. For instance, everyone is familiar with the deep-frying method of cooking, which produces both beneficial and unpleasant molecules. Due to oxidative hydrolytic and polymerization reactions of oils, there is formation of nonvolatile and volatile breakdown products such as hydroperoxides, sterol derivatives, oxidized triacylglycerol monomer, conjugated dienes, polymers, trans-configuration compounds, nitrogen- and sulfur-containing compounds, and acrylamide to name a few (Nawar 1998; Zaribi 2015, 2016). Research has revealed that 400 diverse chemical compounds that had been recognized in depreciated frying oils, though 19 of these combinations are present in traces (Gere 1982). Around 220 are volatile degradation products among them (White 1991).

The aesthetic, perceptual, and health quality of fried products, as well as its shelf life, are all influenced by the degraded volatile chemicals, monomeric compounds, and polymeric compounds (Warner 1999; Juárez et al. 2011). Whenever food is fried properly, several of these components are actually accountable to give it its savory flavor, pleasing taste, usual crispness, and golden color. Free radicals, conjugated linoleic acids, trans-fatty acids, and some oxidized volatile products (acrolein and other unsaturated aldehydes), which are frequently produced during the degradation of edible oils, are known to be the cause of the off-flavor, which decreases the shelf life of edible oils and may also have negative health effects (Warner 1999; Fullana et al. 2004a, b).

13.3.1 Volatile Decomposition Products

Aromatic compounds, ketones, aldehydes, acids, alcohols, hydrocarbons, esters, and lactones are volatile decomposition products produced while frying (Figure 13.4). Basically all the volatile decomposition products are evaporated via steam during frying (Stevenson et al. 1984). However, there is some evidence that describes that few of such products are reserved in foodstuff and cooking medium. Furthermore, these compounds can be inhaled by the person performing deep frying and as a result can cause them ill health effects.

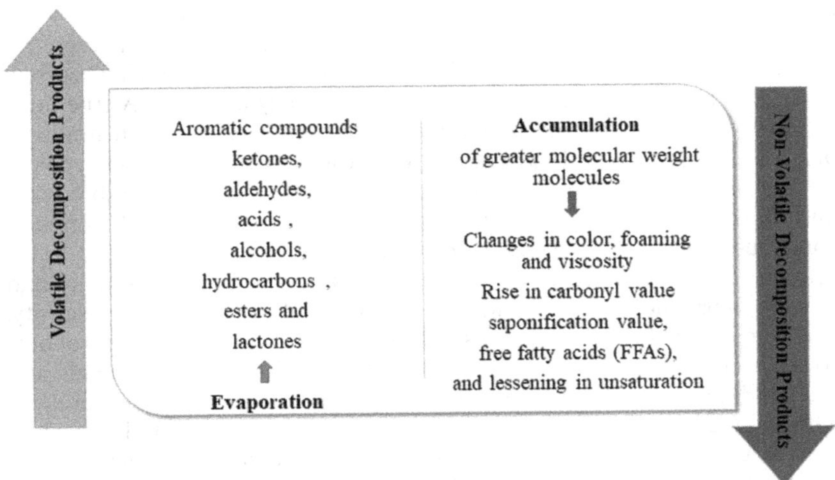

Figure 13.4 List of volatile and nonvolatile decomposition products during frying.

13.3.2 Nonvolatile Decomposition Products

The subsequent class of yields from deep-frying include nonvolatile putrefaction foodstuffs that are produced simultaneously with volatile breakdown products. Polymerization and thermal oxidation of the unsaturated fatty acids in the fatty cooking medium are the main reasons of their formation. It is a severe issue that these chemicals remain in the oil, contribute to further deterioration, are later absorbed by fried food, and are afterwards consumed (Paradis and Nawar 1981; Stevenson et al. 1984; Gertz 2000). High-molecular-weight molecules accumulate steadily and exhibit little volatility, making them more accurate indications of oil abuse. The formation and accumulation of nonvolatile decomposition products are responsible for physical changes in the frying oil (e.g., escalation in color, foaming and viscosity etc. Physical alterations in the frying oil are mainly caused due to formation and accumulation of nonvolatile decomposition products. Apart from this chemical changes, rise in carbonyl value, saponification value, and free fatty acids and lessening in unsaturation are due to consequential rise of the development of high-molecular-weight products. In practice, several different types of reactions occur, and they are all accountable for these changes in the class of frying oil. Solubilization of colored chemicals and fatty components in the food being fried may change the color and composition of the frying medium. Perkins (1967) and White (1991) stated that the elevated temperatures employed in deep frying speed up oxidation process. Oil first gains mass as oxygen is absorbed during air heating, and then its peroxide value may also rise. As the heating process proceeds, the peroxides break down and scission products begin to distil off, resulting in a net weight loss. Three different types of further degradation of hydroperoxides are possible: (i) fission or breaking up to form acids, aldehydes, alcohols, and hydrocarbons, which similarly contributes to the blackening of frying oils and flavor changes; (ii) dehydration to form ketones; and (iii) free-radical development of dimers and trimers to produce polymers, which entirely results in the rise of viscosity. The frequency of viscosity surges and polymer synthesis are synchronous. Artman (1969) reported that due to the conjugation of double bonds and the buildup of oxygenated products, the oil absorbs more ultra violet light. The production of new unsaturated linkages during the initial phases of heat oxidation may allow the oil's iodine value to increase; however, when the dual bonds are used up in subsequent reactions, the value of the oil decreases. The frying procedure predominantly affects unsaturated fatty acids (Gupta 2005). The rate of oxidation is roughly inversely proportionate to how unsaturated the fatty acids are. Linolenic acid (three double bonds) is thus much more vulnerable than oleic acid (one double bond). Hydrolysis caused by moistness in the items being fried additionally produces FFAs, monoglycerides, diglycerides, and glycerol. Due to the breakage and oxidation of double bonds, FFAs may also develop during oxidation (Perkins 1967). The "steam blanket effect" that moisture produces in the fryer cuts interaction with the air and aids in removing peroxides, tastes, and aromas that would otherwise build up in the frying oil. Thus, moisture also has a favorable impact on food quality (Stevenson et al. 1984). The heat that speeds up the polymerization of dimers and cyclic compounds causes the last phase of alteration in the quality of frying oil. Large molecules created by carbon-to-carbon and/or carbon-to-oxygen-to-carbon bridges between a number of fatty acids are the compounds that come from this process, irrespective of the fact that the mechanisms are extremely complicated and still not fully understood (Landers and Rathman 1981) and increasing the amount of polar compounds. Substantial improvements in these substances cause oil viscosity to increase, as well as foaming, "gum" deposition, and color darkening (Stevenson et al. 1984).

During lipid oxidation, more carbonyl groups are produced such as aldehydes and ketones that later react with asparagine and forms acrylamide, even in the absence of reducing sugar. Yasuhara et al. (2003) and Ehling et al. (2005) also opinioned that acrylamide could be developed via the formation of acrolein. Acrolein, which is one of the oil byproducts, further oxidizes to acrylic acid and consequently reacts with ammonia to form acrylamide.

13.3.2.1 Acrylamide

The chemical name for acrylamide (ACR) is C_3H_5NO, and it is an organic substance. Other names for acrylamide include ethylene carboxamide, acrylic amide (50%), propenamide, vinyl amide 2-propenamide, and propenamide (Stadler and Goldmann 2008). Acrylamide is a substance that is frequently employed in the trade to create polyacrylamide, a polymer that is used in a variety of items in the current society. Monomers (single units) and polymers both are formed in acrylamide. A single unit of acrylamide is considered hazardous and may cause human cancer. Due to the Maillard reaction, ACR was found in several foods cooked at high temperatures, above 120°C. The amount of acrylamide present in different types of food is influenced by the food's content, the cooking method, the temperature, and the time of cooking process (baking, frying, and grilling) (Yaylayan et al. 2003).

13.3.2.1.1 Properties of Acrylamide The two possible states of acrylamide (ACR) are liquid and crystalline solid. The monomer is a solid, colorless crystal which dissolves in polar solvents (such as acetone, methanol, and dimethyl ether), and water but not nonpolar ones (e.g., benzene, heptane, and carbon tetrachloride). The melting point of acrylamide is 84.5°C, and its boiling point is 125.5°C at a pressure of 25 mm Hg (Stadler and Goldmann 2008). In both soil and water, it is highly mobile. It does not build up in the soil, where it may be quickly removed. It is possible to biodegrade polyacrylamide (Smith et al. 1996; Altunay et al. 2016; Zamani et al. 2017).

Figure 13.5 illustrates chemical structures of polyacrylamide and acrylamide. ACR is used for the manufacturing of polyacrylamide which is made for diverse purposes. The specific gravity of acrylamide is 1.122 g/cm^3 at 30°C, and it can be freely polymerized if exposed to ultraviolet radiation or heated to melting point. The key spot of reaction is sulfhydryl groups enclosed on proteins.

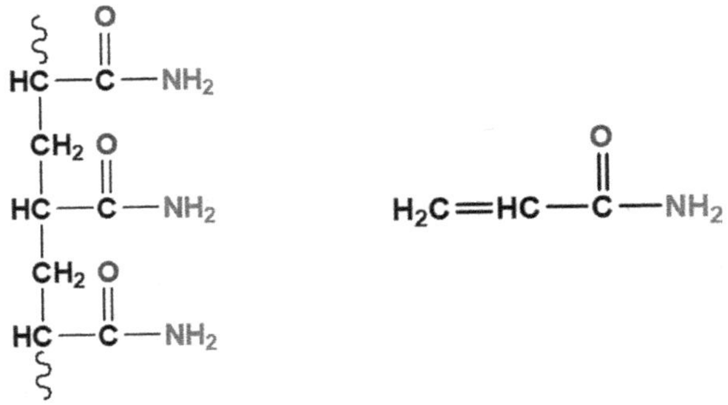

Figure 13.5 Chemical structure of polyacrylamide and acrylamide.

13.3.2.1.2 Acrylamide in Food The majority of the acrylamide found in food is produced during heating (baking, broiling, frying, or roasting) when the amino group of the free amino acid asparagine reacts with the carbonyl group of reducing sugars like glucose in food (baking, broiling, frying, roasting). The majority of exposure to these precursors comes from plant sources like potatoes and grains (barley, rice, and wheat), but not from animal sources like chicken, meat, and fish. Potato chips, French fries, tortilla, bread crust, crisp bread, other baking items, cereals, and coffee are common processed foods with elevated quantities of acrylamide (Mousavi Khaneghah et al. 2022) because observed acrylamide levels vary greatly even within a single meal, such as potatoes, thus the source of the food is crucial. However, other factors can be just as significant.

Wide variations in precursor levels, which is present both naturally in some cultivars and as a result of postharvest handling, as well as processing conditions (e.g., pH, temperature, time, nature of food matrix) appear to be the causes of the wide variations in acrylamide levels found in diverse food categories and brands of the same food category (e.g., cereals, potato chips, French fries, coffee). The food supply is frequently contaminated with acrylamide. The temperature at which food is cooked directly affects its acrylamide levels. Foods that had been cooked or the unheated control did not contain any acrylamide; the level that was discovered was typically below 5 mg/kg. The quantity of acrylamide in heated protein-rich foods ranged from 5 to 50 mg/kg, whereas carbohydrate-rich foods such beets, potatoes, some heated crisp bread, and processed potato products had higher concentrations. Potato products alone are responsible for 50% of individual ACR exposure. The detected amount ranged from 150 to 4,000 mg/kg. Consumption patterns show that the investigated acrylamide levels of heated meals can result in a daily dose of few more tens of micrograms (Tareke et al. 2002). The use of ACR in food packaging results in additional indirect exposure to ACR (Zhang et al. 2007).

13.3.2.1.3 Mechanism of Acrylamide Toxicity The small chemical compound ACR has an electrophilic vinyl group and is susceptible to nucleophilic assault (Afifi et al. 2020). In rats, it has been demonstrated that this substance absorbs quickly and completely through the digestive tract. It is also absorbed through the skin (Zamani et al. 2017). Studies on human toxicokinetics revealed that ACR has a terminal half-life of 2.4–7.0 hours (Matoso et al. 2019).

An imbalance between the biological oxidant and antioxidant can lead to oxidative stress. It might also be the first sign of a lot of different diseases (Mojtahedzadeh et al. 2014). Free radicals are continuously generated in vivo. For mitigating the detrimental effects of free radicals, the body has defense mechanisms such as antioxidant enzymes glutathione S-transferase (GST), superoxide dismutase (SOD), glutathione peroxidase

(GPx), and catalase (Shaki et al. 2012). Recent research has indicated that oxidative stress has been brought on by ACR. Reactive oxygen species (ROS) are produced, which have an impact on the cellular redox chain. It undergoes oxidation to become glycidamide and is subsequently coupled with GSH. A group of nucleophiles can interact with ACR and glycidamide in cells (such as $-NH_2$, –SH, or –OH). Under the action of hepatic GSH-S transferases, ACR directly reacts with GSH (Bucur et al. 2018). GST and SOD activities rise when ACR concentration rises, while GSH concentration falls as ACR concentration rises (Chen et al. 2014). Additionally, it has been noted that ACR can cause apoptosis because of oxidative stress (Yilmaz et al. 2017). Chronic dietary ACR intake may make people more susceptible to oxidative damage.

13.3.2.1.4 Neurotoxicity of Acrylamide According to the findings of several investigations, ACR is a powerful neurotoxin. The lowest observed adverse effect level (LOAEL) for ACR neurotoxicity in animal experiments is 2 g/kg/day, whereas the no-observable adverse effect level (NOAEL) is in the range between 0.2 and 0.5 g/kg/day. The only hazardous effects that have been seen in both humans and animals are those related to ACR neurotoxicity. Numerous lab animal species, including cats, rodents, rats, guinea pigs, rabbits, and monkeys, have been the subject of studies showing the triad effects of extended routine exposure to ACR, which include ataxia, back legs foot splay, and skeletal tissues weakening (Exon 2006). The neurotoxic effects of ACR in the workplace have been measured by a number of research (LoPachin 2004). Another study revealed that ACR disrupted the nervous system by preventing human glioblastoma and neuroblastoma cells from differentiating into their mature forms (Chen and Chou, 2015).

13.3.2.1.5 Carcinogenicity and Genotoxicity of Acrylamide ACR is classified as a "probable human carcinogen" by the International Agency for Research on Cancer (IARC 1994; Pelucchi et al. 2006). Epidemiological studies and human occupational use have not revealed any evidence of the carcinogenic consequences of ACR. ACR is structurally related to cancer-causing substances like vinyl carbamate and acrylonitrile. However, extending the lives of animals exposed to significant concentrations of contaminants in their water causes both male and female species to produce many tumors at various places (Exon 2006).

The carcinogenicity of ACR has been proven by a great deal of research. The findings suggested that ACR may raise the risk of several malignancies. For instance, exposure to ACR was quite linked to breast and kidney malignancies in postmenopausal women. However, some research contests these results (Pelucchi et al. 2017). While there is no relationship between dietary ACR and the threat to bladder or prostate cancer, there was a positive correlation with dietary ACR and renal cell carcinoma

(Hogervorst et al. 2008). The genotoxicity of ACR as well as its primary metabolite, glycidamide, have been examined in a number of investigations. The potential to cause genetic harm is one of the key factors in assessing carcinogenicity (Exon 2006). For instance, research has demonstrated that ACR at doses of 10, 20, and 30 mg/kg significantly increased DNA toxicity when associated with the control group (Alzahrani 2011). In vitro tests and in vivo animal models were used in another work to show that ACR had genotoxic effects in cell culture (Exon 2006).

13.3.2.1.6 Reproductive Toxicity of Acrylamide Animals obtaining high doses of ACR from a lab also had their reproductive toxicity assessed. There is no evidence that ACR is hazardous to human reproduction. For reproductive toxicity in rats, the no-observed adverse effect level (NOAEL) was calculated to be 2–5 g/kg/day. The dose is 2,000 times higher than anticipated dietary exposures and at least tetra fold over the acrylamide neurotoxicity predicted by doses (Tyl and Friedman 2003). In comparison to the control group, ACR treatment of 0.5–10 mg/kg also slowed rat development and reduced epididymal sperm stores. Histopathologic abnormalities were also found in the testes of rats receiving treatment (Wang et al. 2010). In one study, testosterone and prolactin levels in male rats who received intraperitoneal injections of ACR dosages (20 mg/kg) dropped in a dose-dependent manner (Zamani et al. 2017). According to a different study, ACR may harm the mouse female reproductive system. It was demonstrated that oral doses of ACR dramatically reduced body weights, organ weights, and corporal lutea counts, and serum progesterone levels fell as a result of ACR depending on dosage (Wei et al. 2014).

13.3.2.1.7 Immunotoxicity of Acrylamide There is less information on ACR's immunotoxicity than its neuro and reproductive toxicity. ACR's immunotoxicity was seen in female BALB/c mice during a survey. According to the study, ACR caused a drop in lymphocyte counts, spleen and thymus weights, and ultimate body weight. Additionally, pathological changes in the lymph nodes, thymus, and spleen have been noted. Furthermore, T lymphocytes and natural killer (NK) cells have been shown to significantly decrease under the influence of ACR (Fang et al. 2014).

13.3.3 Polar Compounds

In recent years, a sizeable portion of the world's oils and fats are used for the frying business. The quality and safety of frying oil are a problem as the demand for fried meals rises. It is widely known that in order to produce fried food with distinctive organoleptic properties, frying requires heating the oil to a temperature of up to 190°C (Aydınkaptan and Barutçu Mazı 2017). The overheating of the oil during frying in the presence of air and humidity causes a wide variety of chemical reactions. Intricacy of these reactions heavily depends on the food's structure and matrix (Karoui et al. 2011), frying processes, and the composition of the frying oil (Olivero-David et al. 2014) and conditions for the process (such as frying time, temperature, and pressure) (Aladedunye and Przybylski 2008). As fried food absorbs a lot of oil, deterioration of frying oil quality would not only have an adverse effect on the food's sensory qualities but also produce unintended breakdown byproducts that could be harmful to consumer health (Gertz and Behmer 2014; Zribi et al. 2014). Total polar components (TPCs), polyglycerol esters, polycyclic aromatic hydrocarbons, 3-chloropropanol esters, and trans-fatty acids are among the hazardous compounds that are created during the high-temperature processing through oxidation, polymerization, decomposition, and hydrolysis (Tynek et al. 2001). A large amount of TPCs are found in frying oils. During continual frying, peroxides and hydroperoxides are converted into TPCs, whose polarity is greater than that of triglycerides (Xu et al. 1999). According to Kassama and Ngadi (2005), these are the main causes of the quality deterioration of deep-fried foods since they provide the finished items or fried oil an unfavorable odor, an unpleasant hue, and a high viscosity (Choe and Min 2006). Excess TPCs have a negative impact on human health and nutritional function. As a result, many nations mandate that the TPC content of frying oil not be greater than 27% (Gil et al. 2004). Li et al. (2016) fed Kunming mice a high-fat diet including TPCs as a dietary intervention to confirm the effect of TPCs on animal health. Later it was revealed that the mice's body and liver weight had grown after 12 weeks, and their ability to tolerate glucose dramatically worsened. After receiving TPC treatment for 18 months, Billek (2000) demonstrated that rats' body weight decreased, but their liver weight and inflammation increased. The results in terms of hepatic lipid buildup are consistent; however, the impact of TPCs on obesity is still debatable. Oxidative stress and mitochondrial dysfunction are linked to liver lipid buildup and degeneration (Das et al. 2008; Wei et al. 2009). Although it has been demonstrated that the TPCs formed during the frying process have a negative impact on liver function in mice, the precise mechanism underlying these effects is still unclear.

13.4 Conclusion

Culinary world's most widely prevalent cooking method is frying. In the present scenario, fried food products are made with an attitude to attract consumers by satisfying their heart, mind, and body. Most of these fried foods are easy to serve and highly processed by various means. Despite the warning of medical practitioners and nutritionist regarding the calories, fat and cholesterols contents, and toxic compounds, fried foods are gaining popularity. The main reason behind this is they are sensory bombs. Trans-fat acids, polar compounds and acrylamide are main substances created during the frying process that poses as a serious hazard to public health. Therefore, it is crucial to recognize and evaluate their impacts on human health as well as develop strategies to reduce the oil intake and creation of these chemicals without diminishing the flavor of fried foods.

References

Afifi, Mohamed. "A Review on the New Trends of Acrylamide Toxicity." *Biomedical Journal of Scientific & Technical Research* 27, no. 2 (April 27, 2020). https://doi.org/10.26717/bjstr.2020.27.004480.

Aladedunye, Felix A., and Roman Przybylski. "Degradation and Nutritional Quality Changes of Oil during Frying." *Journal of the American Oil Chemists' Society* 86, no. 2 (December 10, 2008): 149–56. https://doi.org/10.1007/s11746-008-1328-5.

Altunay, Nail, Ramazan Gürkan, and Ulaş Orhan. "A Preconcentration Method for Indirect Determination of Acrylamide from Chips, Crackers and Cereal-Based Baby Foods Using Flame Atomic Absorption Spectrometry." *Talanta* 161 (December 2016): 143–50. https://doi.org/10.1016/j.talanta.2016.08.053.

Alzahrani, Hind Abdullah Seed. "Protective Effect of L-Carnitine against Acrylamide-Induced DNA Damage in Somatic and Germ Cells of Mice." *Saudi Journal of Biological Sciences* 18, no. 1 (January 2011): 29–36. https://doi.org/10.1016/j.sjbs.2010.07.004.

Artman, N.R. "The Chemical and Biological Properties of Heated and Oxidized Fats." In *Advances in Lipid Research*, ed. R. Paoletti, 245–350. New York: Academic Press, 1969.

Aydınkaptan, E., and I. Barutçu Mazı. "Monitoring the Physicochemical Features of Sunflower Oil and French Fries during Repeated Microwave Frying and Deep-Fat Frying." *Grasas y Aceites* 68, no. 3 (September 8, 2017): 202. https://doi.org/10.3989/gya.1162162.

Bastida, S., and F.J. Sánchez-Muniz. "Thermal Oxidation of Olive Oil, Sunflower Oil and a Mix of Both Oils during Forty Discontinuous Domestic Fryings of Different Foods." *Food Science and Technology International* 7, no. 1 (February 1, 2001): 15–21. https://doi.org/10.1177/108201301772662644.

Ben Hammouda, Ibtissem, Akram Zribi, Amir Ben Mansour, Bertrand Matthäus, and Mohamed Bouaziz. "Effect of Deep-Frying on 3-MCPD Esters and Glycidyl Esters Contents and Quality Control of Refined Olive Pomace Oil Blended with Refined Palm Oil." *European Food Research and Technology* 243, no. 7 (December 26, 2016): 1219–27. https://doi.org/10.1007/s00217-016-2836-4.

Billek, Gerhard. "Health Aspects of Thermoxidized Oils and Fats." *European Journal of Lipid Science and Technology* 102, no. 8–9 (September 2000): 587–93. http://doi.org/10.1002/1438-9312(200009)102:8/9<587::aid-ejlt587>3.0.co;2-#.

Blumenthal, Michael M. "A New Look at the Chemistry and Physics of Deep-Fat Frying." *Food Technology* 45 (1991): 68–71.

Boskou, George, Fotini N. Salta, Antonia Chiou, Elena Troullidou, and Nikolaos K. Andrikopoulos. "Content of Trans, Trans-2,4-decadienal in Deep-fried and Pan-fried Potatoes." *European Journal of Lipid Science and Technology* 108, no. 2 (February 2006): 109–15. https://doi.org/10.1002/ejlt.200500236.

Bucur, Madalina-Petruta, Bogdan Bucur, and Gabriel-Lucian Radu. "Simple, Selective and Fast Detection of Acrylamide Based on Glutathione S-Transferase." *RSC Advances* 8, no. 42 (2018): 23931–36. https://doi.org/10.1039/c8ra02252f.

Budžaki, Sandra, and Bernarda Šeruga. "Moisture Loss and Oil Uptake during Deep Fat Frying of 'Kroštula' Dough." *European Food Research and Technology* 220, no. 1 (November 5, 2004): 90–95. https://doi.org/10.1007/s00217-004-1058-3.

Chen, Jong-Hang, and Chin-Cheng Chou. "Acrylamide Inhibits Cellular Differentiation of Human Neuroblastoma and Glioblastoma Cells." *Food and Chemical Toxicology* 82 (August 2015): 27–35. https://doi.org/10.1016/j.fct.2015.04.030.

Chen, Wei, Yang Shen, Hongming Su, and Xiaodong Zheng. "Hispidin Derived from Phellinus Linteus Affords Protection against Acrylamide-Induced Oxidative Stress in Caco-2 Cells." *Chemico-Biological Interactions* 219 (August 2014): 83–89. https://doi.org/10.1016/j.cbi.2014.05.010.

Choe, Eunok, and David B. Min. "Chemistry and Reactions of Reactive Oxygen Species in Foods." *Journal of Food Science* 70, no. 9 (May 31, 2006): R142–59. https://doi.org/10.1111/j.1365-2621.2005.tb08329.x.

Chung, J., J. Lee, and E. Choe. "Oxidative Stability of Soybean and Sesame Oil Mixture during Frying of Flour Dough." *Journal of Food Science* 69, no. 7 (May 31, 2006): 574–78. https://doi.org/10.1111/j.1365-2621.2004.tb13652.x.

Costa, R.M., and F.A.R. Oliveira. "Modelling the Kinetics of Water Loss during Potato Frying with a Compartmental Dynamic Model." *Journal of Food Engineering* 41, no. 3–4 (August 1999): 177–85. https://doi.org/10.1016/s0260-8774(99)00095-3.

Cuesta, C., F.J. Sánchez-Muniz, C. Garrido-Polonio, S. López-Varela, and R. Arroyo. "Thermoxidative and Hydrolytic Changes in Sunflower Oil Used in Fryings with a Fast Turnover of Fresh Oil." *Journal of the American Oil Chemists Society* 70, no. 11 (November 1993): 1069–73. https://doi.org/10.1007/bf02632144.

Dana, Dina, Michael M. Blumenthal, and I. Sam Saguy. "The Protective Role of Water Injection on Oil Quality in Deep Fat Frying Conditions." *European Food Research and Technology* 217, no. 2 (August 1, 2003): 104–09. https://doi.org/10.1007/s00217-003-0744-x.

Das, S.K., V. Balakrishnan, S. Mukherjee, and D.M. Vasudevan. "Evaluation of Blood Oxidative Stress-related Parameters in Alcoholic Liver Disease and Non-alcoholic Fatty Liver Disease." *Scandinavian Journal of Clinical and Laboratory Investigation* 68, no. 4 (January 2008): 323–34. https://doi.org/10.1080/00365510701673383.

Ehling, Stefan, Matt Hengel, and Takayuki Shibamoto. "Formation of Acrylamide from Lipids." Formation of Acrylamide from Lipids | SpringerLink, 2005. Accessed October 24, 2022. https://link.springer.com/chapter/10.1007/0-387-24980-X_17.

Exon, J.H. "A Review of the Toxicology of Acrylamide." *Journal of Toxicology and Environmental Health, Part B* 9, no. 5 (December 2006): 397–412. https://doi.org/10.1080/10937400600681430.

Fang, J., C.L. Liang, X.D. Jia, and N. Li. "Immunotoxicity of Acrylamide in Female BALB/c Mice." *Biomedical and Environmental Sciences: BES* 27, no. 6 (June 2014): 401–09. https://doi.org/10.3967/bes2014.069.

Farkas, B.E., R.P. Singh, and T.R. Rumsey. "Modeling Heat and Mass Transfer in Immersion Frying. I, Model Development." *Journal of Food Engineering* 29, no. 2 (August 1996): 211–26. https://doi.org/10.1016/0260-8774(95)00072-0.

Frega, N., M. Mozzon, and G. Lercker. "Effects of Free Fatty Acids on Oxidative Stability of Vegetable Oil." *Journal of the American Oil Chemists' Society* 76, no. 3 (March 1999): 325–29. https://doi.org/10.1007/s11746-999-0239-4.

Fullana, Andres, Angel A. Carbonell-Barrachina, and Sukh Sidhu. "Comparison of Volatile Aldehydes Present in the Cooking Fumes of Extra Virgin Olive, Olive, and Canola Oils." *Journal of Agricultural and Food Chemistry* 52, no. 16 (July 16, 2004a): 5207–14. https://doi.org/10.1021/jf035241f.

Fullana, Andres, Angel A. Carbonell-Barrachina, and Sukh Sidhu. "Volatile Aldehyde Emissions from Heated Cooking Oils." *Journal of the Science of Food and Agriculture* 84, no. 15 (2004b): 2015–21. https://doi.org/10.1002/jsfa.1904.

Gere, A. "Studies of the Changes in Edible Fats during Heating and Frying." *Food / Nahrung* 26, no. 10 (1982): 923–32. https://doi.org/10.1002/food.19820261025.

Gertz, Christian. "Chemical and Physical Parameters as Quality Indicators of Used Frying Fats." *European Journal of Lipid Science and Technology* 102, no. 8–9 (September 2000): 566–72. http://doi.org/10.1002/1438-9312(200009)102:8/9<566::aid-ejlt566>3.0.co;2-b.

Gertz, Christian, and Dagmar Behmer. "Application of FT-NIR Spectroscopy in Assessment of Used Frying Fats and Oils*." *European Journal of Lipid Science and Technology* 116, no. 6 (June 2014): 756–62. https://doi.org/10.1002/ejlt.201300270.

Gil, Bogim, Yong Jin Cho, and Suk Hoo Yoon. "Rapid Determination of Polar Compounds in Frying Fats and Oils Using Image Analysis." *LWT - Food Science and Technology* 37, no. 6 (September 2004): 657–61. https://doi.org/10.1016/j.lwt.2004.02.006.

Gupta, M.K. "Frying Oils." In *Vol. 4 of Bailey's Industrial Oil & Fat Products*, 6th ed., ed. F. Shahidi, 1–31. Canada: Wiley-Interscience, 2005.

Hogervorst, Janneke G., Leo J. Schouten, Erik J. Konings, R. Alexandra Goldbohm, and Piet A. van den Brandt. "Dietary Acrylamide Intake and the Risk of Renal Cell, Bladder, and Prostate Cancer." *The American Journal of Clinical Nutrition* 87, no. 5 (May 1, 2008): 1428–38. https://doi.org/10.1093/ajcn/87.5.1428.

Houhoula, Dimitra P., Vassiliki Oreopoulou, and Constantina Tzia. "The Effect of Process Time and Temperature on the Accumulation of Polar Compounds in Cottonseed Oil during Deep-Fat Frying." *Journal of the Science of Food and Agriculture* 83, no. 4 (2003): 314–19. https://doi.org/10.1002/jsfa.1314.

IARC. *IARC Monographs on the Evaluation of Carcinogenic Risks to Humans*, Vol. 60, 389–433. Lyon: IARC, 1994.

Juárez, María Daniela, Cibele Cristina Osawa, María Elina Acuña, Norma Sammán, and Lireny Aparecida Guaraldo Gonçalves. "Degradation in Soybean Oil, Sunflower Oil and Partially Hydrogenated Fats after Food Frying, Monitored by Conventional and Unconventional Methods." *Food Control* 22, no. 12 (December 2011): 1920–27. https://doi.org/10.1016/j.foodcont.2011.05.004.

Karoui, Iness Jabri, Wissal Dhifi, Meriam Ben Jemia, and Brahim Marzouk. "Thermal Stability of Corn Oil Flavoured with Thymus Capitatus under Heating and Deep-Frying Conditions." *Journal of the Science of Food and Agriculture* 91, no. 5 (January 6, 2011): 927–33. https://doi.org/10.1002/jsfa.4267.

Kassama, L.S., and M.O. Ngadi. "Pore Development and Moisture Transfer in Chicken Meat during Deep-Fat Frying." *Drying Technology* 23, no. 4 (April 1, 2005): 907–23. https://doi.org/10.1081/drt-200054239.

Kim, I.-H., C.-J. Kim, and D.-H. Kim. "Physicochemical Properties of Methyl Linoleate Oxidized at Various Temperatures." *Korean Journal of Food Science and Technology* 31 (1999): 600–05.

Krokida, M.K., V. Oreopoulou, and Z.B. Maroulis. "Water Loss and Oil Uptake as a Function of Frying Time." *Journal of Food Engineering* 44, no. 1 (April 2000): 39–46. https://doi.org/10.1016/s0260-8774(99)00163-6.

Landers, R.E., and D.M. Rathmann. "Vegetable Oils: Effects of Processing, Storage and Use on Nutritional Values." *Journal of the American Oil Chemists' Society* 58, no. 3, Part1 (March 1981): 255. https://doi.org/10.1007/bf02582352.

Li, Xiaodan, Xiaoyan Yu, Dewei Sun, Jinwei Li, Yong Wang, Peirang Cao, and Yuanfa Liu. "Effects of Polar Compounds Generated from the Deep-Frying Process of Palm Oil on Lipid Metabolism and Glucose Tolerance in Kunming Mice." *Journal of Agricultural and Food Chemistry* 65, no. 1 (December 23, 2016): 208–15. https://doi.org/10.1021/acs.jafc.6b04565.

LoPachin, Richard M. "The Changing View of Acrylamide Neurotoxicity." *Neuro Toxicology* 25, no. 4 (June 2004): 617–30. https://doi.org/10.1016/j.neuro.2004.01.004.

Mallikarjunan, P.K., Ngadi, M.O., and Chinnan, M.S. *Breaded Fried Foods*. Boca Raton, FL: CRC Press, 2010.

Matoso, Viviane, Paula Bargi-Souza, Fernanda Ivanski, Marco A. Romano, and Renata M. Romano. "Acrylamide: A Review about Its Toxic Effects in the Light of Developmental Origin of Health and Disease (DOHaD) Concept." *Food Chemistry* 283 (June 2019): 422–30. https://doi.org/10.1016/j.foodchem.2019.01.054.

May, William A., Robert J. Peterson, and Stephen S. Chang. "Chemical Reactions Involved in the Deep-Fat Frying of Foods: IX. Identification of the Volatile Decomposition Products of Triolein." *Journal of the American Oil Chemists' Society* 60, no. 5 (May 1983): 990–95. https://doi.org/10.1007/bf02660214.

Mellema, M. "Mechanism and Reduction of Fat Uptake in Deep-Fat Fried Foods." *Trends in Food Science & Technology* 14, no. 9 (September 2003): 364–73. https://doi.org/10.1016/s0924-2244(03)00050-5.

Meltzer, J.B., E.N. Frankel, T.R. Bessler, and E.G. Perkins. "Analysis of Thermally Abused Soybean Oils for Cyclic Monomers." *Journal of the American Oil Chemists' Society* 58, no. 7 (July 1981): A779–84. https://doi.org/10.1007/bf02887322.

Mojtahedzadeh, M., A. Ahmadi, A. Mahmoodpoor, M.T. Beigmohammadi, M. Abdollahi, Z. Khazaeipour, F. Shaki, B. Kuochaki, and N. Hendouei. "Hypertonic Saline Solution Reduces the Oxidative Stress Responses in Traumatic Brain Injury Patients." *Journal of Research in Medical Sciences: The Official Journal of Isfahan University of Medical Sciences* 19, no. 9 (September 2014): 867–74. PMID: 25535502.

Moreira, Rosana, M. Elena Castwell-Perez, and Maria A. Barrufet. *Deep-Fat Frying. Fundamentals and Applications*. Gaithersburg, MD: Aspen Publication, 1999. https://doi.org/10.1604/978083421321010.1007/b114124.

Mousavi Khaneghah, Amin, Yadolah Fakhri, Amene Nematollahi, Fatemeh Seilani, and Yasser Vasseghian. "The Concentration of Acrylamide in Different Food Products: A Global Systematic Review, Meta-Analysis, and Meta-Regression." *Food Reviews International* 38, no. 6 (August 18, 2022): 1286–304. https://doi.org/10.1080/87559129.2020.1791175.

Nawar, Wassef W. "Thermal Degradation of Lipids." *Journal of Agricultural and Food Chemistry* 17, no. 1 (January 1969): 18–21. https://doi.org/10.1021/jf60161a012.

Nawar, Wassef W. "Chemical Changes in Lipids Produced by Thermal Processing." *Journal of Chemical Education* 61, no. 4 (April 1984): 299. https://doi.org/10.1021/ed061p299.

Nawar, W. W. "Volatile Components of the Frying Process | Grasas y Aceites." Volatile Components of the Frying Process | Grasas y Aceites, August 30, 1998. https://doi.org/10.3989/gya.1998.v49.i3-4.727.

Naz, Shahina, Rahmanullah Siddiqi, Hina Sheikh, and Syed Asad Sayeed. "Deterioration of Olive, Corn and Soybean Oils Due to Air, Light, Heat and Deep-Frying." *Food Research International* 38, no. 2 (March 2005): 127–34. https://doi.org/10.1016/j.foodres.2004.08.002.

Oke, E. K., M. A. Idowu, O. P. Sobukola, S. A. O. Adeyeye and A. O. Akinsola. "Frying of Food: A Critical Revie." Journal of Culinary Science & Technology, 16, no. 2 (April 2018): 107–127, DOI: 10.1080/15428052.2017.1333936

Olivero-David, Raul, Carmen Mena, M. Angeles Pérez-Jimenez, Blanca Sastre, Sara Bastida, Gloria Márquez-Ruiz, and Francisco J. Sánchez-Muniz. "Influence of Picual Olive Ripening on Virgin Olive Oil Alteration and Stability during Potato Frying." *Journal of Agricultural and Food Chemistry* 62, no. 48 (November 20, 2014): 11637–46. https://doi.org/10.1021/jf503860j.

Paradis, A.J., and W.W. Nawar. "Evaluation of New Methods for the Assessment of Used Frying Oils." *Journal of Food Science* 46, no. 2 (March 1981): 449–51. https://doi.org/10.1111/j.1365-2621.1981.tb04882.x.

Paul, S., G.S. Mittal, and M.S. Chinnan. "Regulating the Use of Degraded Oil/Fat in Deep-fat/Oil Food Frying." *Critical Reviews in Food Science and Nutrition* 37, no. 7 (November 1997): 635–62.

Peers, Kenneth E., and Peter A.T. Swoboda. "Deterioration of Sunflower Seed Oil under Simulated Frying Conditions and during Small-Scale Frying of Potato Chips." *Journal of the Science of Food and Agriculture* 33, no. 4 (April 1982): 389–95. https://doi.org/10.1002/jsfa.2740330414.

Pelucchi, Claudio, Carlotta Galeone, Fabio Levi, Eva Negri, Silvia Franceschi, Renato Talamini, Cristina Bosetti, Attilio Giacosa, and Carlo La Vecchia. "Dietary Acrylamide and Human Cancer." *International Journal of Cancer* 118, no. 2 (January 15, 2006): 467–71. https://doi.org/10.1002/ijc.21336.

Pelucchi, C., V. Rosato, P.M. Bracci, D. Li, R.E. Neale, E. Lucenteforte, D. Serraino, et al. "Dietary Acrylamide and the Risk of Pancreatic Cancer in the International Pancreatic Cancer Case–Control Consortium (PanC4)." *Annals of Oncology* 28, no. 2 (February 2017): 408–14. https://doi.org/10.1093/annonc/mdw618.

Perkins, E.G. "Non-volatile Decomposition Products in Heated Fats and Oils." *Food Technology* 21 (1967): 611–16.

Pokorny, J. "Flavor Chemistry of Deep Fat Frying in Oil." In *Flavor Chemistry of Lipid Foods*, ed. David B. Min, and Thomas H. Smouse. AOCS Honored Scientist Ser., 1989. https://doi.org/10.1604/9780935315240.

Romero, Antonio, Carmen Cuesta, and Francisco J. Sánchez-Muniz. "Effect of Oil Replenishment during Deep-Fat Frying of Frozen Foods in Sunflower Oil and High-Oleic Acid Sunflower Oil." *Journal of the American Oil Chemists' Society* 75, no. 2 (February 1998): 161–67. https://doi.org/10.1007/s11746-998-0028-5.

Sánchez-Muniz, F.J., C. Cuesta, and C. Garrido-Polonio. "Sunflower Oil Used for Frying: Combination of Column, Gas and High-Performance Size-Exclusion Chromatography for Its Evaluation." *Journal of the American Oil Chemists' Society* 70, no. 3 (March 1993): 235–40. https://doi.org/10.1007/bf02545301.

Shaki, Fatemeh, Mir-Jamal Hosseini, Mahmoud Ghazi-Khansari, and Jalal Pourahmad. "Toxicity of Depleted Uranium on Isolated Rat Kidney Mitochondria." *Biochimica et Biophysica Acta (BBA) - General Subjects* 1820, no. 12 (December 2012): 1940–50. https://doi.org/10.1016/j.bbagen.2012.08.015.

Smith, Eldon A., Susan L. Prues, and Frederick W. Oehme. "Environmental Degradation of Polyacrylamides. 1. Effects of Artificial Environmental Conditions: Temperature, Light, and pH." *Ecotoxicology and Environmental Safety* 35, no. 2 (November 1996): 121–35. https://doi.org/10.1006/eesa.1996.0091.

Stadler, Richard H., and Till Goldmann. "Chapter 20 Acrylamide, Chloropropanols and Chloropropanol Esters, Furan." In *Comprehensive Analytical Chemistry*, ed. Yolanda Picó, Vol. 51, 705–32. Elsevier, 2008. https://doi.org/10.1016/S0166-526X(08)00020-2.

Stevenson, S.G., M. Vaisey-Genser, and N.A.M. Eskin. "Quality Control in the Use of Deep Frying Oils." *Journal of the American Oil Chemists' Society* 61, no. 6 (June 1984): 1102–8. https://doi.org/10.1007/bf02636232.

Stier, Richard F. "Chemistry of Frying and Optimization of Deep-Fat Fried Food Flavour—An Introductory Review." *European Journal of Lipid Science and Technology* 102, no. 8–9 (September 2000): 507–14. http://doi.org/10.1002/1438-9312(200009)102:8/9<507::aid-ejlt507>3.0.co;2-v.

Takeoka, Gary R., Gerhard H. Full, and Lan T. Dao. "Effect of Heating on the Characteristics and Chemical Composition of Selected Frying Oils and Fats." *Journal of Agricultural and Food Chemistry* 45, no. 8 (August 1, 1997): 3244–49. https://doi.org/10.1021/jf970111q.

Tareke, Eden, Per Rydberg, Patrik Karlsson, Sune Eriksson, and Margareta Törnqvist. "Analysis of Acrylamide, a Carcinogen Formed in Heated Foodstuffs." *Journal of Agricultural and Food Chemistry* 50, no. 17 (July 17, 2002): 4998–5006. https://doi.org/10.1021/jf020302f.

Tompkins, Carol, and Edward G. Perkins. "Frying Performance of Low-Linolenic Acid Soybean Oil." *Journal of the American Oil Chemists' Society* 77, no. 3 (March 2000): 223–29. https://doi.org/10.1007/s11746-000-0036-2.

Tseng, Yi-Chang, Rosana Moreira, and X. Sun. "Total Frying-Use Time Effects on Soybean-Oil Deterioration and on Tortilla Chip Quality." *International Journal of Food Science and Technology* 31, no. 3 (May 1996): 287–94. https://doi.org/10.1046/j.1365-2621.1996.00338.x.

Tyl, R. W., and M. A. Friedman. "Effects of Acrylamide on Rodent Reproductive Performance." *Reproductive Toxicology* 17, no. 1 (January 2003): 1–13. https://doi.org/10.1016/s0890-6238(02)00078-3.

Tynek, Maria, Zdzisllawa Hazuka, Roman Pawlowicz, and Marta Dudek. "Changes in the Frying Medium During Deep-Frying of Food Rich in Proteins and Carbohydrates." *Journal of Food Lipids* 8, no. 4 (December 2001): 251–61. https://doi.org/10.1111/j.1745-4522.2001.tb00200.x.

Varela, G. "Current Fact about the Frying Of Foods." In *Frying of Food, Principle, Changes, New Approaches*, ed. G. Vesrala, A.E. Bender, and I.D. Morton, 9–25. Chichester: Ellis Horwood Ltd, 1988.

Wang, Hao, Pan Huang, Tietao Lie, Jian Li, Reinhold J. Hutz, Kui Li, and Fangxiong Shi. "Reproductive Toxicity of Acrylamide-Treated Male Rats." *Reproductive Toxicology* 29, no. 2 (April 2010): 225–30. https://doi.org/10.1016/j.reprotox.2009.11.002.

Warner, K. "Impact of High Temperature Food Processing on Fats and Oils." In *Impact of Processing on Food Safety. Advance in Experimental Medicine and Biology*, ed. L.S. Jackson, M.G. Knize, and J.N. Morgan, Vol. 459. Boston, MA: Springer, 1999. https://doi.org/10.1007/978-1-4616-4853-9_5.

Wei, Quanwei, Jian Li, Xingmei Li, Lei Zhang, and Fangxiong Shi. "Reproductive Toxicity in Acrylamide-Treated Female Mice." *Reproductive Toxicology* 46 (July 2014): 121–28. https://doi.org/10.1016/j.reprotox.2014.03.007.

Wei, Yongzhong, Suzanne E. Clark, John P. Thyfault, Grace M.E. Uptergrove, Wenhan Li, Adam T. Whaley-Connell, Carlos M. Ferrario, James R. Sowers, and Jamal A. Ibdah. "Oxidative Stress-Mediated Mitochondrial Dysfunction Contributes to Angiotensin II-Induced Nonalcoholic Fatty Liver Disease in Transgenic Ren2 Rats." *The American Journal of Pathology* 174, no. 4 (April 2009): 1329–37. https://doi.org/10.2353/ajpath.2009.080697.

White, Pamela J. "Methods for Measuring Changes in Deep-Fat Frying Oils." *Food Technology* 45, no. 2 (1991): 75–80. http://pascalfrancis.inist.fr/vibad/index.php?action=getRecordDetail&idt=5404188.

Wu, Chung-May, and Shu-Yueh Chen. "Volatile Compounds in Oils after Deep Frying Or Stir Frying and Subsequent Storage." *Journal of the American Oil Chemists' Society* 69, no. 9 (September 1992): 858–65. https://doi.org/10.1007/bf02636333.

Xu, Xin-Qing, Viet Hung Tran, Martin Palmer, Keith White, and Philip Salisbury. "Chemical and Physical Analyses and Sensory Evaluation of Six Deep-Frying Oils." *Journal of the American Oil Chemists' Society* 76, no. 9 (September 1999): 1091–99. https://doi.org/10.1007/s11746-999-0209-x.

Yasuhara, Akio, Yuuka Tanaka, Matt Hengel, and Takayuki Shibamoto. "Gas Chromatographic Investigation of Acrylamide Formation in Browning Model Systems." *Journal of Agricultural and Food Chemistry* 51, no. 14 (June 10, 2003): 3999–4003. https://doi.org/10.1021/jf0300947.

Yaylayan, Varoujan A., Andrzej Wnorowski, and Carolina Perez Locas. "Why Asparagine Needs Carbohydrates to Generate Acrylamide." *Journal of Agricultural and Food Chemistry* 51, no. 6 (February 11, 2003): 1753–57. https://doi.org/10.1021/jf0261506.

Yilmaz, B.O., N. Yildizbayrak, Y. Aydin, and M. Erkan. "Evidence of Acrylamide- and Glycidamide-Induced Oxidative Stress and Apoptosis in Leydig and Sertoli Cells." *Human & Experimental Toxicology* 36, no. 12 (January 8, 2017): 1225–35. https://doi.org/10.1177/0960327116686818.

Yoon, S.H., M.Y. Jung, and D.B. Min. "Effects of Thermally Oxidized Triglycerides on the Oxidative Stability of Soybean Oil." *Journal of the American Oil Chemists' Society* 65, no. 10 (October 1988): 1652–56. https://doi.org/10.1007/bf02912571.

Zamani, E., M. Shokrzadeh, M. Fallah, and F. Shaki. "A Review of Acrylamide Toxicity and Its Mechanism." *Pharmaceutical and Biomedical Research* 3, no. 1 (February 2017): 1–7. https://doi.org/10.18869/acadpub.pbr.3.1.1.

Zhang, Y., J. Chen, X. Zhang, X. Wu, and Y. Zhang. "Addition of Antioxidant of Bamboo Leaves (AOB) Effectively Reduces Acrylamide Formation in Potato Crisps and French Fries." *Journal of Agricultural and Food Chemistry* 55, no. 2 (January 2007): 523–28. https://doi.org/10.1021/jf062568i.

Zribi, Akram, Hazem Jabeur, Felix Aladedunye, Ahmed Rebai, Bertrand Matthäus, and Mohamed Bouaziz. "Monitoring of Quality and Stability Characteristics and Fatty Acid Compositions of Refined Olive and Seed Oils during Repeated Pan- and Deep-Frying Using GC, FT-NIRS, and Chemometrics." *Journal of Agricultural and Food Chemistry* 62, no. 42 (October 13, 2014): 10357–67. https://doi.org/10.1021/jf503146f.

Zribi, Akram, Hazem Jabeur, Guido Flamini, and Mohamed Bouaziz. "Quality Assessment of Refined Oil Blends during Repeated Deep Frying Monitored by SPME-GC-EIMS, GC and Chemometrics." *International Journal of Food Science & Technology* 51, no. 7 (May 21, 2016): 1594–603. https://doi.org/10.1111/ijfs.13129.

Zribi, Akram, Hazem Jabeur, Bertrand Matthäus, and Mohamed Bouaziz. "Quality Control of Refined Oils Mixed with Palm Oil during Repeated Deep-Frying Using FT-NIRS, GC, HPLC, and Multivariate Analysis." *European Journal of Lipid Science and Technology* 118, no. 4 (July 7, 2015): 512–23. https://doi.org/10.1002/ejlt.201500149.

Chapter 14

Biodiesel

An Emerging Fuel Derived from Used Cooking Oil

Abhirup Mitra
Invertis University

Anupriya Mazumder
St. Joseph's University Bengaluru

Winny Routray
NIT, Rourkela

DOI: 10.1201/9781003329244-14

14.1 Introduction

The demand for fuel is an insatiable requirement of the modern world. Fuel is a necessity to run a wide array of generic processes. Hence, industries and government bodies are searching for new sources of fuel and technologies to meet the requirement. Biodiesel has eventually become a vital fuel commodity, and different blends and original forms are currently being explored. Biodiesel quality in terms of fuel efficiency and yield obtained through different processes and from various raw materials are important criteria of research for finding economically viable options for producing biodiesel. Biodiesel is derived from various feedstocks including animal sources, plant sources, and microbial sources (Yaashikaa et al., 2022). Recently, a lot of research groups have focused on the generation of biodiesel from waste resources such as fishery waste (Yahyaee et al., 2013) and non-edible oil (Fadhil et al., 2017). Waste cooking oil is a potential feedstock for the production of biodiesel (Hajjari et al., 2017; Kulkarni & Dalai, 2006), which is being increasingly explored for obtaining economical produce. The production of waste cooking oil has increased in the last few years, which will further be increasing with the increase in food junctions at different setups, such as food courts at universities and shopping complexes. This can be attributed to the interest in accessibility to different cuisines and busy work schedule leading to lack of time to cook.

Figure 14.1 demonstrates the increasing research interest in different aspects of biodiesel production from cooking oil. The figure was prepared from the data collected on year-wise publication of articles on the topic of "biodiesel" and "biodiesel AND cooking oil" and has been plotted using Tableau Public 2022.2. It can be observed that with the increased energy consciousness, the research on biodiesel and allied topics has increased.

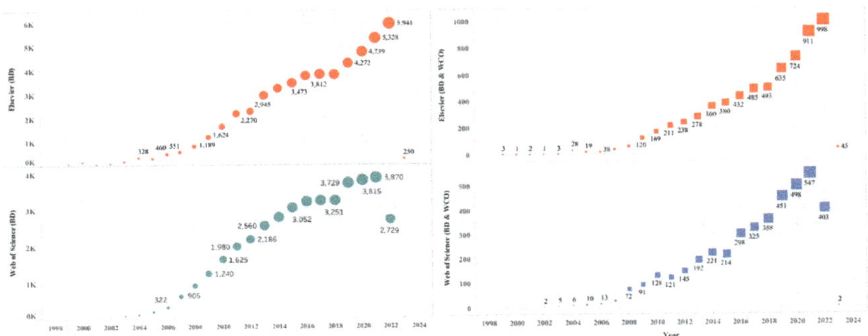

Figure 14.1 Summarization of the number of manuscripts published on the topic "biodiesel" (BD) and "biodiesel AND waste cooking oil" (BD & WCO) according to the databases Science Direct and Web of Science.

This chapter briefly discusses the different aspects of biodiesel and the properties of cooking oil that make it a viable feedstock for production, along with the future prospects of this fuel source.

14.2 Biodiesel

Biodiesel is defined by ASTM as "a fuel comprised of mono-alkyl esters of long-chain fatty acids derived from vegetable oils or animal fats, designated B100" (Hoekman et al., 2012). The US Congress has added an additional requirement to the definition of 'biomass-based diesels'. The baseline lifecycle of biomass-based diesel is at least 50% less than other fuel sources and are produced mainly from canola oil, soybean oil, or other vegetable sources, waste grease, and tallow (Fu et al., 2016). It has been primarily used in the ground transportation where pure biodiesel forms and many blends (blends with 5% or 20% biodiesel with petroleum diesel) have been explored. Generally, it is composed of lower alkyl esters of fatty acids, wherein the free fatty acids are esterified for production of biodiesel. The most commonly present fatty acid methyl ester compounds are summarized in Table 14.1. Biodiesel is considered an eco-friendly fuel. It is considered non-toxic and biodegradable, contains low emission profiles, and is sustainable. Biodiesel production is encouraged throughout the world for several concerns, including (i) global warming and climate change, (ii) greenhouse gas emission, (iii) depletion of fossil fuels, and (iv) interest in developing domestic and secure supply of fuels (Sissine, 2007; Ulusoy et al., 2004).

Table 14.1 Common Fatty Acids and Methyl Esters Observed in Biodiesel

Common Name	*Formal Name*	*CAS No.*	*Molecular Formula*	*Molecular Weight*
Methyl palmitate	Hexadecanoic acid, methyl ester	112-39-0	$C_{17}H_{34}O_2$	270.45
Methyl stearate	Octadecanoic acid, methyl ester	112-61-8	$C_{19}H_{38}O_2$	298.50
Methyl oleate	9-Octadecenoic acid (9Z)-, methyl ester	112-62-9	$C_{19}H_{36}O_2$	296.49
Methyl linoleate	9,12-Octadecadienoic acid (Z, Z)-, methyl ester	112-63-0	$C_{19}H_{34}O_2$	294.47
Methyl linolenate	9,12,15-Octadecatrienoic acid (Z, Z, Z)-, methyl ester	301-00-8	$C_{19}H_{32}O_2$	292.45

Note: The table has been adapted from the article "Storage and oxidation stabilities of biodiesel derived from waste cooking oil" by Fu et al. (2016) with due copyright permission.

Biodiesel prepared from soya or canola oil has been currently used in diesel engines without any further modification of the machinery. These vegetable feedstocks can be used as fuel in any compression-ignition (CI) engine that usually operates on diesel. Various research done on biodiesel suggested that due to the high lubrication capacity of biodiesel, it reduces engine damage by up to 30%. Thus, 2% inclusion of biodiesel in normal diesel will help achieve excellent results in terms of engine life (Agarwal et al., 2003). Biodiesel yields 220% more energy than it requires for production, transportation, and distribution. Higher octane rating, excellent lubrication, a significant reduction in toxic emission, optimum horsepower, and torque can be achieved by mixing 20% biodiesel with diesel. In the USA, soybean oil is mostly used for biodiesel production. Apart from the various benefits mentioned, biodiesels have certain limitations, which should be taken care of before scaling up the application. These limitations include (i) unavailability in large quantities, (ii) a 15% increase in NO^x that contributes to the generation of smog, (iii) increased viscosity of biodiesel at ambient temperatures, and (iv) additives required for lowering the fuel's gel point.

These disadvantages have been attributed to the presence of unsaturated bonds in the molecules obtained from the parent feedstock. This affects the quality of biodiesel during long-term storage and subsequent handling, and has been categorized as auto-oxidation (main reason) and photo-oxidation (Fu et al., 2016). Oxidized B100 biodiesel samples have been observed to have inferior low-temperature properties as compared to the unoxidized samples, which could affect its utilization in places with colder climates (Fu et al., 2016). Antioxidants such as pyrogallol have been added and explored for enhancing the characteristics and functioning of biodiesel (Avase et al., 2015). Advanced analytical techniques are being explored for real-time assessment of the storage quality, including application of electronic nose (Vidigal et al., 2021), and statistical approaches are being followed for prediction of oxidation stability (Çamur & Al-Ani, 2022).

Different methods of production and required properties of biodiesel have been discussed in later sections.

14.3 Characteristics of Used or Waste Cooking Oil

As mentioned before, used or waste cooking oil (WCO) is generally derived from sunflower oil, canola oil, olive oil, vegetable oil, and other animal-based oils such as butter and lard after cooking. The WCO has been categorized into two variations, which are known as yellow grease and brown grease. This categorization is based on the availability of free fatty acids (FFA). The yellow grease has <15% FFA content, whereas brown grease has >15% FFA content (Kulkarni & Dalai, 2006). Usually, WCO consists of 95% of triglycerides that further divide into glycerol and FFA. These fatty acids consist of large number of carbon chains of C_{12}–C_{22}. The FFA present in C_{12}–C_{22} chains include lauric, oleic, palmitic, and linoleic acids.

The used non-edible oil or WCO is appealing as a starting material for biodiesel production for its easy usage. However, there are various problems associated with fat content of vegetable and animal oils like (i) engine fouling, (ii) insufficient combustion, (iii) inadequate atomization of fuel, and (iv) toxicity due to high viscosity of the fuel. Hence, many studies demonstrate the better functioning of biodiesel blend as compared to 100% biodiesel (Fu et al., 2016).

Important properties considered for the evaluation of biodiesel and often being studied include “viscosity, density, peroxide value, heat of combustion, acid number, and phase behavior” (Fu et al., 2016). The properties of biodiesel should be close to pure fossil fuels in terms of its combustion and flash points to perform well in various diesel-based applications. To commercialize the biodiesel prepared using WCO, it should have a kinematic viscosity of 3.0–6.0 mm^2/s at 40°C and an acidity index of 0.5 mg KOH/g. The WCO should achieve a yield of 83% of conversion for pure biodiesel at the end of all treatments. A comparative analysis of linseed oil–derived biodiesel with fossil fuel showed that compression and engine performance are optimum in all modes of operation if the low-load conditions are maintained. The viscosity of oil can be reduced using microemulsion, pyrolysis, oil blending, and esterification. The low-load condition reduced the formation of particulate matter while diesel fuel mixtures are fed. The blended oil has shown comparatively lower emission than fossil fuels in high-load engines (Veinblat et al., 2018). Palm oil blends and methyl esters were investigated as alternative fuels by El-Araby et al. (2018), wherein the simulation outcomes of palm oil blends performed better as compared to correlation predictions. In a different study focused on "evaluation of performance, combustion and emission characteristics of B100 biodiesel and blends", higher brake-specific fuel consumption at lower engine speed and fractional load conditions were observed for biodiesel blends. A marginally shorter ignition delay and lower peak heat release rate were observed in the case of biodiesel as compared to diesel. Furthermore, HC and NOx have been observed to decrease with the use of biodiesel. However, lower engine speed has been observed to have the opposite trend with a substantial effect on the engine combustion and emission formation processes (An et al., 2013).

The physical properties of WCO are also very crucial in terms of ease of application. The WCO is dark in color and highly viscous, with high acidity level, specific heat, and foul smell (da Silva César et al., 2017). The frequent consumption of used or WCO can be hazardous to human health, often leading to very serious health conditions like cancer. Thus, the production of biodiesel using hazardous WCO as feedstock requires preliminary chemical analyses to avoid accumulation toxicity (Doğan, 2016). Furthermore, to efficiently use WCO for biodiesel production, the oil must undergo various pre-treatments to meet the specific criteria for various reactions required to produce biodiesel. Initially filtering and settling are done to remove solid particle, which is followed by further heating to remove water content (Montefrio et al., 2010). The remaining liquid content or the feedstock of the biodiesel is further subjected to the similar treatment as employed for the fossil fuels.

The reprocessing of WCO for industrial uses is difficult attributed to the presence of high FFA content, especially for biodiesel production. Hence, acid pre-treatment or esterification and catalyzation by acid catalyst like sulfuric or hydrochloric acid are employed to convert the free fatty acid into biodiesel. The other method for accelerating the forward reaction of esterification is through addition of sulfuric acid and methanol to WCO, thus lowering the FFA level below 3%. Chai et al. (2014) found that addition of 5% of sulfuric or hydrochloric acid is sufficient if FFA range varies between 15% and 35%. The ideal ratio of methanol to FFA has been observed around 19.8:1 for 15%–25% FFA content.

14.4 Methods Adapted for Production of Biodiesel from Used Cooking Oil

Biodiesel production majorly includes oil generation/extraction and subsequent transesterification for biodiesel formation. In the case of conversion of waste oil to biodiesel, a wide array of catalytic methods

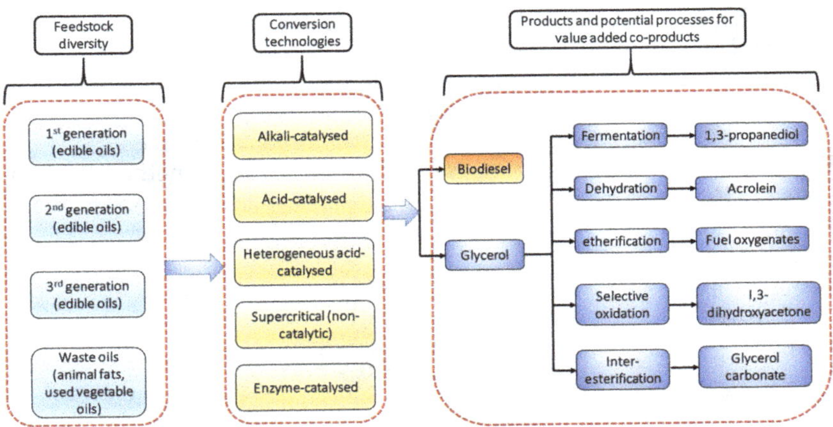

Figure 14.2 Different feedstocks and technologies used for preparing biodiesel along with the by-products and their applications. The figure has been adapted from the article "An overview to process design, simulation and sustainability evaluation of biodiesel production" by Pasha et al. (2021).

are applied as summarized in Figure 14.2. Transesterification is a preferred method as compared to other methods including dilution, micro-emulsification, or pyrolysis, wherein the reaction of a fat or oil (triacylglycerol) with an alcohol leads to yielding biodiesel (fatty acid methyl ester) and glycerol. Catalysts including chemical alkalis or acids, biological enzymes, and microorganisms as a source of enzymes have been explored for the transesterification process. Transesterification has been suggested and observed as a better option as the corresponding process significantly reduces the high viscosity of oil, wherein a biodiesel with the similar physical properties as petroleum diesel fuel is obtained (Dave et al., 2022). Factors affecting the biodiesel process include molar ratio of alcohol, catalyst characteristics and concentration, free fatty acid and water content, reaction temperature and time, and availability of agitation/mixing during the process (Mathew et al., 2021). Application of alkali catalysts is highest, which accounts for about 100% in commercial sector, among which KOH or NaOH is used together with an alcohol (methanol or ethanol). However, in some cases depending on FFA and water content, esterification using acid-catalyzed alcoholysis might precede transesterification (Rizwanul Fattah et al., 2020). Catalysts have been broadly categorized as homogeneous catalysts, heterogeneous catalysts, biocatalysts, and nanocatalysts. One of the reported advantages of heterogeneous catalysts over homogeneous catalysts is that these can be easily separated and in certain cases reused (Rizwanul Fattah et al., 2020). Comprehensive discussion about biodiesel production from various sources can be found in other reviews and chapters, which can be further referred for information on biodiesel production (Babadi et al., 2022; Changmai et al., 2020; Mohiddin et al., 2021; Rizwanul Fattah et al., 2020; Zhong et al., 2020).

Techniques adapted to produce, purify, and utilize biodiesel from WCO can be further categorized as physical and chemical methods as mentioned in Figure 14.3. Some of these specific methods are discussed in the following sections.

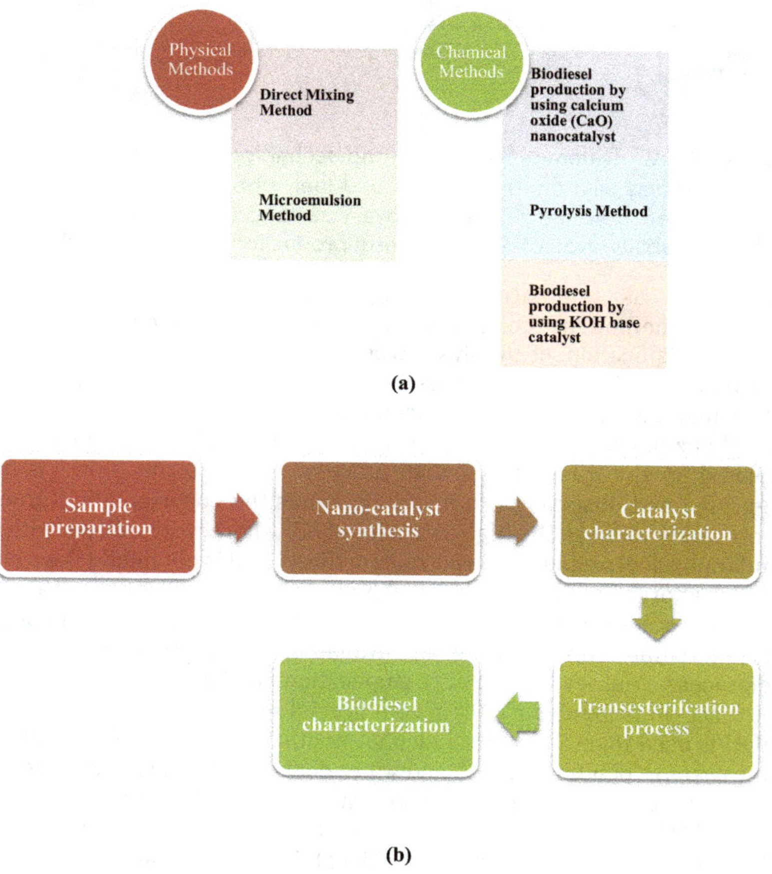

Figure 14.3 (a) Popular methods for production of biodiesel from waste oil, and (b) biodiesel production by using nanocatalyst/catalyst.

14.4.1 Chemical Methods

Chemical methods are majorly based on application of various catalysts based on acids, alkalis, and heterogeneous acid catalysts. Many specific catalysts have been characterized for the efficient production of biodiesel. Although homogeneous catalysts are preferred, heterogeneous catalysts with specific characteristics and recoverability are being explored and encouraged. Nano- and bifunctional catalysts have emerged as very efficient catalysts attributed to the available large surface area (Changmai et al., 2020).

14.4.1.1 Biodiesel Production Using Base and Acid Catalysts

Base catalysts including alkaline metal-based hydroxides (Na and K), methoxides, and carbonates are used during transesterification (Meng et al., 2008; Zhang et al., 2003). Esterification process is catalyzed using various acids including sulfonic, sulfuric, and hydrochloric acids (Rizwanul Fattah et al., 2020). As observed in the cases of oil and catalysts selection, the original oils from which waste oil has been derived affect the choice of different parameters including the selection of catalyst and alcohol (Zhang et al., 2003). In several studies, alkali metal–supported surfaces have been employed as solid catalysts, as observed in the case of transesterification over rice husk silica supporting alkali metal and zeolite nanocomposite trapping alkali (AbuKhadra et al., 2020; Hindryawati et al., 2014).

14.4.1.1.1 Description of a General Method of Alkaline Catalysis of Waste Cooking Oil Aly et al. (2019) in their research showed that when the free fatty acid was lower than 5%, transesterification of the raw material could be done using alkalis as catalysts. During this study, the production of biodiesel from used cooking oil was initiated with mixing the raw materials of used cooking oil, which was then precipitated for 24 hours. After pretreatment, a preliminary analysis of the used cooking oil was carried out, which tested the FFA level (max. 5%). The transesterification process was carried out by mixing KOH (1% by weight of oil) with methanol (methanol:oil ratio = 6:1). The used cooking oil was then heated to 65°. After reaching the temperature, a solution of methanol and KOH was slowly added while pumping (stirring). Heating and mixing were done uniformly at 65°C with a variation of 30, 45, 60, 75 and 90 minutes. After heating, the mixture was allowed to stand for more than 1 hour. After the precipitation, the separation process was carried out by first removing the lower part (glycerol) and then the upper liquid (biodiesel). Cleaning of biodiesel was done by washing with hot water (temperature 70°C) twice as much washing. The ratio between the volume of biodiesel and water for washing was 1:1 in this study. The biodiesel was then heated to 110°C for 10 minutes using a hot plate to remove moisture. Qualitative analysis of biodiesel characteristics was also carried out (Aly et al., 2019).

14.4.1.2 Production Using Nanosized Solid Catalysts along with a Case Study

Various studies have employed a wide array of catalysts for the production of biodiesel, which has provided appropriate results in terms of quality of the product and yield. Studies have also demonstrated the difference in action based on the size of the catalyst.

During a research, Degfie et al. (2019) demonstrated the production of biodiesel and the characterization of used cooking oil collected from various cafes, eating places, and street food vendors in city named Addis Ababa. They showed the whole process in five steps which includes characterization, as shown in Figure 14.3.

During the sample preparation stage of the study, waste palm cooking oil was collected from cafes, eating places, and fast food street vendors in the city of Addis Ababa, obtained after frying of different food commodities. WCO was settled for 4–6 days at ambient temperature and pressure and later filtered through 100 nm sieves to remove all suspended food particles and inorganic residues, followed by heating at 110°C for water elimination. Degfie et al. (2019) demonstrated the preparation of CaO nanocatalyst by the thermal decomposition method according to the procedure of Tang et al. (2013). The nitrate solution was prepared by mixing 11.81 g of calcium nitrate tetrahydrate ($Ca(NO_3)_2.4H_2O$) dissolved in 25 mL of ethylene glycol solution, and 2.10 g of sodium hydroxide was added to the above mixture with vigorous stirring. In order to obtain nanoparticles of uniform size, after stirring for 10 minutes, the gel solution was kept in a static state for about 5 hours, followed by washing with distilled water and then drying under vacuum. Finally, different sizes of CaO nanoparticles were obtained after calcination at 500°C. Subsequently, catalyst characterization was carried out for detailed understanding. The properties of the synthesized catalyst were characterized by X-ray diffraction (XRD) to identify the main components and to determine the crystallite size. XRD analysis was performed with a Mini Flex 600 XRD system with Ni-filtered CuKα radiation at λ = 0.154 nm, and a JSM-IT300 LV scanning electron microscope (SEM) was used to study the morphology of the synthesized catalyst (Degfie et al., 2019). Subsequently, transesterification process was carried out. Biodiesel is produced from triglycerides in the presence of alcohol with a catalyst through a transesterification reaction. Production of biodiesel from WCO with methanol in the presence of nanocatalysts-sized calcium oxide was carried out on a laboratory scale. WCO was preheated to the desired reaction temperature before methanol and catalyst were added to the reaction flask. The calculated amount of methanol to oil ratio was poured into the reactor. CaO catalyst was added in the range between 0.5% and 5% by weight with respect to the weight of WCO, and then the resulting reaction mixture was stirred for 10 minutes. 100 mL of WCO was added, and the temperature of the mixture was adjusted from 30°C to 70°C, with an interval of 5°C. The transesterification took place with constant stirring of the reaction mixture for the required time. All transesterification reactions were carried out at atmospheric pressure with a stirring speed of 1,500 rpm. A thermometer was inserted into the flask to monitor the reaction temperature. After the reaction was complete, the mixture was transferred to a separatory funnel and allowed to stand overnight. Three phases were formed due to solid catalyst and higher density of glycerol than biodiesel (Degfie et al., 2019). Biodiesel characterization is an essential step to study the properties relevant for a highly efficient fuel. The separated biodiesel was heated above the boiling point of methanol (64.7°C) to remove excess unreacted methanol. In addition, very little suspended solid catalyst was removed by settling for 2–3 days, after which biodiesel viscosity, specific gravity, water and sediment, total acidity, ash content, sulfur content, flash point, and cloud point were checked according to the American Society for Testing and Materials (ASTM D 6751).

14.4.1.3 Pyrolysis Method

Pyrolysis method has been used for biodiesel production from various substrates such as animal fats and WCO (Ito et al., 2012; Shahruzzaman et al., 2018). Trabelsi et al. (2018) demonstrated the process of thermal cracking of WCO by using pyrolysis method in a lab-scale fixed-bed reactor. They observed by analyzing the final temperature of pyrolysis, which was maintained between 550°C and 800°C. Along with that, the heating rate was also taken into account and maintained at 5, 15, 20, and 25°C/min, respectively. The product produced after the pyrolysis was also investigated, and it was observed that the maximum yield of biodiesel (80 wt%) was obtained at the temperature of 800°C by maintaining heating rate at 15°C/min. The properties of the biodiesel showed pyrolytic oil with the highest caloric value (HHV around 8,843 kg/Kcal) supporting its high demand as fuel. However, few other properties like high acidity index around 126.8 mg KOH/g sample and high viscosity about 8.95 cSt need some upgradation. The GC/MS characterization was also done of the biodiesel which showed high molecular complexity supporting its use as a source of chemical products or active molecules. It was also demonstrated that the syngas heating value, which was approximately 8 MJ/kg, was perfect for its use as an energy source for the reactor. The leftover biochar was also perfect for the application as fertilizer because of its richness of iron and organic carbons. The stoichiometric model of WCO pyrolysis was also established based on the pyrolysis of products yielded, the CHNS-O composition of the raw material, the syngas chemical composition, and remaining biochar.

14.4.2 Biological Methods

Enzymatic hydroesterification is a recently much explored method which has been gaining popularity. This method has been reported as effective compared to conventional methods as it demonstrates faster reaction rate and prevents soap formation. Also, it has lower energy consumption and is efficient in conversion of low-quality feedstock as well. This process consists of two parts: enzymatic hydrolysis and enzymatic esterification (Pourzolfaghar et al., 2016). Lipase is a common enzyme used for biodiesel production and has been used for conversion of waste oil to biodiesel (Patchimpet et al., 2020). Mainly lipases of microbial origin are preferred as compared to plant and animal sources. Also, enzyme formulations consisting of a proportion of various enzymes and multi-enzyme systems are also being used (Babaki et al., 2017; Pourzolfaghar et al., 2016). Furthermore, for increasing the efficiency of the process and reusability of the enzymes, immobilized enzymes prepared through different substrates are being explored. Biodiesel production from WCO was improved using single or mixed lipases on polyhydroxy alkanoate by Binhayeeding et al. (2020), magnetic nanocomposites by Parandi et al. (2022), and genipin cross-linked chitosan beads by Khan et al. (2020). Sometimes enzyme-catalyzed methods are combined with acid-catalyzed processes for higher efficiency of the overall process (Saifuddin et al., 2009). However, this method has not been commercialized due to higher enzyme costs (Pourzolfaghar et al., 2016).

14.4.3 Physical Methods

In this section, the final product derived using various methods have been explained, and information about the direct application and the corresponding consequences is provided.

14.4.3.1 Direct Mixing Method

Few biodiesels produced from WCO are liquid at ambient conditions due to the high unsaturated fatty acids concentration, which can be directly used in diesel engines or combined with pure diesel without further treatment. The advantages of this method include low cost, easier operation, and high thermal efficiency (grease lubricant has a high heating value). However, there are lubricants in biodiesel that result in its high density, high viscosity, and low volatility; these deficiencies have a poor impact on overall engine combustion performance, together with negative atomization effects, plugging of injector nozzles, and increased corrosion. These issues can lead to more serious problems, including insufficient fuel combustion; reduction of engine power; reduced engine life; increased carbon monoxide, hydrocarbon, particulate matter; and dangerous exhaust emissions. Because there are numerous problems with this approach, it serves as a basis for comparing other biodiesel production and application methods using optimized biorefinery concepts and methods which improve biodiesel efficiency.

14.4.3.2 Microemulsion Method

The microemulsion technique is used to improve the excessive viscosity, low fluidity, and other drawbacks of biodiesel fuels. The microemulsion method involves mixing animal and vegetable oils with solvents and microemulsions or surfactants to make microemulsion biodiesel fuel. The microemulsion method is characterized by means of a simple and direct reduction of biodiesel viscosity. However, when the engine burns fuel produced via the microemulsion approach for a long time, problems such as increased carbon deposition, incomplete combustion, and increased viscosity of lubricating oils may occur.

14.5 Characterization of Physiochemical Properties of the Biodiesel Produced from Used Cooking Oil

The cooking oil generally has different properties that depend on several factors, including the type of oil source commodity, the time of use, the fried food ingredients, and the frying temperature. The main characteristics of used cooking oil are relatively high levels of density and viscosity of free fatty acids (FFA). The different properties of biodiesel analyzed during the studies include density @ 15°C, kinematic viscosity @ 40°C, cetane number, distillation temperature, ester content, ash content (max), sulfur content (max), flash point, cloud point, cold filter plugging point, lubricity (max), water and sediment (max), water content (max), acid value (max), copper strip corrosion (max), carbon residue, oxidation stability, and iodine value (max) (Rizwanul Fattah et al., 2020). In the following sections, specific studies done extensively on waste oil–derived biodiesel have been discussed.

14.5.1 Infrared Spectroscopy Analysis

FTIR analysis has become a popular method for the characterization of the biodiesel. Aly et al. (2019) noted the presence of the several characteristic peaks of biodiesel including OH, =CH, and C=O absorption peaks, observed at 3,446, 3,004, and 1,744 cm^{-1}, respectively. Asymmetric and symmetric stretching vibrational peaks of the $-CH_2$ group were located at 2,929 and 2,859 cm^{-1}. Antisymmetric and symmetric C–O–C stretching vibrational absorption peaks were found at 1,024 and 1,169 cm^{-1}. The C=C symmetric stretching vibrational peak was at 1,633 cm^{-1}.

14.5.2 Viscosity

Viscosity is also an important property often measured for the property analyses of the biodiesel. Aly et al. (2019) demonstrated the measurement of kinematic viscosity of biodiesel by using Ubbelohde viscometer. It was 4.75 mm^2/s (cSt) for biodiesel prepared from WCO and 4.458 mm^2/s (cSt) for biodiesel prepared from fresh oil. Absolute viscosity was also estimated using the following equation:

$$v = \eta/\rho \tag{14.1}$$

where ν is the kinematic viscosity, η is the dynamic viscosity, and ρ is the density. For biodiesel from waste oil, η has been reported as 3.9×10^{-3} Pa.s, whereas η is 0.04×10^{-3} Pa.s for biodiesel obtained from fresh oil.

14.5.3 Density

Density of the prepared biodiesel samples is also an important property often measured using a hydrometer, at 15°C and under atmospheric pressure. It was observed as 0.888 g/cm^3 for the sample made from WCO and 0.9 g/cm^3 for the sample made from fresh oil (Aly et al., 2019).

14.5.4 Flash Point

The flash point of biodiesel is used as a criterion to limit the level of unreacted alcohol remaining in the finished fuel. The flash point is important in relation to legal requirements and for safety measures associated with fuel handling and storage. The flash point has been measured for both biodiesel samples using Koehler Instruments K14670 TAG Closed Cup flash point meter, 220–240 V. Aly et al. (2019) observed it as 170°C for both samples made from WCO and fresh oil.

Table 14.2 Comparison on Few Characteristics of Biodiesel Produced from Fresh or Recycled Oil and Used Cooking Oil

Characteristic Property	*Biodiesel from Waste Cooking Oil*	*Biodiesel from Fresh Oil/ Recycled Oil*	*Sources*
Density (g/cm^3) at 15°C	0.888	0.9	Aly et al. (2019)
Kinematic viscosity(cSt) at 40°C	4.75	4.458	Aly et al. (2019)
Flash point (°C)	>101	>110	Aly et al. (2019)
Calorific value (MJ/Kg)	37.3	37.4	Aly et al. (2019)
Acid number	4.48	0.14	Aly et al. (2019)
Viscosity (Pa.S)	0.032	0.02775	Aly et al. (2019)

14.5.5 Calorific Value

The amount of thermal energy produced by burning a unit weight or volume of fuel is called its calorific value. The calorific value of solid and liquid fuels can be determined using a bomb calorimeter. The basic principles used to determine the calorific value of a fuel are that a known amount of fuel is burned, the released thermal energy is transferred to a medium of known mass and specific heat, and the temperature rise of this medium is measured. Calorific value was 37.3 MJ/kg for biodiesel from waste oil and 37.4 MJ/kg for biodiesel from fresh oil. These values are 9% lower than for regular gasoline (Aly et al., 2019) (Table 14.2).

14.6 Efficiency of Biodiesel Produced from Used Cooking Oil

The various processes used for converting WCO into biodiesel include direct blending of used or WCO with biodiesel, pyrolysis or catalytic cracking, micro-emulsification, followed by transesterification (Verma & Sharma, 2016). The efficient conversion of WCO to biodiesel depends on the use of different homogenous and heterogeneous catalysts. As mentioned before, the following steps are involved: alkali- or acid-catalyzed transesterification, two-step transesterification process using acid and alkali catalyzation steps, enzyme-assisted catalyzed transesterification, and non-catalytic conversion technologies (Mendecka et al., 2020). Different by-products are formed in the transesterification process of WCO including glycerol, soaps, polymeric pigments, and unreacted alcohols (Banerjee & Chakraborty, 2009; Fereidooni & Mehrpooya, 2017). Hence, to make the process economical, further biorefineries are encouraged which will lead to production of a wide array of products and benefit the overall businesses.

Various research studies were conducted to access the economics, environmental impact, and production efficiency of transesterification of WCO to biodiesel. The most common method used worldwide is alkali-catalyzed transesterification. Jegannathan et al. (2011) concluded that alkali-catalyzed transesterification process is best in the production of biodiesel from virgin oils in terms of manufacturing cost. They also stated that the best performance is achieved when enzyme-catalyzed process is applied. Zhang et al. (2003) concluded in his research that acid-catalyzed transesterification process of WCO was more economic compared to other technologies. Several advantages are found in alkali-catalyzed transesterification process: (i) it has lower alcohol to oil ratio, (ii) catalyzation reaction can occur in low pressure and temperature, (iii) higher rate of reaction, (iv) improved corrosion resistance, and (v) relatively smaller equipment size is required. The disadvantages associated with alkali catalysis are as follows: (i) for treating WCO, a pre-treatment unit is required, and (ii) if pre-treatment of WCO is excluded, the biodiesel yield is low due to by-product formation. Acid-catalyzed transesterification can occur in high water and FFA composition. It does not require pre-treatment unit. The effect of acid treatment is less hazardous and thus has lower environmental issues. Acid-catalyzed transesterification has several drawbacks. The rate of reaction is low and requires expensive stainless-steel equipment. It requires higher alcohol to oil ratio and operates at high-temperature and high-pressure conditions.

In individual cases, depending on the feedstock source of raw material, the economic and technological analyses are basic requirements and will have to be adapted and analyzed in each case.

14.7 Future Perspectives and Conclusion

The heterogeneous catalysts are being increasingly explored for biodiesel production, which are not only economical but also reusable. Also one of the biggest challenges of the industry is consumption of high amount of water for the removal of homogenous catalysts, which can be overcome as heterogeneous catalysts are in different phases (Rizwanul Fattah et al., 2020). Currently, chemical-based transesterification is still the most widely used technology for obtaining biodiesel. However, future techniques should concentrate on economic and technical viability of the enzymatic transesterification to decrease the carbon footprint of the process and make it cleaner. Furthermore, application of statistics in various aspects of biodiesel production should be further practiced at larger scale as well to avail maximum inference with minimum expenditure of time and resources (Al-Hamamre & Yamin, 2014; Naidoo et al., 2016; Omar & Amin, 2011). Future practices should also follow biorefinery setups for maximum utilization of this feedstock which is a waste. This will also allow maximum utilization of the by-products and wastes generated during biodiesel production. Overall, this resource has been explored recently, and each sample of the WCO feedstock depends on the original sample of oil used (such as canola, coconut, or sunflower oil); hence, further research is required for devising a method that would be applicable universally for most of the cases.

References

AbuKhadra, M. R., Basyouny, M. G., El-Sherbeeny, A. M., El-Meligy, M. A., & Abd Elatty, E. (2020). Transesterification of commercial waste cooking oil into biodiesel over innovative alkali trapped zeolite nanocomposite as green and environmental catalysts. *Sustainable Chemistry and Pharmacy, 17*, 100289.

Agarwal, A. K., Bijwe, J., & Das, L. (2003). Effect of biodiesel utilization of wear of vital parts in compression ignition engine. *Journal of Engineering for Gas Turbines and Power, 125*(2), 604–611.

Al-Hamamre, Z., & Yamin, J. (2014). Parametric study of the alkali catalyzed transesterification of waste frying oil for Biodiesel production. *Energy Conversion and Management, 79*, 246–254.

Aly, S. T., Farouk, S. M., Mokhtar, K. H., Yasmine, E., Mai, E., Nada, E., Menna, A., & Fatma, R. (2019). Synthesis and characterization of biodiesel from waste cooking oil and virgin oil. *International Journal of Engineering Research and Technology, 8*(12). DOI : 10.17577/IJERTV8IS120112.

An, H., Yang, W. M., Maghbouli, A., Li, J., Chou, S. K., & Chua, K. J. (2013). Performance, combustion and emission characteristics of biodiesel derived from waste cooking oils. *Applied Energy, 112*, 493–499. DOI: 10.1016/j.apenergy.2012.12.044.

Avase, S. A., Srivastava, S., Vishal, K., Ashok, H. V., & Varghese, G. (2015). Effect of pyrogallol as an antioxidant on the performance and emission characteristics of biodiesel derived from waste cooking oil. *Procedia Earth and Planetary Science, 11*, 437–444.

Babadi, A. A., Rahmati, S., Fakhlaei, R., Barati, B., Wang, S., Doherty, W., & Ostrikov, K. (2022). Emerging technologies for biodiesel production: processes, challenges, and opportunities. *Biomass and Bioenergy, 163*, 106521.

Babaki, M., Yousefi, M., Habibi, Z., & Mohammadi, M. (2017). Process optimization for biodiesel production from waste cooking oil using multi-enzyme systems through response surface methodology. *Renewable Energy, 105*, 465–472.

Banerjee, A., & Chakraborty, R. (2009). Parametric sensitivity in transesterification of waste cooking oil for biodiesel production—A review. *Resources, Conservation and Recycling, 53*(9), 490–497.

Binhayeeding, N., Klomklao, S., Prasertsan, P., & Sangkharak, K. (2020). Improvement of biodiesel production using waste cooking oil and applying single and mixed immobilised lipases on polyhydroxyalkanoate. *Renewable Energy, 162*, 1819–1827.

Çamur, H., & Al-Ani, A. M. R. (2022). Prediction of oxidation stability of biodiesel derived from waste and refined vegetable oils by statistical approaches. *Energies, 15*(2), 407.

Chai, M., Tu, Q., Lu, M., & Yang, Y. J. (2014). Esterification pretreatment of free fatty acid in biodiesel production, from laboratory to industry. *Fuel Processing Technology, 125*, 106–113.

Changmai, B., Vanlalveni, C., Ingle, A. P., Bhagat, R., & Rokhum, S. L. (2020). Widely used catalysts in biodiesel production: a review. *RSC Advances, 10*(68), 41625–41679.

da Silva César, A., Werderits, D. E., de Oliveira Saraiva, G. L., & da Silva Guabiroba, R. C. (2017). The potential of waste cooking oil as supply for the Brazilian biodiesel chain. *Renewable and Sustainable Energy Reviews, 72*, 246–253.

Dave, D., Pohling, J., & Routray, W. (2022). Marine Oils as Biodiesel. In *Bailey's Industrial Oil and Fat Products*, Fereidoon Shahidi (Ed.) (pp. 1–20). Wiley. DOI: 10.1002/047167849X.bio097.

Degfie, T. A., Mamo, T. T., & Mekonnen, Y. S. (2019). Optimized biodiesel production from waste cooking oil (WCO) using calcium oxide (CaO) nano-catalyst. *Scientific Reports, 9*(1), 1–8.

Doğan, T. H. (2016). The testing of the effects of cooking conditions on the quality of biodiesel produced from waste cooking oils. *Renewable Energy, 94*, 466–473.

El-Araby, R., Amin, A., El Morsi, A., El-Ibiari, N., & El-Diwani, G. (2018). Study on the characteristics of palm oil–biodiesel–diesel fuel blend. *Egyptian Journal of Petroleum, 27*(2), 187–194.

Fadhil, A. B., Al-Tikrity, E. T., & Albadree, M. A. (2017). Biodiesel production from mixed non-edible oils, castor seed oil and waste fish oil. *Fuel, 210*, 721–728.

Fereidooni, L., & Mehrpooya, M. (2017). Experimental assessment of electrolysis method in production of biodiesel from waste cooking oil using zeolite/chitosan catalyst with a focus on waste biorefinery. *Energy Conversion and Management, 147*, 145–154.

Fu, J., Turn, S. Q., Takushi, B. M., & Kawamata, C. L. (2016). Storage and oxidation stabilities of biodiesel derived from waste cooking oil. *Fuel, 167*, 89–97. DOI: 10.1016/j.fuel.2015.11.041.

Hajjari, M., Tabatabaei, M., Aghbashlo, M., & Ghanavati, H. (2017). A review on the prospects of sustainable biodiesel production: a global scenario with an emphasis on waste-oil biodiesel utilization. *Renewable and Sustainable Energy Reviews, 72*, 445–464.

Hindryawati, N., Maniam, G. P., Karim, M. R., & Chong, K. F. (2014). Transesterification of used cooking oil over alkali metal (Li, Na, K) supported rice husk silica as potential solid base catalyst. *Engineering Science and Technology, an International Journal, 17*(2), 95–103.

Hoekman, S. K., Broch, A., Robbins, C., Ceniceros, E., & Natarajan, M. (2012). Review of biodiesel composition, properties, and specifications. *Renewable and Sustainable Energy Reviews, 16*(1), 143–169.

Ito, T., Sakurai, Y., Kakuta, Y., Sugano, M., & Hirano, K. (2012). Biodiesel production from waste animal fats using pyrolysis method. *Fuel Processing Technology, 94*(1), 47–52.

Jegannathan, K. R., Eng-Seng, C., & Ravindra, P. (2011). Economic assessment of biodiesel production: comparison of alkali and biocatalyst processes. *Renewable and Sustainable Energy Reviews, 15*(1), 745–751.

Khan, N., Maseet, M., & Basir, S. F. (2020). Synthesis and characterization of biodiesel from waste cooking oil by lipase immobilized on genipin cross-linked chitosan beads: a green approach. *International Journal of Green Energy, 17*(1), 84–93.

Kulkarni, M. G., & Dalai, A. K. (2006). Waste cooking oil an economical source for biodiesel: a review. *Industrial & Engineering Chemistry Research, 45*(9), 2901–2913.

Mathew, G. M., Raina, D., Narisetty, V., Kumar, V., Saran, S., Pugazhendi, A., Sindhu, R., Pandey, A., & Binod, P. (2021). Recent advances in biodiesel production: challenges and solutions. *Science of the Total Environment, 794*, 148751.

Mendecka, B., Lombardi, L., & Kozioł, J. (2020). Probabilistic multi-criteria analysis for evaluation of biodiesel production technologies from used cooking oil. *Renewable Energy, 147*, 2542–2553.

Meng, X., Chen, G., & Wang, Y. (2008). Biodiesel production from waste cooking oil via alkali catalyst and its engine test. *Fuel processing technology, 89*(9), 851–857. DOI: 10.1016/j.fuproc.2008.02.006.

Mohiddin, M. N. B., Tan, Y. H., Seow, Y. X., Kansedo, J., Mubarak, N., Abdullah, M. O., San Chan, Y., & Khalid, M. (2021). Evaluation on feedstock, technologies, catalyst and reactor for sustainable biodiesel production: a review. *Journal of Industrial and Engineering Chemistry, 98*, 60–81.

Montefrio, M. J., Xinwen, T., & Obbard, J. P. (2010). Recovery and pre-treatment of fats, oil and grease from grease interceptors for biodiesel production. *Applied Energy, 87*(10), 3155–3161.

Naidoo, R., Sithole, B., & Obwaka, E. (2016). Using response surface methodology in optimisation of biodiesel production via alkali catalysed transesterification of waste cooking oil. *Journal of Science and Industrial Research, 75*, 188–193.

Omar, W. N. N. W., & Amin, N. A. S. (2011). Optimization of heterogeneous biodiesel production from waste cooking palm oil via response surface methodology. *Biomass and Bioenergy, 35*(3), 1329–1338.

Parandi, E., Safaripour, M., Abdellattif, M. H., Saidi, M., Bozorgian, A., Nodeh, H. R., & Rezania, S. (2022). Biodiesel production from waste cooking oil using a novel biocatalyst of lipase enzyme immobilized magnetic nanocomposite. *Fuel, 313*, 123057.

Pasha, M. K., Dai, L., Liu, D., Guo, M., & Du, W. (2021). An overview to process design, simulation and sustainability evaluation of biodiesel production. *Biotechnology for Biofuels, 14*(1), 129. DOI: 10.1186/s13068-021-01977-z.

Patchimpet, J., Simpson, B. K., Sangkharak, K., & Klomklao, S. (2020). Optimization of process variables for the production of biodiesel by transesterification of used cooking oil using lipase from Nile tilapia viscera. *Renewable Energy, 153*, 861–869.

Pourzolfaghar, H., Abnisa, F., Daud, W. M. A. W., & Aroua, M. K. (2016). A review of the enzymatic hydroesterification process for biodiesel production. *Renewable and Sustainable Energy Reviews, 61*, 245–257. DOI: 10.1016/j.rser.2016.03.048.

Rizwanul Fattah, I., Ong, H., Mahlia, T., Mofijur, M., Silitonga, A., Rahman, S., & Ahmad, A. (2020). State of the art of catalysts for biodiesel production. *Frontiers in Energy Research, 8*, 101.

Saifuddin, N., Raziah, A., & Farah, H. (2009). Production of biodiesel from high acid value waste cooking oil using an optimized lipase enzyme/acid-catalyzed hybrid process. *E-journal of Chemistry, 6*(S1), S485–S495.

Shahruzzaman, R. M. H. R., Ali, S., Yunus, R., & Yun-Hin, T. Y. (2018). Green biofuel production via catalytic pyrolysis of waste cooking oil using malaysian dolomite catalyst. *Bulletin of Chemical Reaction Engineering & Catalysis, 13*(3), 489–501.

Sissine, F. (2007). Energy Independence and Security Act of 2007: a summary of major provisions.

Tang, Y., Xu, J., Zhang, J., & Lu, Y. (2013). Biodiesel production from vegetable oil by using modified CaO as solid basic catalysts. *Journal of Cleaner Production, 42*, 198–203.

Trabelsi, A. B. H., Zaafouri, K., Baghdadi, W., Naoui, S., & Ouerghi, A. (2018). Second generation biofuels production from waste cooking oil via pyrolysis process. *Renewable Energy, 126*, 888–896.

Ulusoy, Y., Tekin, Y., Cetinkaya, M., & Karaosmanoglu, F. (2004). The engine tests of biodiesel from used frying oil. *Energy Sources, 26*(10), 927–932.

Veinblat, M., Baibikov, V., Katoshevski, D., Wiesman, Z., & Tartakovsky, L. (2018). Impact of various blends of linseed oil-derived biodiesel on combustion and particle emissions of a compression ignition engine–A comparison with diesel and soybean fuels. *Energy Conversion and Management, 178*, 178–189.

Verma, P., & Sharma, M. (2016). Review of process parameters for biodiesel production from different feedstocks. *Renewable and Sustainable Energy Reviews, 62,* 1063–1071.

Vidigal, I. G., Siqueira, A. F., Melo, M. P., Giordani, D. S., da Silva, M. L., Cavalcanti, E. H., & Ferreira, A. L. (2021). Applications of an electronic nose in the prediction of oxidative stability of stored biodiesel derived from soybean and waste cooking oil. *Fuel, 284,* 119024.

Yaashikaa, P. R., Kumar, P. S., & Karishma, S. (2022). Bio-derived catalysts for production of biodiesel: a review on feedstock, oil extraction methodologies, reactors and lifecycle assessment of biodiesel. *Fuel, 316,* 123379. DOI: 10.1016/j.fuel.2022.123379.

Yahyaee, R., Ghobadian, B., & Najafi, G. (2013). Waste fish oil biodiesel as a source of renewable fuel in Iran. *Renewable and Sustainable Energy Reviews, 17,* 312–319.

Zhang, Y., Dube, M., McLean, D., & Kates, M. (2003). Biodiesel production from waste cooking oil: 1. Process design and technological assessment. *Bioresource Technology, 89*(1), 1–16.

Zhong, L., Feng, Y., Wang, G., Wang, Z., Bilal, M., Lv, H., Jia, S., & Cui, J. (2020). Production and use of immobilized lipases in/on nanomaterials: a review from the waste to biodiesel production. *International Journal of Biological Macromolecules, 152,* 207–222.

Index

Note: Bold page numbers refer to tables and *italic* page numbers refer to figures.

INDEX

INDEX

INDEX

INDEX

For Product Safety Concerns and Information please contact our EU representative GPSR@taylorandfrancis.com
Taylor & Francis Verlag GmbH, Kaufingerstraße 24, 80331 München, Germany

www.ingramcontent.com/pod-product-compliance
Lightning Source LLC
Chambersburg PA
CBHW052010230326
41598CB00078B/2370
* 9 7 8 1 0 3 2 3 5 8 9 9 4 *